T0391701

Engendering Agricultural Development

a division of
NIPA GENX ELECTRONIC RESOURCES & SOLUTIONS P. LTD.
New Delhi-110 034

About the Editors

Prof. Binoo P Bonny

She is currently the Professor and Head, Department of Agricultural Extension and Project Coordinator, Centre for Gender Studies in Agriculture and Farm Entrepreneurship Development, Kerala Agricultural University. She has done PhD in Agricultural Extension from Indian Agricultural Research Institute (IARI), New Delhi with Jawaharlal Nehru Trust Scholarship. She is also trained in the Advanced Certificate Course at Centre for Development Innovations, Wageningen University, Netherlands under NUFIC Fellowship. She started her research career as Scientist (Agrl. Extension) in the Agriculture Research Service at ICAR-IIHR, Bangalore and later joined Kerala Agricultural University as Assistant Professor. Involved in teaching, research and extension for more than 20 years, she has published more than 100 research papers and has guided many students in their PG and PhD research. She also serves as the Associate Editor of Journal of Tropical Agriculture and Associate Director of Research (Monitoring and Evaluation) of Kerala Agricultural University.

Prof. K P Sudheer

Started the career as scientist in Agro Processing Division, CIAE (ICAR), Bhopal, Professor Sudheer has put in more than two decades of illustrious service as teacher, researcher, and extension scientist. With a PhD in Agricultural Engineering from IARI New Delhi and Post doctorate from KU Leuven Belgium, he is presently heading the Department of Agricultural Engineering, RAFTAAR Agri Business Incubator, and Centre of Excellence in Post-harvest Technology at College of Agriculture, Thrissur, KAU. He has secured research funding from Ministry of Food Processing Industries, NAHEP- ICAR, Kerala State Council for Science, Technology & Environment, Ministry of Rural Development, Ministry of MSME, NABARD, RKVY, Food Corporation of India, and Government of Kerala, and has to his credit 300 research publications, six textbooks and several bulletins in the field of Post-harvest Technology. Professor Sudheer is the recipient of several prestigious fellowships for international training including the Norman E Borlaug Fellowship by United States Dept. of Agriculture (USA), NUFFIC Fellowship from Netherlands, VLIR- UDC Fellowship from Belgium, CINADCO Fellowship from Israel and ERASMUS MUNDUS Fellowship from Sweden. His contributions to post harvest sector of Kerala have been recognized with the Krishi Vigyan Award for the best Agricultural Scientist, by the Government of Kerala. He has also received the 'Bhakshyamithra Award' for the best food processing scientist for promoting food processing entrepreneurship in the State. He is the recipient of the prestigious title 'ICAR National Fellow' for his research on 'Safety and quality of Non-thermal Processed Fruits and Vegetables'. Recently he has also received the Best Teacher Award in Agricultural higher education (by ICAR) of Kerala Agricultural University, Commendation Medal and Best Reviewer Award of Indian Society of Agricultural Engineers, New Delhi.

Dr. Smitha S.

Smitha S is a doctoral degree holder in the discipline of Agricultural Extension from National Dairy Research Institute (NDRI), Karnal. She is working as Assistant Professor (Agril Extension) in Kerala Agricultural University since 2019. She is the recipient of fellowship from University Grants Commission and NDRI-Merit Fellowship for her doctoral research. She had published many research papers in various Indian journals. Her research interests include Gender studies, ICT and women in agriculture.

Engendering Agricultural Development Dimensions and Strategies

Editors

Binoo P Bonny

Professor & Head
Department of Agricultural Extension &
Centre for Gender Studies in Agriculture and
Farm Entrepreneurship Development
College of Agriculture, Kerala Agricultural University
Thrissur, Kerala-680 656

K P Sudheer

(ICAR National Fellow)
Professor & Head
Department of Agricultural Engineering &
KAU RAFTAAR Agribusiness Incubator
College of Agriculture
Kerala Agricultural University, Thrissur, Kerala-680 656

Smitha S

Assistant Professor
Centre for Gender Studies in Agriculture and
Farm Entrepreneurship Development
College of Agriculture
Kerala Agricultural University, Thrissur, Kerala-680 656

CRC Press is an imprint of the
Taylor & Francis Group, an **informa** business

a division of

NIPA GENX ELECTRONIC RESOURCES & SOLUTIONS P. LTD.

New Delhi-110 034

First published 2023
by CRC Press
4 Park Square, Milton Park, Abingdon, Oxon, OX14 4RN

and by CRC Press
6000 Broken Sound Parkway NW, Suite 300, Boca Raton, FL 33487-2742

CRC Press is an imprint of Informa UK Limited

British Library Cataloguing-in-Publication Data
A catalogue record for this book is available from the British Library

ISBN: 9781032395005 (hbk)
ISBN: 9781032395012 (pbk)
ISBN: 9781003350002 (ebk)

DOI: 10.4324/9781003350002

Typeset in Times New Roman
by NIPA, Delhi

Dedicated to

All Women in Agriculture

Message

Gender research is to remain relevant and important, as the gender inequalities persist, despite all efforts and resistance. So, I am only happy to learn that Kerala Agricultural University scientists under its Centre for Gender studies in Agriculture has been successful in bringing out a publication on *Engendering Agricultural Development: Dimensions and Strategies*. It is gratifying to find it as a collaborative work that has brought together the leading researchers working in gender in agriculture from across the country and abroad. I understand that the right mix of experience and youth among the authors has added vigour and relevance to the critical discourses on the topics. Moreover, it has brought better credentials to the systematic analysis of the meaning of gender, evolution of gender theory, policy and practices detailed from all the specialized sectors in agriculture.

I am hopeful that the publication will fill the void in post liberalized gender literature by its focus on empirical knowledge from research on gender roles and vulnerabilities, gendered programmes and perspectives in agriculture. The overall range, the acuity, and the liveliness of contemporary thoughts in gender research covered by the content is sure to get wide acceptance in the academic and development circles. This will enable streamlining academic and research thoughts that will guide development trajectories in practically significant ways.

I congratulate the team and wish that they will inspire to bring the best from all in supporting, advocating for, and living the broader cause of gender in development, especially agriculture.

Dr. R Chandra Babu
Vice Chancellor
Kerala Agricultural University

Foreword

Gender concerns in agriculture has occupied centre stage in development discourses especially after early 1990s. The book on *Engendering Agricultural Development: Dimensions and Strategies* enriches this by providing a concise account of the significant transitions over the years. While the book documents the vastness, diversity and vibrancy of research works in gender in agriculture documented from different parts of India inspiring several young researchers to further explore this area to gain new insights, it also presents the continuing plight of undernourished, underpaid and marginalized women who remain mostly invisible in terms of recognition and compensation.

The compendium of writing by leading scholars involved in gender related studies located across the country and abroad has been well organized under three sections viz. *Gender in agriculture development (Part I), Gender in allied sectors of agriculture (Part II) and Data, Tools and approaches in gender analysis (Part III).* It will definitely be appreciated as a genuine attempt to provide insights on the initiatives and outcome of feminization in agriculture on the development of countries like India which holds global relevance. The content covered tries to highlight gender roles and vulnerabilities, importance and impact of gendered programmes and perspectives in agriculture and allied sectors emerging from different regions of the country in the context of concerns such as globalization, urbanization, climate change including adverse effects of Covid 19 on life and livelihoods. Foresight on depiction of gender as a cross cutting theme to work towards UN Sustainable Development Goals (SDG) is highly commendable.

The tireless efforts of the editors in bringing on board the best minds and their works in the field is worth mentioning. I also compliment the contributors for their engaging communication skills in operationalizing the concepts of gender in agricultural development with corroborating evidences from literature and primary research data. The content provides a deft analysis and evidences of relationship between gender and historical doctrines, socio-cultural, behavioural and political theories as applied to the context of present-day concerns on gender empowerment and research and extension. Food security, nutrition and health, ownership rights on land and other resources, partnership and collective efforts, capacity development, gender sensitive monitoring and evaluation of women empowerment programs are also discussed along with data, tools and strategies for gender analysis and gender vulnerability measurement.

I hope and wish the publication will serve as a useful reference handbook for gender researchers, academics and policy makers world over providing the much-needed impetus for planned, inclusive, accelerated growth and sustainable agricultural development.

Dr. Mruthyunjaya
Formerly National Director
National Agricultural Innovation Project
ICAR, New Delhi
Formerly Director ICAR-National Institute on
Agricultural Economics and Policy Research, New Delhi

Preface

As in all major sectors of the economy, in agriculture also, gender differentiates the roles, responsibilities, resources, constraints and opportunities of women and men. This calls for the recognition of gender as a social characteristic that cuts across caste, class, occupation, age and ethnicity. As such, precise gender information is the need of the day.

Gendered dimensions of agriculture and food security are predominated by the key roles played by women in agricultural production, food processing and marketing. They play a decisive role in dietary diversity and are responsible for nutrition at home. Women comprise 20 to 50 percent of the agricultural labour force in developing countries and 79 percent in least developed countries. Their roles vary considerably between and within regions and are changing rapidly in many parts of the world, where economic and social forces are transforming the agricultural sector. This book is an attempt to comprehend and compile the history, present status and future trends of gender roles in agriculture.

The book comprises of three divisions *viz.*, Gender in agriculture development (Part I), Gender in allied sectors of agriculture (Part II) and Data, Tools and approaches in gender analysis (Part III), that explicates the prevalent gendered relegations. It provides insights on the gender dimensions in Indian agriculture, including initiatives, policy reforms and mends the literature gap in gender roles in the sector. The gender roles and impacts from different cultural and geographical horizons of Indian agricultural and allied sectors in the emerging contexts of globalization, urbanization, climate change and the Covid19 pandemic is gathered in this book. The experience of bringing together the ideas of eminent academics including Professors, Scientists and Research scholars from across the country and abroad has been a difficult but a rewarding learning experience. We gratefully appreciate the sincere efforts of all authors in giving the best. The message and foreword by eminent agricultural scholars have brought better credentials to the publication.

The readership of the book covers academicians, researchers, students and social workers who strive towards a gender-neutral world. We are hopeful that this book will be successful in providing better insights into scientific approach towards gender studies.

Editors

Contents

Part-II: Gender in Allied Sectors of Agriculture

Part-III: Tools and Approaches in Gender Analysis

List of Contributors

Abha Manohar K., *Assistant Professor, Department of Forestry, Centurion University of Technology and Management, Forest Park, Bhubaneswar, Odisha-751009.*

Aiswarya T. Pavanan, *Assistant Professor, College of Climate Change and Environmental Science (CCCES), Kerala Agricultural University, Vellanikkara Thrissur-680656.*

Akhil Ajith, *Research Scholar, Department of Agricultural Extension, College of Agriculture Kerala Agricultural University, Vellanikkara, Thrissur-680656.*

Allan Thomas, *Assistant Professor (Agricultural Extension) and Programme Coordinator Krishi Vigyan Kendra, Kerala Agricultural University, Wayanad-673593.*

Ann Annie Shaju, *Project Assistant, NAHEP-CAAST Project, Agri Business Incubator, Kerala Agricultural University, Vellanikkara, Thrissur-680656.*

Anu Susan Sam, *Assistant Professor, Department of Agricultural Economics, Regional Agricultural Research Station, Kumarakom, Kottayam-686563.*

Aparna Radhakrishnan, *Assistant Professor (Agril Extension), Agricultural Information and Sales Centre, Kerala Agricultural University, Vengeri, Kozhikode-673010.*

Archana Bhatt, *Scientist, Community Agrobiodiversity Centre (CAbC), M.S. Swaminathan Research Foundation- Community Agrobiodiversity Centre (MSSRF), Community Agrobiodiversity Centre Puthoorvayal, Meppadi, Wayanad-673577.*

Archana Raghavan Sathyan, *Assistant Professor, Department of Agricultural Extension, College of Agriculture, Vellayani, Thiruvananthapuram-695522.*

Ayisha Mangat, *Research Scholar, Department of Farm Machinery and Power Engineering, Kelappaji College of Agricultural Engineering and Technology, Tavanur Malappuram-679573.*

Binoo P. Bonny, *Professor and Project Coordinator, Centre for Gender Studies in Agriculture and Farm Entrepreneurship Development (CGSAFED), College of Agriculture, Kerala Agricultural University, Vellanikkara, Thrissur-680656.*

Chitra Parayil, *Associate Professor, Department of Agricultural Economics, College of Agriculture, Kerala Agricultural University, Vellanikkara, Thrissur, 680656.*

De, H.K., *Principal Scientist, (Agricultural Extension), ICAR-Central Institute of Freshwater Aquaculture, Kausalyaganga, Bhubaneswar, Odisha-751002.*

Devi Soumyaja, *Assistant Professor, School of Management Studies, Cochin University of Science and Technology (CUSAT), Cochin-682022.*

Hema M., *Assistant Professor, Department of Agricultural Economics, College of Agriculture Kerala Agricultural University, Vellanikkara, Thrissur-680656.*

Hymavathi T.V., *Professor and Head, Department of Food and Nutrition, Professor Jayashankar Telangana State Agricultural University, Rajendranagar, Hyderabad, Telangana-500030.*

Jayasree Krishnankutty, *Professor and Head, Communication Centre, Kerala Agricultural University, Mannuthy, Thrissur-680651.*

Jyotsna C., Ph.D. *Research Scholar, Tata Institute of Social Sciences, V.N. Purav Marg Mumbai-400088.*

Linipriya Vasan, *Assistant Professor, Government Law College, Ayyanthole, Thrissur-680003.*

Lokesh S., *Senior Research Fellow, Agribusiness Incubator, Kerala Agricultural University Vellanikkara, Thrissur-680656.*

Mathura Swaminathan, *Professor, Economic Analysis Unit, Indian Statistical Institute, Mysore Road, Bangalore-560059.*

Nimisha Mittal V., *Lead Researcher, Centre for Research on Innovations and Science Policy (CRISP), Banjara Hills, Hyderabad-Telangana-500034.*

Niyati S., S*enior Research Fellow, Economic Analysis Unit, Indian Statistical Institute, Mysore Road, Bangalore-560059.*

Parvathy Venugopal, *Survey Coordinator, National Bat Monitoring Programme, Bat Conservation Trust, SE11 5RD, London.*

Ponnuswamy K., *Principal Scientist, Dairy Extension Division, ICAR-National Dairy Research Institute, Karnal, Haryana-132001.*

Prema A, *Professor and Head, Department of Agricultural Economics, College of Agriculture Kerala Agricultural University, Vellanikkara, Thrissur-680656.*

Rasheed Sulaiman, V. *Director, Centre for Research on Innovations and Science Policy (CRISP) Banjara Hills, Hyderabad- Telangana-500034.*

Reeja George P, *Associate Professor, Department of Veterinary and Animal Husbandry Extension College of Veterinary and Animal Sciences, Kerala Veterinary and Animal Sciences University Mannuthy, Thrissur-680651.*

Ruchira Bhattacharya, *Assistant Professor, Centre for Gender Studies and Development, National Institute of Rural Development and Panchayati Raj, Rajendranagar, Hyderabad-500030.*

Sabita Mishra, *Principal Scientist, ICAR-Central Institute for Women in Agriculture (CIWA) Plot No. 50-51, Mouza-Jokalandi, Baramunda, Bhubaneswar-751003, Odisha.*

Saha, G.S., *Principal Scientist (Agricultural Extension) ICAR-Central Institute of Freshwater Aquaculture, Kausalyaganga, Bhubaneswar, Odisha-751002.*

Shilpa Karat, *Research Fellow, Communication Centre, Kerala Agricultural University, Mannuthy Thrissur-680651.*

Shinoji K.C., *Scientist (Agril. Extension), ICAR-Indian Institute of Soil Science, Bhopal 462038*

Sivaraman I., *Scientist (Fisheries Extension) ICAR-Central Institute of Freshwater Aquaculture Kausalyaganga, Bhubaneswar, Odisha-751002.*

Smitha S., *Assistant Professor, Centre for Gender Studies in Agriculture and Farm Entrepreneurship Development (CGSAFED), College of Agriculture, Kerala Agricultural University, Vellanikkara, Thrissur-680656.*

Sreelakshmi K. *Unni, Business Manager, KAU RAFTAAR Agri Business Incubator, Kerala Agricultural University, Vellanikkara, Thrissur-680656.*

Sreeram V., *Assistant Professor, Department of Agricultural Extension, Regional Agricultural Research Station, Kerala Agricultural University, Wayanad- 673593.*

Srivastava, S.K., *Former Director (Acting) and Principal Scientist (Agril. Entomology) ICAR- Central Institute for Women in Agriculture (CIWA), Plot No.50-51, Mouza-Jokalandi Baramunda, Bhubaneswar, Odisha-751003.*

Sudheer, K.P., *ICAR National Fellow and Head, Agri-business Incubator (ABI) and Department of Agricultural Engineering, College of Agriculture, Kerala Agricultural University Vellanikkara, Thrissur-680656.*

Sulaja O.R., *Assistant Professor, Department of Agricultural Extension, College of Agriculture Kerala Agricultural University, Vellanikkara, Thrissur-680656.*

Suma Nair, *Assistant Professor (Agricultural Engineering), Krishi Vigyan Kendra, Kerala Agricultural University, Vellanikkara, Thrissur-680656.*

Suresh Appukuttan, *Assistant Professor, Department of Management, Amrita Vishwa Vidyapeetham, Amritapuri Campus, Kollam-690525.*

Swain, S.K., *Director (Acting) and Principal Scientist, ICAR-Central Institute of Freshwater Aquaculture, Kausalyaganga, Bhubaneswar, Odisha-751002.*

Tanuja S., *Scientist (Fish Processing Technology), ICAR-Central Institute for Women in Agriculture (CIWA), Plot No.50-51, Mouza-Jokalandi, Baramunda, Bhubaneswar, Odisha-751003.*

Part-I

Gender in Agriculture Development

1

Gender in Agriculture: A Development Perspective

Binoo P. Bonny*, Akhil Ajith and Lokesh S.

1. Emergence and history of the concept of gender in development

In a chronological point of view it was the period of late 1960s to mid1970s that marked a turning point in development that saw the emergence of gender as an issue of concern. It was during this period that many important publications on psychology of women and gender roles started its manifestation. However, the recurring theme in most of these publications focussed the white male bias in research and research interpretation. Yet, these research publications served as a platform for a thorough discussion of the biological and social factors that contributed to the understanding of psychology of women. The book entitled, *The Development of Sex Differences*, edited by Maccoby (1966), can be seen as a pivotal moment in the field of gender psychology. The book with several chapters that lay the foundations of children's gender development focused on theories of gender development. The publication of the book *Man and Women, Boy and Girl* in the 1970s put forth an advanced theory about gender identity and suggested that social factors played a more significant role in gender identity and gender roles. This sparked the debate of nature - nurture issues which are still significant to this very date. It's during the same decade that scholars started to challenge the conceptualisations of masculinity and femininity and migrate from the unidimensional simplistic models of bipolar opposites.

Contributions of Carl Gustav Jung (originally Karl Gustav Jung) also had a great role in this migration. Carl Jung's school of analytical psychology, proposed the concepts of *anima* and *animus*. These were part of his theory called collective unconscious which suggests that every woman has an unconscious masculine realm known as *animus* and every man has an unconscious feminine side known as *anima*. He postulated that this personal unconscious symbolized prototypical masculine or feminine principles and not an aggregate of either father or mother. It had the potential to transcend to the personal psyche and influence the behavior.

**Author contact: binoo.pb@kau.in*

The results of the study on the content categories and construct covered by the gender articles by Ruble *et al.* (2006) is presented in Table 1. The table results showed that more articles were written on the two out of six areas of the content matrix developed. These areas of focus were activities and interests which included toys and occupations and personal-social attributes which measured roles and abilities. Among these, the majority of studies were conducted on the constructs of identity, perception and behavioural enactment that used the survey method. Less representation of experimental and observational methods were seen in this regard. Further, use of cross sectional or longitudinal designs for research in gender studies saw a declining trend in terms of use. This over reliance on non-experimental studies set a limit to the goals and questions that could be focused in such studies. Experimental methods that can test the causal directions of gender stereotyped preferences and behaviour is important to address the influence of biological, social and cognitive factors which remained significant with respect to the development of theory.

Table l. Classification of gender based articles in the content matrix

Area of research	**Constructs**				
	A Concepts or beliefs	**B Identity or self perception**	**C Prefer-ences**	**D Behavi-oural enactment**	**Total**
Biological/ Categorical Sex	19	4	4	0	27
Activities and interests	47	13	38	40	138
Personal-social attributes	45	75	5	84	209
Gender-based social relationships	1	6	7	10	24
Styles and symbols	3	23	4	12	42
Values regarding gender	0	5	39	3	47
Total	**115**	**126**	**97**	**97**	**487**

This further led to consistent research interest towards certain categories of gender related topics over time such as gender differences, socialisation, and stereotyping. The decade wise evolution of major areas of gender research over the decades is presented as Table 2. Global scenario as is evident from the results of the table suggest that on an average, more than half of the gender research categories examined the differences between the genders upto the 1990s. A slight decrease was seen in the trend during the post 90 period wherein focus shifted to areas such as socialisation, networks and peer group influence. This also influenced a wide range of studies in the subject of stereotyping and related cognitions and behaviours. Further Hyde (2005) proposed for the gender similarities hypothesis in which he argued that women and men are similar in almost all psychological variables. Further, Hyde (2007) also argued that more theoretical studies were needed to understand gender as a stimulus variable than as an individual variable. Some of these studies investigated how individual and group interaction along with gender segregation influenced the behavioural pattern of a particular gender. Considerable amounts of studies were also carried out for understanding how attitudes of children were influenced by egalitarian gender roles. Further, more self-perceptions of traits, abilities and gender identity were studied based on self-ratings of masculinity and femininity.

Table 2. Topic categories addressed by gender research over the decades

Decade	**Topic categories of gender research**								
	Gender difference	**Sociali-zation**	**Stereo-typing**	**Gender identity**	**Cross cultural**	**Media**	**Indivi-dual differ-ences**	**Inter group process**	**Biol-ogy**
1970s	30	15	14	7	2	2	9	2	0
1980s	103	83	59	35	21	8	23	2	0
1990s	135	71	45	33	19	21	29	2	1
2000 onwards	97	73	38	37	32	20	62	3	0

An increase in attention to cultural differences were seen in gender research post 2000. This led to the adaptation of cultural differences and theories to develop a framework for defining gender differences. Involvement of demographic samples were also noted which indicated the focus on cultural issues along with cross cultural approaches. Content analysis of books and media were used to delineate gender stereotyping involved in the media, also emerged as an area of study over time. The implications of such media have been widely discussed. But less research attention was given to issues with respect to the identification of health and eating disorders. On a public medium these topics have received much attention and have been considered as one of clear indications of the prominence gained by gender in development. In short it can be concluded that the gender research was about social sciences, for most part and has been based on assumptions of gender discriminations indoctrinated over the years from the 19th century.

2. Gender based measures

In 1995, two well being indexes were introduced to the Human Development Index (HDI), the Gender - Related Development Index (GDI) and the Gender Empowerment Measure (GEM). Till then HDI was criticized for its assumption that everyone in the society has reached average achievements for the three components of the index, namely, life expectancy, education and income. Another limitation was that the index did not account for the inter group inequality in a society. These criticisms were identified valid, because it is clear from the practical experience that not everyone in the society will be reaching the average achievements for all the three components of the HDI and there will be inequality among the two prominent genders of the society with respect to these components. Further it was also understood that human development cannot be measured without accounting the disparities among the genders of the society with respect to its components. This formed the basis for the introduction of GDI by Anand and Sen (1995). The purpose of GDI was to penalize the HDI, and to discount the HDI of the prevailing gender inequality. This suggests that the gap between HDI and GDI indicated the difference in achievements of HDI components by men and women and this must be reduced from the HDI to understand the real position of the country or state. It could be inferred that the overall wellbeing of a country is affected by gender inequality. Hence GDI is used to adjust HDI downward.

The gender equity in terms of political and economic participation is measured by GEM. The GEM measurements are based on the following three indicators:

1. Male and female shares of parliamentary seats
2. Male and female shares in administrative, professional, technical and managerial positions.
3. Power over economic resources as measured by earned income.

But several researchers have over the years posted criticisms towards GDI and GEM. The major criticism that GDI faces according to Schuler (2007) is that GDI is a measure prone to misinterpretation. He argues that the most common mistake in calculating GDI is that researchers interpret it as a direct measure of gender inequality. This is not true as GDI is an adjustment made to HDI based on the difference in measures to the components of the HDI with respect to gender. Also, changes in measurement of GDI post 1999 has affected its data integrity. Similarly incorrect interpretation of GEM is also made due to the computational feature of the income component. GEM is said to have more consistency if male and female income shares were used to compute the income component. Further in case of developing countries GDI and GEM has limitations due to lack of sub national data on gender and differences in social structure. For example, the Human Development Report of West Bengal indicates that the state had more participation of women in their administrative portfolios even when the national scenario was not so representative. Also the calculation of income component in GDI and GEM were based on the estimation of non-agricultural income, while in countries like India, the major source of employment for rural women is agriculture. But again the question is in the validity of the indicators used in GDI and GEM. Scientists like Dijkstra and Hanmer (2011) argue that as a first step, assessment of socio-economic gender dimensions of human development, GDI is useful. GDI can be accepted as a relative measure of gender inequality and not an accurate representation since the HDI values have a larger influence on the measurement. Accordingly they suggested a new measure called Relative Status of Women (RSW) which uses the same indicators of HDI but assess the relative position of women compared with men. Again the weak correlation of RSW with GDP, compared to GDI indicates that RSW can bring forth the information about the country's development that is not captured by per capita income.

Similarly, novel attempts were made to formulate national level measures for analysing the extent of gender inequality in the past also. Most of these attempts were made from developing countries. One such earlier attempt can be seen in a human development report of Gujarat, in 2004, which introduced Gender Development Measure-1 (GDM-1) and the Gender Equity Index (GEI). Except for the per capita expenditure, GDM-1 and GEI have all the components of HDI.The Modified Gender-related Development Index (MGDI), suggested by Dholakia (2005) also suggests the widening of GDI by addition of two variables to the original index. These variables are, gender specific percentages in ministerial positions and gender specific economic participation rates. Many papers like Mehta (1996) have also tried to improve upon GEM, by creating constructs of three measures of gender empowerment.

Another noteworthy example comes from the Economic Commission for Africa (2004). The African Gender and Development Index (AGDI) developed by them isa composite index consisting of two parts. The first part Gender Status Index (GSI) contains elements like basic capabilities, economic power and political power. Women's advancement and empowerment and government policy performance regarding that is measured by the second part called the African Women's Progress Scoreboard (AWPS). But again the reliability of these measurements are based on the country and region of development.

3. Gender in agriculture

Women have always constituted the major workforce in agriculture, especially in rural areas. According to FAO (2011) the female share in agriculture has seen a rise from 1980-2010 in most of the agrarian countries of Latin America, Europe, South Asia, North Africa and Sub-Saharan Africa. Female labour force participation rate (LFPR) estimates portray India to have the lowest rate of 24.7 per cent compared to G8 countries (58.2%), Emerging Market economies (GM) (52.2%) and South Asian countries (41.6%) as depicted in Fig. 1. It could be inferred from the results that India has a long way to tread to be a gender-equal and equitable society for women. However, there has been a major concern with respect to these estimates that it does not account for the actual work by women. This is because the official statistics does not consider many unpaid work by women such as gardening and household chores. In most of the regions mentioned above, women also take care of members of the family. Though cultural norms are often cited as the major impediment for a greater female LFPR in India, it needs to be understood and interpreted on multiple dimensions. Moreover, the remarkable progress the country has achieved in women empowerment in the 21st century through its gender friendly policies and programmesneed to be discussed in the light of the prevailing female LFPR trends.

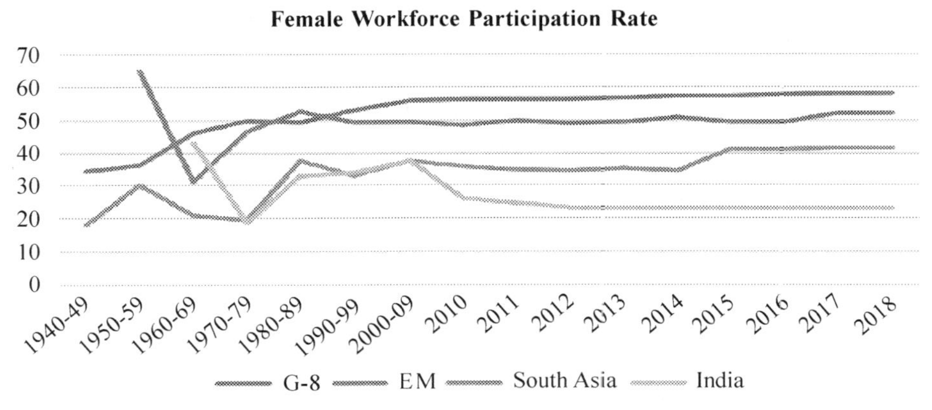

Fig. 1. Comparison of female Labour Work Participation Rate (LWPR) in different countries
Source: Verick, Sher (ILO), 2017

India has made remarkable gains in the female literacy which recorded an increase from the 18 percent in 1947 to 62 per cent as per the estimates from the last census in 2011. Other major indicators of women development such as the maternal mortality rates and fertility rates also has recorded a decline during the period. An estimate by the National Family Health Survey (2017) noted, that compared to countries with similar economic development and income, the percentage of Indian women who said they were in control of their health and reproductive decisions was 8.2 percentage points higher. These implied that the present Indian society is ready to empower women by giving better responsibilities but it mostly remain confined to the limits of the households and family farms as reflected in the current trends in LFPR. This poses the very grave concern as to whether the higher enrolment of women at the school and college levels seen since 2003 will actually result in higher participation of women in the economy.

This is well explained by the U shaped economic function presented as Fig. 2 which provides an implicit frame work to interpret the impact of education level in women's employment across the economic development process as reported by Goldin (1995). The results from the figure indicate that the capability of women to enter into a job market of security and comfort is dependent on the turn of educational level from school to degree and post graduations. The improvements in the initial years of schooling will keep them in low paid unorganized jobs of drudgery and disrespect as indicated by the dip of the U curve. However, the curve rises when the educational level turns into collegiate levels of education which earns them entry into better paid organized job markets. These suggest that the effect of cultural norms on limiting female LFPR cannot be underestimated unless it is forfeited by exceptional gains in their education.

Fig. 2. Female LFP rate per cent as a function of level of education in different locations
Source: Adapted from Goldin (1995)

3.1. Feminization of Agriculture (FoA)

Further due to the increased migration of men from agriculture and various socio-political issues, the agriculture sector has also been witnessing what is called Feminization of Agriculture (FoA). The phenomenon which started during the 1960s is defined as the rising share of farm work undertaken by women as independent producers, as unremunerated family workers, or as agricultural wage workers relative to men (WB 2016). In India alone, agriculture employs 80 per cent of women in which 33 per cent are part of the agricultural workforce and 48 per cent are self-employed farmers. Hence FoA can also be seen as the result of the renewed recognition of agriculture as a pathway out of poverty (World Bank, 2008).

But even with increasing participation of women in multiple roles in agriculture, they are not often recognized as farmers in many Indian policies as is evident from the condition of women in Bihar. In Bihar agriculture is highly feminized with a total workforce constituting 50.1% of women, yet only 7% women have land rights. Therefore, FoA need to be redefined more comprehensively in terms of the extent to which women define, control and enact the social processes of agriculture rather than their mere participation (Pattnaik *et al.*, 2018). Thesesuggest that the concept of the FoAcan be interpreted in two different ways. It can either be the increase in the amount of agriculture work done by women or the extent to which women are at the centre stage controlling the social process of agriculture. The first version encompasses the smallholder production responsibilities and wage worker participation. The second interpretation, addresses the ownership of land, power to make decisions and recognition of contributions in the public sphere. Pattnaik *et al.* (2018) by analyzing the Primary Census Abstracts of 1981, 1991, 2001, and 2011 in India found out that the female workforce grew at twice the rate of population increase. But as a result of the increase in male participation, the proportion established was modest. Examining the four occupational categories in the census namely cultivator, agricultural labourer, household industry and others, an increase in proportion of male and female agricultural workers and decrease in proportion of cultivators were noted in the decade of 2011. Thus with minimal rights to property and declining cultivator status agricultural labour became the marginal occupation of women over time. This, refer to the importance of the first interpretation of FoA as it underlines the fact that feminization not necessarily means empowerment. Even the increased participation of one gender also indicate the decreased participation of another and underpaid tenuous employment which is far from empowering and close to disempowering.

Therefore, true empowerment comes when the control of decision making is with women as showcased in several case studies. The strawberry based agribusiness model developed by Technology mission in Horticulture, Meghalaya, is one such cases of women empowerment. Due to the intervention of the mission, the strawberry production has increased in districts of RiBhoi, East Khasi Hills, West Khasi Hills and Garo Hills. This has brought enhanced participation of the women farmers of RiBhoiStrawberry Growers Association in strawberry cultivation. Moreover, strawberry production techniques flourished as a means of livelihood and profit making business which generates savings in the bank accounts of these women growers.

3.2. Gender and Sustainable Development Goals (SDG)

Gender equity and equality has been considered as a fundamental human right based on democratic values. Ensuring the right has been considered as the best meansto meetthe multitudeof pressing challenges that range from economic crisis to climate change, violence against women and escalating conflicts. Women who form the major victims of these problems are also the custodians of indigenous knowledge and experiential wisdom that offer solutions to these diverse problems. It was in this context the UN adopted its 2030 Agenda for Sustainable Development focussing 17 Sustainable Development Goals (SDGs) adopted in 2015 (replacing the Millennium Development Goals). It symbolize a roadmap for sustainable and inclusive development that provide equitable distribution of resources and opportunities to all. In fact, achievement of gender equality and women's empowerment is integral to each of the 17 goals. The first Sustainable Development Goal, SDG 1 calls for ending poverty in all forms everywhere across the globe. It targets ensuring equal entitlements to economic resources and equal access to basic services, ownership and rights over land, inheritance, natural resources, and technology. The second goal, SDG 2 aims that ending hunger in all forms that persist around the world. It aims to improve food and nutrition security by promoting sustainable agriculture. It stresses the importance of gender equality by securing equal access for them to land, other productive resources and inputs, knowledge, financial services, markets and opportunities for non-farm employment. SDG 4 and SDG 5 have more specific gender targets to achieve such as gender equality in education and gender parity that value unpaid domestic work of women with equal to economic resources and property rights respectively. These indicate the significance gained by gender balanced development in an inclusive and shared future.

4. International agencies in gender development

The 1985 United Nations Third World Conference on Women held in Nairobi served as the first international platform that held a discussion on the concept of gender mainstreaming. A decade later in 1995 Beijing Declaration at the Fourth

United Nations World Conference on Women established gender mainstreaming as a strategy for implementing gender policy. The concept has been used to define the process of assessing and including the concerns of men and women as integral dimensions of design, implementation, monitoring and evaluation of policies and programmes in socio-economic development. Ever since many specialized agencies of UN and agencies from government and non-government sectors have been engaged in gender main streaming processes in the national and international arena. These agencies have been playing important role in ensuring gender sensitive approach in policy making, establishing institutional arrangements to facilitate gender balanced development and to devise tools to delineate gender disaggregated data.A brief description of major international agencies involved in the processes have been included here.

4.1. United Nation Women

The UN Women is an entity of the United Nation working for the betterment and empowerment of women around the globe. It came into existence in 2011 based on the report *Comprehensive Proposal for the Composite Entity for Gender Equality and the Empowerment of Women* submitted by Secretary-General in 2010. However, it had its initiation with the International Research and Training Institute for the Advancement of Women (INSTRAW) established in 1976. UN Women was founded by merging INSTRAW with the different gender divisions of UN *viz.,* Division for the Advancement of Women (DAW); the Office of the Special Adviser on Gender Issues and Advancement of Women (OSAGI), and the United Nations Development Fund for Women (UNIFEM).The main aim of UN Women is to promote gender equality, expand the opportunities for women and to fight against discernments. Currently UN Women is one of the members in the United Nations Development Group.

The important goals of the UN Women are supporting the intergovernmental bodies which are working on status of women, formulation of better policies for women development, global ideals and standards development. The important areas of work of UN Women include the following:

- Leadership and political participation
- Economic empowerment
- Ending violence against women
- Humanitarian action
- Peace and security
- Governance and national planning
- The 2030 agenda for sustainable development
- HIV and AIDS

A chronological list of major achievement milestones of UN Women is presented herewith:

Milestone	Details
Declaration on the Elimination of Violence against Women (1993)	The first international instrument to explicitly address and define forms of violence against women.
ICPD programme of Action (1994)	A 23-year action plan that puts people and their rights at the heart of development and recognizes women's sexual and reproductive health as key to everyone's well-being.
Beijing Declaration and Platform for Action (1995)	A comprehensive framework adopted at the Fourth World Conference for Women with a road map of actions under 12 critical areas to advance women's rights.
UN Security Council Resolution 1325 (2000)	The first UN legal and political framework to recognize that war impacts women differently and to call for women's participation in conflict prevention and resolution.
UN Millennium Declaration (2000)	A set of eight time-bound goals unanimously adopted by world leaders to end poverty with a 15-year deadline. In 2015, the world reflects on progress and gaps, and develops its next transformative agenda: The Sustainable Development Goals.
A mass action for peace (2003)	A relentless civil war impels thousands of Liberian women to form a movement.
The Gulabi Gang: Justice for women (2006)	A "gang" of tens of thousands of women dressed in pink (gulabi) collectively tackle social injustices against women in the UP (India) and are inspiring similar uprisings in the nation.
Challenging the status quo (2011 onwards)	Arabian Peninsula to the capitals of North Africa, streams of women vigorously protest for their rights as part of a broader uprising: the pan-Arab movement.

4.2. OXFAM

As the name indicates, Oxford Committee for Famine Relief, popular as OXFAM, was initiated to assist the famine affected people of Britain in 1942. The institution played a vital role in organizing the food for women and children of Greece when it was occupied by opponents at the time of the Second World War. Oxfam continued to work all over Europe even after the Second World War, until Europe reached a normal phase with respect to hunger and poverty. In 1995 few independent non-governmental organizations joined to form Oxfam Internationals. The main aim has been to increase the working efficiency of Oxfam and to reduce global poverty and injustice.Currently Oxfam International is working in more than 90 countries, to save the lives in emergencies, help people rebuild their livelihoods and campaign for genuine, lasting change, keeping women's rights at the heart of everything. Oxam is currently working on getting equal access for the resources to women, supporting the decision making of women and women leadership. It is working with different partners over 40 countries to stop domestic violence against women and is also challenging the laws of different country where women are not given priority.

4.3. International Center for Research on Women (ICRW)

ICRW was founded in 1976 to make sure the balanced distribution of international development interventions among men and women. It was mainly constituted because women's roles, efforts, requirements, difficulties and their contributions were not considered while countries planned their developmental programs and poverty alleviation programs. ICRW was formed to overcome these disparities made towards women in the world. During the initial years ICRW worked on quantifying women's contribution to the economy being in the various professions and positions. It was mainly focused on how poor households depended on women for their livelihoods. As an impact of ICRW, women led households were also considered for evaluating the effectiveness of poverty relief programs where women led households were not considered earlier. Micro finance programs were initiated as a research effect of ICRW in the 1980's which has become a leading light for millions of poor women worldwide and have demonstrated better repayment tracts than men. IRCW also took the lead in research activities related to women health issues and is the first institute to conduct research and find the impacts of AIDS on women.

ICRW merged with Re:Gender (formerly known as National Council for Research on Women) in 2016 and continue its research, advisories and services to minimize the global poverty level by improving the lives of women and girls in partnership with other organizations.

5. Conclusion

The concept of gender broadly encompasses the roles or relationships between men and women in a given social context which are mostly culturally defined with spatial and temporal dimensions. As such equity and equal access to the benefits of development interventions formed the indispensable precondition for broad-basing the Sustainable Development Goals. All of the 17 SDG goals have integrated gender balance as a cardinal component and define the current development pathways that target 2030.

More significantly, agriculture has been widely recognized as the largest economic activity that involve women in various stages of production, processing, marketing and postharvest decision making. It is extensively reported that gender bias skewed against women pervades the agricultural system that necessitates a different orientation and approach in the efforts of research, extension and education. This has brought to fore the need and value of gender disaggregated data in development and has led to the emergence of different institutional provisions to facilitate gender mainstreaming at grassroots to international arenas of development. Gender Development Index (GDI) and the Gender Empowerment Measure (GEM) are gender sensitive tools based on gender disaggregated data. These have served to provide for inclusive development perspectives and better participation of women by addressing issues related to unequal representation of women in resource entitlements, credit access and even in agricultural extension programmes. This has been instrumental in evolving better gender sensitive and inclusive programmes and strategies that help in the augmentation of crop production, food security and livelihood security of all including women in agriculture. These also have the potential to serve as effective induction pads for gender sensitive programmes and strategies in agricultural development.

References

Anand, S. and Sen, A. 1995. Gender inequality in human development: Theories and measurement. Human Development Occasional Papers (1992-2007) HDOCPA-1995-01, Human Development Report Office (HDRO), United Nations Development Programme (UNDP).

Dholakia, K. 2005. Human Development Index and status of women. Mimeo, The University of Texas at Dallas DOI: http://www.Utdallas.edu/,kruti/HDI_status_%20of_ women_ 072305.pdf.

Dijkstra, A.G. and Hanmer, L.C. 2000. Measuring socio-economic gender inequality: toward an alternative to the UNDP Gender-related Development Index.Feminist Economics. 6(2): 41–75.

Food and Agricultural Organization (FAO), 2011. Women in agriculture closing the gender gap for development, The State of Food and Agriculture 2010-11. FAO, Rome.

Goldin C. 1995. The U-Shaped Female Labor Force Function in Economic Development and Economic History. In: Schultz, T. P. Investment in Women's Human Capital and Economic Development. University of Chicago Press ; pp. 61-90.

Hyde J.S. 2005. The gender similarities hypothesis. The American Psychologist. 60: 581–592. doi: 10.1037/0003-066X.60.6.581.

Hyde J.S. 2007. New directions in the study of gender similarities and differences. New Directions in Psychological Science.16: 259–263. doi: 10.1111/j.1467-8721. 2007.00516.x.

Maccoby, Eleanor E. (Ed.). 1966. The Development of Sex Differences. Stanford: Stanford University Press.

Mehta, A.K. 1996. Recasting indices for developing countries: A gender empowerment measure. Economic and Political Weekly, 31(43): WS80–WS86.

National Family Health Survey, 2017. National Family Health Survey (NFHS-4) 2015-16. Ministry of Health and Family Affairs, India.

Pattnaik, Itishree, Dutt, KuntalaLahiri, Lockie, Stewart and Pritchard,Bill. 2017. The feminization of agriculture or the feminization of agrarian distress? Tracking the trajectory of women in agriculture in India.Journal of the Asia Pacific Economy. DOI: 10.1080 13547860. 2017.1394569

Ruble, D.N., Martin, C.L., & Berenbaum, S.A. 2006. Gender development. In Eisenberg,N.; Damon,W. and Lerner, R. M. (Eds.), Handbook of Child Psychology: Vol. 3, Social, Emotional, and Personality Development.Hoboken: Wiley. pp. 858–932. doi:10.1002/9780470147658.chpsy0314

Verick, Sher, 2017. The paradox of low female labour force participation, Decent Work Team for South Asia and Country Office for India, ILO.

World Bank, 2008. Gender in Agriculture Sourcebook. Washington, DC: World Bank. http://documents.worldbank.org/curated/en/799571468340869508/Gender-in-agriculture-sourcebook.

World Bank, 2016. World Development Indicators. http://data.worldbank.org/indicator/NV.AGR.TOTL.ZS.

2

Gender Disaggregated Data in Agriculture

Mathura Swaminathan and S. Niyati*

1. Introduction

For a proper understanding and analysis of women's position in India's rural economy, we require adequate and good quality data. This chapter examines the limitations of gender-disaggregated data on the rural economy of India with respect to three aspects: the ownership of land and other assets, participation in the rural workforce, and women's share of agricultural incomes and earnings. In approaching the need for better data on gender, the following features of gender statistics as identified by UN organisations should be kept in mind (Corner undated):

a) All statistics on individuals should be collected, collated and presented disaggregated by sex;

b) All variables and characteristics should be analysed by and presented with sex as a primary and overall classification;

c) Specific efforts should be made to identify gender issues and to ensure that data addressing these are collected and made available.

Further, when data have to be gender-disaggregated, it require to follow the following steps as noted in *Engendering Statistics* (Hedman *et al.*, 1996), a publication of Statistics Sweden.

i) Formulation of concepts and definitions used in data collection that adequately reflect the diversity of women and men and capture all aspects of their lives;

ii) Development of data collection methods that take into account stereotypes and social and cultural factors that might produce gender biases.

These points are illustrated in the discussions in this chapter on gender disaggregated data on the rural economy of India.

**Author contact: madhura@isibang.ac.in*

2. National databases

On ownership of land and other assets, there are two major sources of national-level serial data, namely the Land and Livestock Holding Surveys and the All-India Debt and Investment Surveys (AIDIS), both conducted by the National Survey Office (NSO), earlier National Sample Survey Office (NSSO). The limitations of these two sources with respect to data on the ownership of assets by women are discussed.

Secondly labour force statistics, in particular, the Employment and Unemployment Surveys (EUS) and the Periodic Labour Force Surveys (PLFS) of NSSO, are reviewed. These surveys provide information on the participation of women in the rural workforce by occupation and industry. There is a large literature on the limitations of standard labour force surveys in capturing women's work, particularly, in rural areas. The literature is summarised briefly. The role of time-use surveys, which collect information on the actual time spent in different activities, in under standing work done by women is also discussed with reference to the recent national Time Use Survey 2019.

The third area of focus is women's role in crop production and women's earnings from farming. The Comprehensive Cost of Cultivation of Principal Crops (CCPC) scheme under the supervision of the Ministry of Agriculture and Farmers' Welfare, and is the primary source of information on costs of crop cultivation and incomes for 25 crops in 19 states. Lastly, the Situation Assessment Surveys of 2002 and 2013, conducted by the NSO, provide data on incomes of farming and otherallied activities for cultivator households.

2.1. Ownership of land and other assets

It is well recognized in the literature that enhancing women's status in Indian society requires changes in ownership or control over property and other assets. In the rural economy, land is considered as the asset par excellence. So, it is important for researchers and policy makers to obtain data on women's ownership of land and other assets.The decennial Land and Livestock Holding surveys (LHS) of the NSO collect data on area of land owned and operated by households. Estimates of agricultural land and homestead can be separately obtained in the survey. The All India Debt and Investment surveys (AIDIS) provide estimates of the net value of land owned by household. However, both these surveys consider a 'household' as the primary unit. Hence, there is no information on area or value of land legally owned by female members of a household.

The only estimate of relevance to women that can be computed from unit-level data is the land area owned and operated and value of land and other assets owned by female-headed households. This categorization, of male and female-headed households, cannot be considered a useful analytical category for looking at women's control over land or other assets. To elaborate on the limitations of the concept of female-headed households, Ramachandran and Swaminathan (2001) argue, that there is ambiguity in the definition of head of household taking with respect to the case of the Census of India. The Instruction Manual of the Census of India (2000) states "the head of household for census purposes is a person who is recognised as such by the household. She or he is generally the person who bears the chief responsibility for managing the affairs of the household and takes decisions on behalf of the household. The head of household need not necessarily be the oldest male member or an earning member, but may be a female member or a younger member of either sex. It may be remembered that there are female-headed households and in such a case the name of the female head should be recorded at serial number 1 (Ramachandran and Swaminathan, 2001). They further point out that "there are indeed many households in which a single person manages the affairs of the household, takes decisions on its behalf and is recognised by other members of the household as being its head." In such cases, the head of household is easy to identify.

However, in many households, a person-often a woman-in the household "bears the chief responsibility for managing the affairs of the household" and/or "takes decisions on behalf of the household" without being recognised by other members of the household-or by the community-as being the head of household." This is often the case, as evidence from village data show, when the concerned person is a woman. For example, if a man is disabled or unemployed or migrant for the large part of the year, the women or spouse may be the major earner and take major decisions but is usually not recognised as head of household. As a Census enumerator or an investigator in a large scale survey such as by the NSS has to collect information on a large number of parameters, "they cannot engage in a discussion on the identification of a head of household, and finally record the person recognised by the household as its head (Census of India 2000, ibid., cited in Ramachandran and Swaminathan, 2001). These example suggest that often the woman is managing the household or taking major decisions but the male (spouse or father or father-in-law) may be recognised as the head of household. Thus, it could be inferred that the data on land and asset ownership by female headed households does not really capture women's control over land. It only identifies households without any adult males such as those headed by widows or single mothers. Widows constitute about 10 per cent of the female population and female headed households are approximately 11 per cent of all households in rural India (Census of India, 2000).

The other important source of data on land is the World Agricultural Census. The Agricultural Census uses data consolidated from the land revenue surveys of most States, and considers operational holdings as the primary unit. Here too, there is no disaggregated data for male and female title holders, as all land held by any member of a household constitutes a single operational unit. Even though it is possible to obtain data on land titles held by women from official land records, there has been no effort to consolidate or publish such data. It can be an extremely difficult task since land records for the whole nation are not yet digitized. In the land title recording system in India the responsibility of mutation and registration of land titles lie with the land holder. Land records are often outdated and inaccurate when land holders do not report transfers in land title. These suggest that there do exist a clear need to improve the data on land registration. One of the outcomes of such improvement can be availability of data on land titles held by women.

2.2. Ownership of Livestock

Despite the prominent role of women in livestock rearing, very less is known about the extent of women's ownership of livestock assets. The Land and Livestock Holding Survey provides information on the type and the extent of livestock holdings of a household. The AIDIS provides information on the value of livestock assets held by a household. As with land holdings, the primary unit of analysis is the household. Therefore, there is no gender-disaggregated information on intra-household ownership and control of livestock. The only information available separately for women is that for female-headed households.

3. Statistics on work and employment

3.1. Labour Force Surveys

The Employment and Unemployment Surveys (EUS) and the Periodic Labour Force Surveys (PLFS) of the National Sample Survey Organisation (NSSO) are the official sources for estimates of women's work participation rates (WPR) in India. The quinquennial surveys on employment and unemployment situation in India began in 1972-73, and nine rounds were completed in 2011-12. In 2017-18, considering the importance of frequent rounds of data on employment and unemployment situation, NSSO constituted periodic labour force surveys (PLFS) based on National Statistical Commission's recommendations. Currently, we have completed two rounds of PLFS (2017-18, 2018-19). The surveys provide information on aggregate labour force participation by 4-digit industry and sector.

In the labour forces surveys, a person engaged in producing goods and services for market or production of primary goods for own consumption and own-account production of fixed assets for a major time during the reference year is defined as a "worker". A person is termed a worker on usual principal status (UPS) if she engaged in economic activity for a relatively long time during the year. A person seeking work or out of the labour force for a relatively long period and who pursued work for a shorter time during the year is termed a worker on usual subsidiary status (USS) (NSSO 2016). Typically, workers are identified by combining principal and subsidiary activity status (UPSS) as this is a more inclusive definition.

The most important official statistic on rural women available is that of low and declining female work participation rates. The WPR for rural women of age 15 years and above was 48 per cent in 2004-05, which recorded a fall to 35 per cent in 2011-12, and further to 25 per cent in 2017-18.

There is an extensive literature on errors in measuring women's work participation and reasons for not capturing women's work adequately (Jose, 1989; Banerjee, 1999; Hirway, 1999; Kaspos *et al.*, 2014; Klasen and Pieters, 2015; Mehrotra and Sinha, 2017; Usami, 2018; Swaminathan, 2020a). This concerns raised in this literature have been, first, around the concept and definition of work and its limitations in terms of actually capturing the work done by women. For example, rural women engage in tasks performed within the household, tasks that are unpaid, tasks that are done intermittently throughout the day, and so on. These tasks may be part of production for own use or even production for the market but get left out in labour force surveys.

Secondly, the literature has noted problems in data collection including investigator bias, respondent bias, recall period, etc. Independent studies on a smaller scale that can control better for data quality, have found women's work participation to be much higher than reported in large-scale surveys.

One approach taken to get a better estimate of women workers from labour force surveys has been by recalculating work participation rates by including tasks categorised as domestic duties (activity status 92) or domestic duty and certain other tasks such as collection of firewood (activity status 93). Each activity status is listed in Appendix 1. In codes 92 and 93, additional information on up to 12 specified activities such as maintaining a kitchen garden or care of household dairy and poultry animals was collected up to 2011-12 (Appendix Table 2). It is thus feasible, as shown by Usami (2018) to identify women engaged in some of these specific activities even though categories as out of the work force (code 92, 93) and include them in an augmented work participation rate.

Estimates of work participation using principal, subsidiary, and augmented definitions of work, with the latter including persons classified in code 92 or 93 but engaged in selected specified activities (SA 01 to 04) are presented in Table 1. In 2011-12, by the standard Usual Principal and Subsidiary Status (UPSS) definition, 35 per cent of women aged 15 and above were in the work force, but the proportion nearly doubled to 62 per cent with an augmented definition.

Table 1. Work participation rates (WPR) of rural women above 15 years of age using different definitions of work in 2011-12

Definition	WPR (%)
Usual Principal Status (UPS)	25
Usual Subsidiary Status (USS)	10
Usual Principal+ Subsidiary Status (UPSS)	35
Specified Activities 01-04	27
UPSS + Specified Activities 01-04	62

Source: Usami (2018)

Nevertheless, there is still a question of explaining the "fall" in work participation. The decline is often explained by an increase in time spent in education or rising household incomes leading to women quitting the work force. Evidence from primary studies suggests otherwise. The crisis of regular employment may better explain women's withdrawal from the work force (Kaspos *et al.,* 2014, Swaminathan *et al.,* 2020).

3.2. Time Use Surveys

Feminist scholars have argued for regular time-use surveys to capture the extent of time women actually spend on various activities, particularly in unpaid work (Jain, 1983; Hirway, 1999). A time-use survey typically collects data for a 24 hours period and on time allocated to all activities during that period, be it economic activity or child care or even sleep. This method can reduce the investigator and respondent biases in identifying workers that occur in standard labour force surveys. However, time-use surveys cannot replace the labour force surveys but can complement the existing surveys by broadening our understanding of women's participation in various activities (Hirway, 2020).

The NSSO conducted a pilot time use survey (TUS) in 1998-99 that covered a small sample of households in six States of India-Gujarat, Haryana, Madhya Pradesh, Meghalaya, Odisha, and Tamil Nadu. The first national time use survey was conducted in 2019 and it covered all States (GOI, 2020). The survey collected information on time spent in different activities in the last 24 hours, broken up in to 30-minute time slots, for each respondent above six years. As a first approximation, any woman engaged in economic activity (activities within the production boundary as defined in NSSO 2016) even for 30 minutes during the reference day was counted as a worker. By this approach, 44 per cent of rural women were identified as workers (Vijayamba pers. Comm.). By contrast, the PLFS 2018-19 reported a work participation rate of 20 per cent.

While the national time-use survey (TUS) of 2019 was long-due and adds to our understanding of women workers, there are some serious limitations in the concepts and definitions. Specifically, TUS 2019 failed to collect information on subsidiary activity status and has only identified principal activity status of each respondent. As mentioned earlier, many women who work at selected agricultural tasks (say harvest) for a shorter duration during the year report themselves as subsidiary workers, and may thus not be counted as workers. Furthermore, the information on the main activities of women in rural areas, namely crop cultivation and animal rearing, are collected under TUS at a highly aggregated level (and not for example by crop or by operation or task). Both these factors imply that the TUS is likely to have underestimated the number of women workers in rural areas.

4. Cost of Cultivation Surveys

It is a well-known fact that many rural women in India work in crop production and livestock rearing for livelihood. Around 85 per cent of the rural female agricultural work force is engaged in crop production and 9 per cent in livestock rearing (Vijayamba, 2020). There would have been changes in female labour absorption in these sectors over time and across regions. The lack of disaggregated information in official statistics poses a challenge to understand these changes. From the standard labour force surveys, we can estimate the workers in crop production, but a further disaggregated understanding of crop-wise labour absorption is not possible.

Information on crop-wise input use, including the use of human labour, can be obtained from the cost of cultivation surveys (CoC) conducted by the Ministry of Agriculture, Government of India. These surveys were initiated under the Comprehensive Scheme for Study of Cost of Cultivation of Production of Principal Crops (CCPC) in 1970-71. Under this scheme, the State Agricultural Universities collect data on the inputs in physical and monetary terms to estimate the cost of cultivation per hectare of various crops (GOI, 1980; CSO, 2008).

The cost of cultivation survey uses a three-stage stratified random sampling with tehsil (sub-district) as the first-stage unit, cluster (a group of two to three villages) as the second-stage unit and the operational holdings in each cluster as the third-stage unit. The operational holdings are listed in ascending order of the size of holdings and arranged into five size-classes. Two holdings are selected from each size-classes. Data on the cost of cultivation and production are collected from these holdings. The scheme operates in 19 States; the North-Eastern States other than Assam and the erstwhile state of Jammu and Kashmir are excluded. It covers 28 crops that are seasonal (except coconut and sugarcane); plantation crops and vegetables are excluded. The method of estimation of cost is given in CSO (2008). For the calculation of costs, data on labour use and payments are collected. Gender-disaggregated information on the labour hours and payments by crop operation are collected in this survey, as is information on animal upkeep.

Sen and Bhatia (2004) and Surjit (2017) have examined the Cost of Cultivation surveys and identified problems relating to (i) coverage, (ii) methodology, (iii) collection and quality of data, and (iv) publication of data. The point we wish to highlight is that data on gender-wise labour absorption are neither published in the reports of CCPC nor are the unit data easily available to researchers. Gender-disaggregated data are only provided on special request, if at all. Using unit level gender-disaggregated data on labour use in rice cultivation between 1994 and 2013 in selected States of India, Niyati (2020) found a sharp decline in female labour use in Andhra Pradesh and Karnataka. The decline in female labour use was mainly on account of mechanisation of harvesting and threshing operations. Gender-disaggregated information on work hours and payments by crop-operation and crop need to be available in the public domain, so as to understand the gendered effects of changes in farming practices.

5. Statistics on income from farming

India's statistical system does not conduct regular surveys on household incomes. In response to the agrarian crisis of the late 1990s, the NSO conducted surveys to collect data on incomes of farming households for the first time in 2002-03. A second survey was done 10 years later, in 2012-13. The Situation Assessment Survey of Farmers (59th round) and Situation Assessment Survey of Agricultural Households (70th round) gathered information on receipts and expenses of farm and non-farm activities of households and total income of farming households from all sources (Bakshi *et al.*, 2012; Sarkar, 2017).

There have been criticism of the SAS surveys in respect of definition of a farmer, the sampling size and design, and the concept used to estimate the cost of cultivation. In 2003, the SAS defined farmer household as "a household with at least one person who possessed some land and engaged in agricultural activities on any part of that land during the last 365 days" (NSSO, 2005). In 2013, farmer household was replaced by "agricultural household" which has been defined as "a household receiving a value of produce more than Rs 3000 from agricultural activities and having at least one member self-employed in agriculture either in the principal status or in subsidiary status during last 365 days" (NSSO, 2014). This change has two implications: first, we cannot compare the results from the two rounds of surveys. Second, the definition used in 2013 excludes households with a value of agricultural produce less than Rs. 3000. Despite spending significant time on agricultural activities, households can have a low output value because of crop failure. Thus, by definition the survey exclude many cultivator households.

Secondly, the cost concepts used are significantly different from the standard approach followed in the cost of cultivation surveys, resulting in an under-estimation of costs or over-estimation of net incomes (Sarkar, 2017).

Data from SAS-2013 showed that cultivation (48 per cent) and livestock rearing (12 per cent) contributed significantly to household incomes (NSSO, 2014). Women we know play a crucial role in crop production and livestock rearing. They contribute as cultivators, wage workers, and unpaid family labour. The SAS surveys have no information on women's contribution (say, in terms of labour hours) to crop and livestock production, and thus it is not possible to examine the share of women in incomes generated from crop and livestock production. Again, the only variable available for any gender analysis is that on female-headed households, a category that does not adequately capture women's control over income or decision making or even effort.

6. Summary

In a rural economy, land is the asset par excellence. Women's economic empowerment will ultimately depend on their ownership of assets, land and housing in particular. The existing official sources of data have very little information on women's ownership of assets. The only information pertaining to women is that for female-headed households (that constitute around 11 per cent of households in India). Households that are officially counted as "female-headed" are invariably households headed by widows or those without adult males.

Collection of data on female ownership of assets is difficult but must be attempted. This can be attempted, first, by collating gender-disaggregated information from existing sources, such as data on land titles from the land records registers. Secondly, information on ownership should be recorded for all public policies that involve asset creation (as for example, house pattas issued in state-sponsored housing schemes, or land pattas given through land reform). Thirdly, a pilot survey can be undertaken to collect data on women's ownership of assets, particularly land and livestock.

The existing labour force surveys (EUS and PLFS) do not adequately capture women's work in rural and informal settings. The underestimation of women workers has many reasons including the concepts and definitions used and the method of data collection. Adopting a broader definition of work that includes women's paid and unpaid work is necessary to capture the true extent of economic activity among women. Supplementing labour force surveys with information on actual time spent on various activities using time-use surveys can improve labour force estimates.

Gender-disaggregated information on work hours and payments involved in crop cultivation and livestock rearing is collected in the Cost of Cultivation Surveys (CCPC) of the Ministry of Agriculture and Farmers' Welfare. These data should be available in the public domain.

Lastly, there is very little information available in India on incomes of rural households. The exception was two recently conducted Situation Assessment Surveys: these surveys do not follow the standard method of income estimation nor do they provide any gender-disaggregated data. There is no information on women's share of crop income and livestock income despite the widespread participation of women in these activities. Regular surveys on household incomes with gender-disaggregated information need to be undertaken.

References

Bakshi, Aparajita, Rawal, Vikas, Ramachandran, V.K. and Swaminathan, Madhura (2012), "Household Income Surveys in India: Lacunae and Illustrations from Village Surveys," paper presented at the 32nd General Conference of the International Association for Research on Income and Wealth, Boston, August 6-10.

Banerjee, Nirmala (1999), "Women in the Emerging Labour Market," The Indian Journal of Labour Economics, Vol. 42, No. 4, pp. 543-557.

Corner, Lorraine (2013), "From Margins to Mainstream, From Gender Statistics to Engendering Statistical Systems" Regional Economic Advisor, UNIFEM in Asia-Pacific & Arab Stateshttp://www1.aucegypt.edu/src/engendering/Documents/engendering%20corporate%20governance/Margins2 Mainstreamg engerstatistics. pdf

CSO (Central Statistics Office) (2008), Manual on Cost of Cultivation Surveys, Central Statistics Office, Ministry of Statistics and Programme Implementation, New Delhi.

GOI (Government of India) (2020), Time Use in India-2019, Ministry of Statistics and Programme Implementation, National Statistical Office, available at http://mospi.nic.in/sites/default/files/publication_reports/Report_TUS_2019_0.pdf, viewed on February 26, 2021.

GOI (Government of India) (1980), Report of the Special Expert Committee on Cost of Production Estimates, Department of Agriculture and Cooperation, Ministry of Agriculture, New Delhi.

Hedman, Birgitta, Perucci, Francesca, Sundström, Pehr (1996), Engendering Statistics: A Tool for Change, Statistics Sweden, Sweden, available at https://www.scb.se/contentassets/886d78607f724c3aaf0d0a72188ff91c/engendering-statistics-a-tool-for-change.pdf

Hirway, Indira (1999), "Estimating work force using time use statistics in India and its implications for employment policies," paper presented in the International Seminar on Time Use Studies, United Nations Development Programme (UNDP), December 1999.

Hirway, Indira (2020), "Work, Employment, and Labour Underutilisation: What the ILO Resolution Means for India," in Madhura Swaminathan, Shruti Nagbhushan, and V.K. Ramachandran (eds.), Women and Work in Rural India, Tulika Books, New Delhi, pp. 3-18.

Jain, Devaki (1983), "Co-opting Women's Work into the Statistical System: Some Indian Milestones," Samya Shakti : A Journal of Women's Studies, Vol. 1, No.1, July, pp. 85-99.
Jose, A.V. (1989), "Female Labour Force Participation in India: A Case of Limited Options," in A.V. Jose (ed.), Limited Options: Women Workers in Rural India, Asian Regional Team for Employment Promotion, International Labour Organisation, Geneva.
Kannan, K.P., and Raveendran, G. (2019), "From Jobless to Job-loss Growth: Gainers and Losers during 2012–18," Economic and Political Weekly, Vol.54 (44).
Kaspos, Steven, Silberman, Andrea, and Bourmpoula, Evangelina (2014), "Why is Female Labour Force Participation Declining so Sharply in India?," ILO Research Paper no. 10, International Labour office, Geneva, August.
Klasen, Stephens, and Pieters, Janneke (2015), "What Explains the Stagnation of Female Labor Force Participation in Urban India?," The World Bank Economic Review, Vol.29(3), pp.449-478.
Mehrotra, Santosh, and Sinha, Sharmistha (2017), "Explaining Falling Female Employment during a High Growth Period," Economic and Political Weekly, Vol.2 (39).
National Sample Survey Office (NSSO) (2016), Instructions to Field Staff Volume – I: Design, Concepts, Definitions and Procedures, Periodic Labour Force Survey, Ministry of Statistics and Programme Implementation, Government of India, New Delhi. available at: http://mospi.nic.in/sites/default/files/Instruction_manual_for_ PLFS.pdf
National Sample Survey Organisation (NSSO) (2014), Key Indicators of Situation of Agricultural Households in India, National Sample Survey 70th Round, Ministry of Statistics and Programme Implementation, Government of India, New Delhi.
National Sample Survey Organisation (NSSO) (2005), "Income, Expenditure and Productive Assets of Farmer Households," Situation Assessment Survey of Farmers, National Sample Survey 59th Round, Report No. 497 (59/33/5), Ministry of Statistics and Programme Implementation, Government of India, New Delhi.
Niyati, S. (2020.), "Trends and Patterns in Female Labour Absorption in Rice Cultivation," chapter 2 of unpublished PhD thesis, "Women's Labour in Rice Cultivation," Indian Statistical Institute, Bangalore.
Ramachandran, V. K., Swaminathan, Madhura, and Rawal, Vikas (2001), "Female Headed Households: A Note on Methodology," unpublished manuscript, paper presented at the Annual Conference of the International Association of Feminist Economics, Oslo, June 20-24.
Sarkar, Biplab(2017), "Situation Assessment Surveys: An Evaluation," Review of Agrarian Studies, vol. 7, no. 2., available athttp://ras.org.in/eaa3e4614a54ad9733b 3d6caf96c7447
Sen, Abhijit, and Bhatia, M. S. (2004), "Cost of Cultivation and Farm Income," in State of the Indian Farmer: A Millennium Study, vol. 14, Academic Foundation in association with the Department of Agriculture and Cooperation, Ministry of Agriculture, Government of India, New Delhi.
Siddiqui, Mohammed Zakaria, Lahiri-Dutt, Kuntala, Lockie, Stewart, and Pritchard, Bill (2017), "Reconsidering Women's Work in Rural India: Analysis of NSSO Data, 2004-05 and 2011-12," Economic and Political Weekly, Vol.52, No.1, pp. 45-52.
Surjit, V. (2017), "The Evolution of Farm Income Statistics in India: A Review," Review of Agrarian Studies, vol. 7, no. 2, available at http://ras.org.in/81c0a1fdaf 64210c6a4 569c674b7231d
Swaminathan, Madhura (2013), "Gender Statistics in India: A Short Note with a focus on the Rural Economy," Prepared for the Central Statistical Office (CSO), New Delhi, June 2013, available at https://mospi.gov.in/documents/213904/416697//1600868031036_ Them_paper_Gender.pdf/d7d691f8-766e-ee39-0b3e-70d237d60cb0

Swaminathan, Madhura (2020a), "Measuring Women's Work with Time-Use Data: An Illustration from Two Villages of Karnataka," in Madhura Swaminathan, Shruti Nagbhushan, and V.K. Ramachandran (eds.), Women and Work in Rural India, Tulika Books, New Delhi, pp. 19-39.

Swaminathan, Madhura (2020b), "Time-Use Survey Report 2019: What Do We Learn About Rural Women?," Review of Agrarian Studies, vol. 10, no. 2, available at http://ras.org.in/e0b4f51bf0390e368bdb2159fb53cb99

Swaminathan, Madhura, Nagbhushan, Shruti, and Ramachandran, V.K. (2020), Women and Work in Rural India, Tulika Books, New Delhi.

Usami, Yoshifumi, with Patra, Subhajit, and Kapoor, Abhinav (2018), "Measuring Female Work Participation in Rural India: What Do the Primary and Secondary Data Show?," Review of Agrarian Studies, vol. 8, no. 2, available athttp:// ras.org.in/ 86c3e748cd7 c8263a1f762 aa12 c3e37e

Vijayamba, R. (2020), "Women in Livestock Economy," in Madhura Swaminathan, Shruti Nagbhushan, and V.K. Ramachandran (eds.), Women and Work in Rural India, Tulika Books, New Delhi, pp. 167-186.

Appendix Table 1. *Activity status with codes and description*

Employed

Worked in household enterprise (self-employed) as own account worker (11)

Worked in household enterprise (self-employed) as employer (12)

Worked as helper (unpaid family worker) in household enterprises (self-employed) (21)

Worked as regular salaried/wage employee (31)

Worked as casual wage labour in public works (41)

Worked as casual wage labour in other types of works (51)

Did not work owing to sickness though there was work in household enterprise (61)

Did not work owing to other reasons though there was work in household enterprise (62)

Did not work owing to sickness but had regular salaried/wage employment (71)

Did not work owing to other reasons but had regular salaried/wage employment (72)

Unemployed

Did not work but was seeking and/or available for work (81)

Did not seek but was available for work (82)

Out of Labour Force

Attended educational institution (91)

Attended domestic duties only (92)

Attended domestic duties and was also engaged in free collection of goods (vegetables, firewood, cattle feed, etc.), sewing, tailoring, weaving, etc. for household use (93)

Rentiers, pensioners, remittance recipients, etc. (94)

Not able to work due to disability (95)

Others (including begging, prostitution, etc.) (97)

Source: NSSO (2016)

Appendix Table 2. *List of specified activities asked of women who attend domestic duties (code 92) or engage in domestic duties and other activities for household consumption (code 93) in usual principal status*, 2011-12.

	Description of Activity
1.	Maintenance of kitchen garden
2.	Working for household dairy/poultry
3.	Free collection of eatables
4.	Free collection of fuel for household use
5.	Paddy husking
6.	Grinding foodgrains
7.	Preparation of molasses
8.	Making baskets and mats
9.	Making dung cakes
10.	Sewing, tailoring, or weaving
11.	Tutoring own or other's children free of charge
12.	Fetching water from outside

Source: Vijayamba (2020).

3

Gender Advocacy for Food and Nutritional Security

Ruchira Bhattacharya*

1. Introduction

There is a global evidence of deterioration of nutrition and food security condition following the Pandemic and subsequent economiclockdown (Summerton, 2020; Laborde *et al.*, 2020). In India, the status of nutrition and food security has been a matter of concern for decades. India scored an alarming rankof 102 out of 116 countries in Global Hunger Index (IFPRI, 2019). The problem has also been a gendered one, with an abysmally higher proportion of malnourishment and hunger in women and children in India. As per NFHS 4 (2015-16), every second woman in reproductive age in India was suffering from anaemia, two in five children were suffering from stunting and every tenth child in India was suffering from acute malnourishment (IIPS, 2017). It is not difficult to imagine that this existing malnutrition-crisis has only worsened in the post-lockdown period. There is a consensus among academicians, policy makers and planners that the nutrition crisis needs to be linked to gender-based exclusions & vulnerabilities (Quisumbing *et al.*, 2020). This chapter attempts to discuss some of these complicacies of food-nutrition-gender linkages to put forth gender advocacy for food & nutrition policies.

Evidence of linkages between gender and nutrition are there althoughit is difficult to establish precise empirical connect. Also, the pathways through which gender influences nutrition and food security are multiple and complex. Food/nutrition-policy, therefore, has a long way to go till it is gender-informed. To begin with, the understanding of the concept of gender is limited at the policy-level. Moreover, most of the nutrition-food related policies in India have not mainstreamed gender or are working ina gender-specific manner rather than redistributive manner[1].

[1] The terms are borrowed from March, Smyth and Mukhopadhyay, 1999 (concept from Kabeer 1994). Gender specific refers to programs meant for targeted gender needs; gender-redistributive refers to programs that attempt to change the resource distribution and gender-based power-imbalances in the society.

**Author contact: ruchirab.nird@gov.in*

It is in this pretext the chapter reviewed the available evidence on gender-nutrition linkages to put forth an argument not only in favour of gender-aware nutrition policies, but also to guide as to *how* Gender should be integrated. The chapter is structured in five sections with sections that discusses the concept, i.e., importance of context in policy-level for conceptualising gender-needs. This is followed by a review of existing work from a Gender Needs framework and selected empirical evidence from recent primary study of the author are also presented as another separate section. The relevance of reviewing gender-nutrition linkage after COVID19 Pandemic also find place in the chapter. Finally the summary concludes the discussion on gender advocacy in nutrition.

2. Significance of 'context' in Gender Needs Framework

Despite the evidence of gendered challenges playing a major role in deciding food and nutrition security, the food policies we have are at the most women-specific i.e., targeting women or attempting to include more women beneficiaries. It rarely incorporated a gendered approach in setting the program goals, monitoring the program outputs, and evaluating the impact.

Undoubtedly, the vulnerability to malnutrition and hunger is higher for women due to non-inheritance of land titles (FAO, 2011), unequal or inadequate wages (Ramachandran *et al.* 2010), lower access to productive resources (Agarwal, 2012), lower participation in gainful livelihood (Sinha *et al.*, 2016), or low status in despotic family structures (Ramalingaswami *et al.*, 1996; Rao, 2008). But the social protection programs such as Integrated Child Development Programme (ICDS) or the Public Distribution System (PDS) which intend to reduce malnourishment and hunger mostly mitigate the immediate causes of malnutrition[2] such as inadequate dietary intake, inability to afford market prices of food grains etc. Other than identity-based prioritizing or targeting, little provision is there to overcome the barriers that stem from underlying social determinants such as gender, caste or ethnicity even though these identities have a major share of the malnourished (Bhattacharya, 2020).

The lacuna begins with the very way gender is understood in the policy arena. When asked about the gender-initiatives of a program, most of the program officers invariably respond with the number of women beneficiaries they have included in the scheme. There are also studies that acknowledge the relation of

---

[2] UNICEF distinguished the determinants of nutrition into immediate such as dietary intake and morbidity, underlying such as availability of good quantity and quality of food, quality of caring practices, safe water, sanitation and basic such as socio-political environment, technology et cetera (Smith and Haddad, 2014).

gender with nutrition but only in the sense of beneficiary's sex (male or female) for example, Nithya and Bhavani (2018). While the inclusion is first step to achieving gender-just policy, the gender-awareness stems from answering the question what these schemes have done for changing these women's lives in comparison to their own past and their male counterparts. This relational and more structural analysis of the policy should have been the basis of evaluating a gender-sensitivity of the program. But in most cases even the programs designed only for women, such as National Rural Employment Guarantee Scheme which is a women-self-help-group based livelihood program, rarely have a gendered-evaluation framework and evaluate the program at household level (Hunt and Brouwers, 2003; Espinosa, 2013). A gendered evaluation is also complicated because of the complex nature of the association of gender and nutrition. Livelihood may provide sustainable food security to a household but in the case of women's nutrition, longer working hours only add to the burden of reproduction care-work and household chores.That in turn can result in chronic energy deficiency in women and lower duration of breast-feeding forchildren (Bamji and Thimayamma, 2000; Engle and Pederson, 1989).

2.1. Gender Needs

The beginning of building a gender-aware or gender-informed policy-framework should be founded on identifying the gender needs of the community in question so that the designing, budgeting, monitoring and evaluation of the program can be executed using a gender-lens. The concept of gender needs comes from Moser (1993) who noted that women and men have different gender roles and positions in society, resulting in different gender needs and interests. Moser makes an exclusive category of Practical Gender Needs (PGN) as opposed to Strategic Gender Needs (SGN). PGN defined as basic needs of survival not unique to women and are basic necessities for any human being such as food shelter, clothing and water. They facilitate the reproductive roles of women by improving material conditions, and are achievable in short-term by policies targeted to food, water-supply, cooking fuel supply, health clinics, etc. SGN on the other hand, is to facilitate productive roles of women and refers to socio-economic and political positions of women compared to men. They relate to structures and systems, are long-term and as per Moser, these needs are common to almost all women. Since they stem from the hidden structural exclusion processes they can be addressed by consciousness-raising, increasing self-confidence education, strengthening women's organisations, political mobilisation, etc.

In the context of multi-dimensional vulnerabilities, identifying these gender needs as exclusive categories is not only difficult but also impractical. The designation of a specific 'need' as practical or strategic is fuzzy. Practical Needs has been created through a power-gap emanating from structural exclusion and hence addressing them means addressing Strategic Need. For example, availing iron and protein-rich food may depend on the bargaining power of the woman and improve a woman's health and self-worth within the household. Access to water within premises may save hours of daily time-use and grant women access to leisure and may even get men to share some of the domestic workload-which is an important pillar of empowerment. A gendered analysis of predictors of nutrition must reflect this categorization of Gender Needs.

Accepting that circumstances define the nature of the need does not contradict Moser's categories, neither does it make the binary of needs any less important for gendering planning and policy. But it adds a task to designing gender-informed policy i.e. identification of social-economic-political barriers specific to genders for effective intervention. That is to capture and differentiate the *context* or the ground-level scenario where the gendered planning is taking place and the intervention is supposed to work. For example, the availability of safe potable water at or near dwellings may be a Practical Need if women or girls are not missing out significant time or leisure for collecting water. In a community where women do not eat less food than their male counterparts, or do not eat last in the households, provision of iron rich food in the PDS, local ICDS and MDM may be a Practical Need. In these cases, the Strategic Need would include closing gender gap in access to institutions that deliver these services e.g. in case of MDM, ensuring that girls & boys have same attendance rate in school by either community mobilization, construction of separate toilets at school, or ensuring safe and easy commute to school for girls. But in communities where women eat last and least culturally or eat lesser protein than male members because of societal pressure, the first stage i.e. provision of protein rich food in local PDS/ICDS/MDM will be a Strategic Need-followed by the second stage of ensuring access to these institutions. Any need may be Strategic or Practical given the context in which it is being identified and pathway in which it affects nutrition.

There are many forms of exclusion that create marginal sub-identities within 'women', e.g., women in landless households are observed to fare worse both in terms of BMI and overall food intake compared to women in landed households (Bamji and Thimayamma, 2000). The heterogeneity not only affects women's endowment and access to resources that influence the outcome, but also the process through which these identities influence food and nutrition security.

2.2. Existing Evidence of Gender-nutrition links in a gender needs framework

Several studies have investigated these intricate relations to obtain empirical validation of the magnitude and direction of the influences. In this section we compile selected studies (Table 1) and try to identify the gendered predictors of nutrition from a Gender Needs framework. This section tries to inform nutrition-policy of the significance of *context* in identifying gendered predictors of nutrition.

Few studies have examined the nutritional outcomes of children and linked these outcomes with women's nutrition, education, income and other endowments. Though women-centric in approach, they can be recognized as advocating Practical Needs as the pathway to achieve nutrition is simply by facilitating women to perform their reproductive roles better. For example, educated mothers (at least till secondary level) and higher gender life expectancy ratio have been observed to reduce child under nutrition (Smith and Haddad, 2015). Women's education and life expectancy in this approach to nutrition is a facilitator for improved care giving. Similarly, studies linking women's nutritional knowledge (Fadare *et al.,* 2019), mothers' BMI, short stature and lack of antenatal care (Kader and Perera, 2014), or mother's employment (Nankinga *et al.,* 2019) to child's nutrition or birth weight consider women's education and health as a strategy to improve child's nutrition. Same can be said for the studies that provide evidence of link between women's nutrition with women's socio-demographic endowments (Girma and Genebo, 2002) but do not illustrate the strategic pathways. In the context of these studies,even though they advocate education, income or employment – all of which improve women's productive capability and agency – they work as Practical Needs. On the other end of the spectrum are the studies that examine the complex relation that women have with nutrition through multiple roles (Rao, 2017), lack of control on land and productive resources (Agarwal, 2012), work (Bisgrove and Popkin, 1996), access to alternate support for caregiving (Lamontagne *et al.,* 1998), decision on household purchases (Amugsi*et al.* 2016), status of within households (Ramalingaswami *et al.,* 1996) and time for care giving or dietary energy conservation in peak agricultural seasons (Rao and Raju, 2020). By relating the strategic constraint faced by women in productive sphere with achievements in nutrition for children and for themselves, these studies advocate Strategic Gender Needs. Studies linking knowledge and education of mothers to malnutrition but highlighting the effect of knowledge and education on mother's control on resources and autonomy treat knowledge as a strategic need (Engle *et al..* 1999). Recent research in developing countries linking women's empowerment in agriculture decision making to nutritional outcome (Malapit and Quisumbing, 2015), macro-nutrient intake (Tsiboe *et al.,* 2018), dietary diversity index (Sraboni *et al.,* 2014) or mitigation of dietary crisis from low production diversity (Malapit *et al.,* 2015) highlight decision making as a strategic need for nutrition.

There are studies where the line between strategic and practical is blurred. Studies examining nutrient intake in pregnant women (Sahoo and Panda, 2006), young age of mother (Fall *et al.*, 2015) and linking maternal morbidity to nutrient deficiency (Villar *et al.*, 2003), perinatal risks to short pregnancy interval and repeated pregnancies (Conde-Agudelo *et al.*, 2006) all advocate improved health and nutritional intake which are Practical Needs in a non-discriminatory and equal society. But in the society where women have no control on their body and health related choices, reproductive rights stand far from Practical Needs. Women's access to better nutrient in pregnancy, age of first pregnancy, number of and gap between pregnancies – all depend on complex intra-household and intra-community power dynamic between men and women of different ages and attributes. Therefore, programs and policies ensuring women's reproductive rights address strategic needs in the societal context.

Table 1. Selected studies on nutrition gender linkages inGender Need Framework.

S.No.	Study	Location	Outcome	Gender Predictor	Effect	Strategic or Practical?
1.	Sraboni, *et al.*2014	Rural Bangladesh	Per capita calorie availability, household dietary diversity, and adult Body Mass Index	Women Empowerment in Agriculture Index	+ve association	By linking nutritional outcome to a gendered power-difference within households and community these studies treat empowerment or decision-making power as strategic needs for nutrition
2.	Malapit and Quisimbing 2015	Ghana	Infant and young child feeding practices ; child's anthropometric measures; and women's dietary diversity score Body Mass Index.	Key Domains of WEAI	Decision making in Credit +ve for girl's weight-for-height & the women's dietary diversity; no association with women's BMI. Gender parity gap +ve for weight-and height-for-ageof boys	
3.	Malapit *et al.* 2015	Rural Nepal	Health outcomes and dietary diversity	WEAI Improved Production Diversity Index	Women Empowerment mitigates ill effect of low Production Diversity and improves nutrition	
4.	Tsiboe *et al.* 2018	Ghana	macro-nutrient intake	Women's Empowerment in Agriculture	+ve association	
5.	Girma and Genebo 2002	Ethiopia	BMI (Adult women) Height and Weight for age and weight for height z-scores (children)	Region of residence, household economic status, employment, control on earnings,	+ve association	Socio-economic predictors in this context are treated as a practical need for nutritional improvement as the study

				women's health, demographic attributes		does not go into the strategic effects of these predictors
6.	Nithya and Bhavani 2018	Rural Maharashtra, India	BMI	Dietary Diversity Score +Gender control variable	-ve effect of Gender in adults, +ve in adolescents	Gender as a control variable: Needs not highlighted
7.	Rao and Raju 2020	India	Energy Expenditure	Gendered Time Use	Weight-loss higher in women of margina-lised tribes during peak season work stress combined with care work; time-stress affects child nutrition	Time-use or leisure in this study is a strategic need as it demands redistribution of labour and gender roles. The study also highlights the importance of context in reading the time-use-nutrition associations
8.	Rao 2017	India	Nutritional security	Gender difference Pathways	Nutrition security improves with equal opportunities for women in their roles as producer consumer and caregiver	Identifies women's roles and situates gender-equality as a strategic predictor of nutrition
9.	Arokiasamy 2004	India	Mortality and Morbidity indicators	Gender & Birth Order	-Ve association with female of Higher Birth Order	By highlighting the relation of malnutrition with reproductive practices and social constraint associated with a girl-child, reproductive health can be treated as a Strategic need
10.	Sahoo & Panda 2006	India	Exploratory Study	Pregnant Women's nutrition	Nutrient intake in Pregnant women lower than recommended level	Access to nutrient in Pregnancy is part of reproductive health and should be considered Strategic Need in a Patriarchal society which gives least importance of women's health and well-being

11	Kader and Perera 2014	India	Low Birth Weight Children	Mother's low education level, BMI <18.5, short stature (height <145cm)and lack of antenatal visits (<4 visits)	Significant association of Low Birth Weight and mother's lack of education, Health, & Antenatal Care	Education, BMI and Ante-natal care is Practical Need in this study as women's well-being is considered in the context of child's nutrition
12	Villar *et al.* 2003	Systematic Review	Maternal Morbidity and Preterm delivery	Nutrient intake of women	Supplemental nutrition may reduce maternal morbidity	Although women's nutrition is considered a facilitator for lower preterm birth, the review discusses the complications of context in which the supplementation may not work for women. Nutritional supplement in this study is a Strategic Need.
13.	Conde-Agudelo *et al.* 2006	Systematic Review	Perinatal Outcomes	Interpregnancy intervals	Interpregnancy intervals significantly associated with adverse perinatal outcomes	Gap between pregnancies is a Reproductive Choice and represent a Strategic Need
14.	Fall *et al.* 2015	Cross-country	Low birthweight, preterm birth, stunting	Mother's Age	Children of young mother's more likely to have low birth weight, preterm birth and stunting	Delaying the age of pregnancy is a Strategic Need as it advocates control of women on reproductive decision
15.	Engle *et al.* 1999	Conceptual Review	Childcare practices	Women's education / knowledge	Women's knowledge and education associated with care in multiple pathways including higher	Although women's knowledge is considered a facilitator for caregiving, the review goes deeper into the pathways and highlights strategic benefit of

					autonomy and control on resources	better education and knowledge for women
16.	Fadare *et al.* 2019	Nigeria	child under-nutrition	Mother's nutrition-related knowledge	+ve association of HAZ and WHZ scores with mother's knowledge	Knowledge here is a Practical Need as it is only to improve care giving practices
17.	Smith and Haddad 2015	Cross-country	child under-nutrition	Women's Gross Second. enrollment; Life Expectancy Ratio Men/Women	-ve effect on child undernutrition	Better education and Health (Life Expectancy) in this study are Practical Needs as they facilitateimproved child nutrition.
18.	Ramalin-gaswamy *et al.* 1996	India	Malnutrition	Status of women	Lower status of women associated with High malnutrition	Advocates a gender redistributive and role changing policy
19.	Lamontagne *et al.* 1998	Nicaragua	Childcare practices	Mother's employment	Employment improves nutrition through income NOT care giving practice	Employment in this context is a strategic need as it advocates alternate caregiving means & redistribution of roles for employed women
20.	Agarwal 2012	India	Food Security	Control on Land	Control on land & resources through group farming can improve food security conditions of women	Access to land, resources and institutional mechanisms are Strategic Needs to overcome constraints of women in their multiple role
21.	Bisgrove and Popkin 1996	Philippines	Women's nutrition	Women's work	women's work improve the quality of diets - more so for lower-income women	In this study 'Work' is a Strategic Need as it improves diets through complex pathways such as increased autonomy in low-income groups

22.	Amugsi *et al.* 2016	Ghana	Dietary Diversity	Control on Household Purchases	+ve association of higher decision in household purchase and Dietary Diversity	Access to decision making- especially regarding expenditures is a Strategic Need as it requires higher agency of women
23.	Nankinga *et al.* 2019	Uganda	Stunting and Wasting	Mother's employment	-ve effect of mother's non-manual work on stunting but higher wasting in childen of women employed by nonfamily	Employment in this context is a practical need as it seeks no redistribution of roles and treats 'women's work' as a predictor of children's malnutrition
24.	Hamad *et al.* 2015	Peru	Women's psychological and physical health	Micro-credit	+ve association with psychological health but no association with physical health	Access to credit here is a Strategic Need as the study acknowledges societal status of women as a context in which Micro-credits work.
25.	Dehingia *et al.* 2019	India	Maternal Health Service Utilization	Microcredit	+ve link between Service Utilization and Micro-credit partcipation through better paying capacity & awareness	Access to credit in this context is a Strategic Need as it is a facilitator for Institutional Service indicating women's autonomy
26.	Moseson *et al.* 2014	Peru	Food security and childhood anemia	Longer participation in microcredit	+ve association of longer participation and food security	Women's access to credit in this context is a Practical Need as it is only a facilitator for household and child's welfare.
27.	Ojha *et al.* 2020	India	Children's Weight for Height	Credit intervention to women	+ve association of credit and WHZ	

Source: Compiled by author from various sources

Similarly, participation in self-help groups have been observed to improve women's nutrition and healthcare-access (Doocy *et al.*, 2005; Kumar, 2006). But there are contestations from the low-income countries that SHG-based lending has either no significant impact (Banerjee *et al.*, 2014), has led women into indebtedness (Brett, 2006; Kabeer, 2005) with control exercised by more powerful Front-line workers (Ferroglioand Nisbett, 2018) and the returns to these financial inputs going to the households, ergo, the men. A dichotomous classification of access to credit into either Strategic and Practical will be misleading at the least and may even be gender blind.

At the very least, a distinction should be made between studies that reveal multiple pathways or socioeconomic status of women as key pathway through which credit may improve women's health (Hamad, 2015; Kumar *et al.*, 2018) and women's autonomy of availing institutional services (Dehingia *et al.*, 2019) from the studies that relate micro-finance as direct means of reducing household food-insecurityand child malnourishment (Moseson *et al.*, 2014; Ojha *et al.*, 2020). While financial empowerment may be a strategic improvement in the sense it may increase women's bargaining power, but to what end the financial empowerment is advocated as means (through improved Household income or greater bargaining power of women?) is what distinguishes the studies in a Gender Needs framework.

It is imperative to clarify here that this emphasis on identifying Strategic Needs from Practical does not discredit the immense significance of evidence advocating fulfilment of Practical Gender Needs. Attainment of gender equality starts with fulfilment of Practical Needs. In fact, any improvement of wellbeing in any dimension is a gain for human development. Also, a fully exclusive distinction between Practical and Strategic will be misleading in the context of nutrition as Gender shows complicated relation with development outcomes. All the predictors marked as Strategic Needs, given a changed context in the same study will be Practical and vice versa. But the purpose of this section is to nudge nutrition-policy discourse towards a higher order by bringing *context* in the question of determinants so that the policies and programs for nutrition can move forward to seeking redistributive gender justice.

3. Evidence of gendered access to food security from a primary study

Evidence from our primary research at field level revealed a systematic relation between women empowerment and nutritional outcome. From a cross-sectional a survey of 1148 Primary Adult Household members (575 men and 573 women) in 574 rural households in 2019 funded by the National Institute of Rural Development and Panchayati Raj the status of women was measured with the index of women empowerment in agriculture index (Ad Hoc) (Bhattacharya *et al.*, 2020). The variables that contributed to the index are listed as follows.

- Nutritional Outcome i.e. low BMI (Height to Weight Ratio <18.5),
- Low dietary diversity index score (IDDS < 4)
- Disease prevalence in last 15 days for individuals, and
- Food Insecurity Score (combining responses to 8 questions relating to household food sufficiency in past fortnight and 6 months for the households)

These outcomes calculated were plotted (Figure 2) against Gender Parity Status in Deprivation or Disempowerment Score. Disempowerment score was created combining responses to questions in 5 Domains of Empowerment viz. Production, Income, Credit and Savings, Leadership and Leisure (Deprivation coded as 1). Households where Deprivation Score of women was higher than men were marked as No Gender Parity, otherwise Gender Parity.

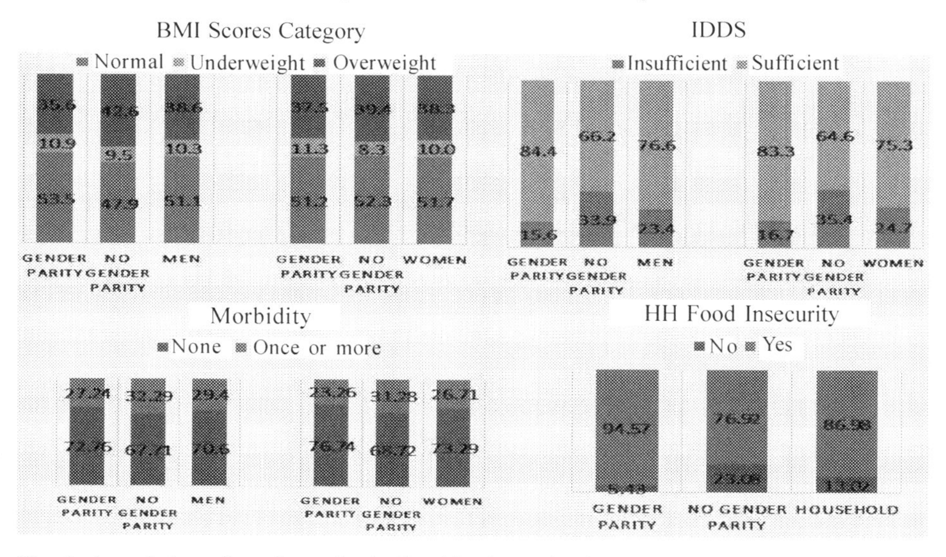

Fig. 1. Association of gender parity in Combined Deprivation Score and selected outcomes of nutrition, health and hunger

Source: Author's own calculations

Results showed that Gender Parity at household has mixed results with nutrition outcomes at individual and household level. As opposed to men, per centof adult women at normal BMI levels (18.5-25) was marginally higher at households without Gender Parity. Men seem to benefit in gender parity household in terms of BMI. A possible explanation for this can be found through time-use studies of women. Women who take less part in the production process and only work at domestic level lose less energy and weight compared to men (Rao and Raju, 2020). With dietary diversity, however, Gender Parity shows a clear result as higher per cent of women received adequate diet in the Gender Parity households

(83%) compared to women in non-gender parity household (64.6%). This result corroborates findings of different studies such as Saborni *et al.* (2014), Malapit *et al.* (2015) that women empowerment translates into Dietary Diversity through a greater control on production, income, purchases and greater knowledge of nutrition.Morbidity seems to be higher for both men and women in No Gender Parity Households compared to the Gender Parity Households. With household food insecurity a reverse relation was observed possibly due to the poorer nature of the households that constituted gender parity category.

4. Relevance of gender advocacy

The recent pandemic followed by a complete economic lockdown has created an unprecedented health, humanitarian, and livelihood crisis (Lancet, 2020). In India, the lockdown created a kind of return-labour-migration which was never observed before. This phenomenon of reverse migration has been the result of the drying up of the centres of growth that made the labourers dependent on themto move back to the peripheries (Iyengar and Jain, 2020).

Post Lock down the sources of seasonal work have dried up with the lockdown juxtaposing with lean season which meant no work for almost two months for women (Swaminathan, 2020). The access to regular employment opportunities, which was already low, further worsened with Anganwadis and schools being closed affecting the independent sources of income of women (Dreze, 2020). It is unavoidable that nutrition, health, education status will also witness varying degrees of setbacks.There is already evidence of decline in household food consumption and dietary diversity especially for women and girl child in the rural households in different primary level studies (FAO, 2020).

To sum up, the crisis has hit women in a disproportionately bigger way with more women vulnerable to losing livelihood and earnings (Genoni *et al.*, 2020; UNFPA, 2020). It also implies that more women aresubjected to unchecked domestic violence (Deshpande, 2020) and further lowering of women's health/ nutrition levels during the period.In order tomitigate these developmental challenges a gender-responsive policy framework is an urgent need of the hour. But what does gender-responsive really mean should be a key question. A simple women-centric or women-targeted transfer may bring palliative relief, but that will not suffice.

5. Conclusion: Context-informed gender lens in nutrition-food security action

The chapter attempted to link a theoretical discussion of Gender Needs to the policy question of gender-mainstreaming in nutrition and hunger. Based on the review of literature, empirical evidence and recent situation it can be concluded that the only way any nutrition policy will be effective in India is when it is 'context-informed'. It implied that it should be flexible and adaptive to underlying economic-socio-political problems faced by different genders. A gender-aware approach to nutrition and food security will be to move beyond palliative and post-facto interventions. It warrants moving ahead from '*What*' the predictors are to '*How*' the predictors work in terms of fulfilling Practical and/or Strategic need. This is not to undermine any palliative measures such as quick cash transfers after pandemic or corrective measures such as providing supplemental nutrition in a malnourished community. But to be honest, what good will an iron tablet do if the woman does not accept her worth? Gendered exclusions and discriminations start early at the household with differential treatments to daughters and sons. In India there is empirical evidence of high levels of domestic violence and societal acceptance of the same (Jejeebhoy, 1998; Ahmad *et al.*, 2019). In fact, prolonged exposure to such treatments results in stress induced malnutrition that a simple supplemental nutrition program cannot overcome (Ackerson and Subramanian, 2008).

In realizing an effective nutrition policy, the multi-dimensional nature of the problem must be dealt by a context-adaptive strategy which can be flexible to the changing nature of the predictors. This alone can ensure its capacity to address the practical need or strategic need involved. In a very crude sense, if the intervention from one policy line-department does not mitigate strategic barriers, there should be ease of access to the interventions from another line-department towards the same goal. Towards that end, Government of India has already started to emphasize 'convergence' across departments in nutrition mission. However, an all-encompassing safety-net around the beneficiaries or in other words a 'convergence' at implementation level has not been the experience (Menon *et al.*, 2019). Therefore, it is imperative that granular base-line data should be collected with comparable methods and materials on indicators of gendered predictors of nutrition. This aids in the realistic evaluation of the needs and leverage a convergence mode to incorporate context specific gendered challenges.

References

Ackerson L.K., Subramanian S.V. 2008. Domestic Violence and Chronic Malnutrition among Women and Children in India. Am J Epidemiol;167:1188–96. Available from: https://academic.oup.com/aje/article/167/10/1188/232214

Agarwal B. 2012. Food Security, Productivity, and Gender Inequality. Report No.: 314.

Ahmad J., Khan N., Mozumdar A. 2019. Spousal Violence Against Women in India: A Social–Ecological Analysis Using Data from the National Family Health Survey 2015 to 2016. Journal of Interpersonal Violence. October. doi:10.1177/0886260519881530

Amugsi D.A., Lartey A., Kimani E., Mberu B.U. 2016 . Women's participation in household decision-making and higher dietary diversity: findings from nationally representative data from Ghana. J Health PopulNutr. May 31 [cited 2020 Nov 27];35(1):16. Available from: http://jhpn.biomedcentral.com/articles/10.1186/s41043-016-0053-1

Arokiasamy, Perianayagam, 2004. "Regional Patterns of Sex Bias and Excess Female Child Mortality in India." Population (English Edition, 2002-) 59 (6): 833–63. https://doi.org/10.2307/3654897.

Bamji M.S., Thimayamma B.V.S. 2000. Impact of women's work on maternal and child nutrition. Ecol Food Nutr . Mar 1; 39(1): 13–31. Available from: https://doi.org/10.1080/03670244.2000.9991602

Banerjee A., Duflo E., Glennerster R., Kinnan C. 2014. The miracle of microfinance? Evidence from a randomized evaluation. Available from: http://www.centre-for-microfinance.org/publications/data/.

Bhattacharya R. 2020. Social identity as a driver of adult chronic energy deficiency: analysis of rural Indian households. J Public Health Policy.

Bhattacharya, Ruchira, Madhuri N.V., Maitra Sudeshna, Md. Sajid. 2020. 'Gender Differences in Nutrition in relation to Women's Access to Food Production in Rural India', Report to National Institute of Rural Development and Panchayati Raj, Hyderabad 500030. [Unpublished Report]

Bisgrove E.Z., Popkin B.M. 1996. Does women's work improve their nutrition: Evidence from the urban Philippines. Soc Sci Med. Nov 1; 43(10): 1475–88.

Brett J.A. 2006. "We Sacrifice and Eat Less": The Structural Complexities of Microfinance Participation. Vol. 65, Organization.

Conde-Agudelo A., Rosas-Bermúdez A., Kafury-Goeta A.C. 2006 .Birth Spacing and Risk of Adverse Perinatal OutcomesA Meta-analysis. JAMA . Apr 19;295(15):1809–23. Available from: https://doi.org/10.1001/jama.295.15.1809

Dehingia N., Singh A., Raj A., McDougal L. 2019. More than credit: Exploring associations between microcredit programs and maternal and reproductive health service utilization in India. SSM - Popul Health. Dec 1; 9. Available from: /pmc/articles/PMC6706634/?report=abstract

Deshpande, Ashwini, 2020. In locked down India, women fight coronavirus and domestic violence. Quartz India. April 16, 2020.

Doocy S., Teferra S., Norell D., Burnham G. 2005. Credit program outcomes: Coping capacity and nutritional status in the food insecure context of Ethiopia. Soc Sci Med. May 1; 60(10): 2371–82.

Dreze J. 2020. Public services like Anganwadis should not have been shut during lockdown. Scroll. Jul 2 [cited: 2020 Nov 27]; Available from: https://scroll.in/article/966207/public-services-like-anganwadis-should-not-have-been-shut-during-lockdown-jean-dreze

Engle P.L., Menon P., Haddad L. 1999. Care and Nutrition: Concepts and Measurement. World Dev. 27(8):1309–37. Available from: http://www.sciencedirect.com/science/article/pii/S0305750X99000595

Engle P.L., Pederson M. 1989. Maternal work for earnings and children's nutritional status in urban Guatemala. Ecology of food and nutrition, 22: 211-223.

Espinosa J. 2013. Moving towards gender-sensitive evaluation? Practices and challenges in international-development evaluation. Evaluation. Apr 12;19(2):171–82. Available from: http://journals.sagepub.com/doi/10.1177/1356389013485195

Fadare O., Amare M., Mavrotas G., Akerele D., Ogunniyi A. 2019.Mother's nutrition-related knowledge and child nutrition outcomes: Empirical evidence from Nigeria. PLoS One. Jan 1;14(2).

Fall CHD, Sachdev H.S., Osmond C., Restrepo-Mendez M.C., Victora C., Martorell R., et al., 2015. Association between maternal age at childbirth and child and adult outcomes in the offspring: a prospective study in five low-income and middle-income countries

(COHORTS collaboration). Lancet Glob Health. Jul 1;3(7):e366–77. Available from: https://doi.org/10.1016/S2214-109X(15)00038-8

FAO, IFAD, UNICEF, WFP and WHO, 2020. The State of Food Security and Nutrition in the World 2020. Transforming food systems for affordable healthy diets. Rome, FAO.

FAO, 2011: The State of Food and Agriculture. Women in Agriculture: Closing the gender gap for development. Rome; 2011. Available from: http://www.fao.org/catalog/inter-e.htm

Feruglio F., Nisbett N. 2018. The challenges of institutionalizing community-level social accountability mechanisms for health and nutrition: a qualitative study in Odisha, India; Available from: https://doi.org/10.1186/s12913-018-3600-1

GenoniAfsana M.E., Khan I., Krishnan N. 2020.Losing Livelihoods. The Labor Market Impacts of COVID-19 in Bangladesh.

Girma W., Genebo T. 2002. Determinants of the nutritional status of mothers and children in Ethiopia 2002. Available from: https://dhsprogram.com/publications/publication-fa39-further-analysis.cfm

Hamad R., Fernald LCH. 2015. Microcredit participation and women's health: Results from a cross-sectional study in Peru. Int J Equity Health. Aug 5 ;14(1):62. Available from: https://equityhealthj.biomedcentral.com/articles/10.1186/s12939-015-0194-7.

Hunt J., Brouwers R. Review on Gender and Evaluation, 2003. Available from: https://www.oecd.org/dac/evaluation/dcdndep/31736413.pdf

IFPRI, 2019. Global food policy report. 2019. Washington, DC: International Food Policy Research Institute (IFPRI). https://doi.org/10.2499/9780896293502

IIPS, 2017. National Family Health Survey (NFHS-4) 2015-16 India A. 2017. Available from: http://www.rchiips.org/nfhs

Iyengar K.P., Jain V.K. 2020. COVID-19 and the plight of migrants in India. Postgrad Med J . 2020 Aug 12, Available from: http://orcid.org/0000-0002

Jejeebhoy S.J. 1998. Wife-Beating in Rural India A Husband s Right-Evidence from Survey Data/ : | Economic and Political Weekly. Econ Politcal Weekly. 33(15).

Kabeer N. 2005. Is Microfinance a 'Magic Bullet' for Women's Empowerment? Analysis of Findings from South Asia. Econ Polit Weekly. Oct; 4709–18.

Kabeer N. 1994. Reversed Realities: Gender Hierarchies in Development Thought. London: Verso.

Kader M., Perera N.K. 2014. Socio-economic and nutritional determinants of low birth weight in India. N Am J Med Science. Jul;6(7):302–8. Available from: https://www.ncbi.nlm.nih.gov/pubmed/25077077

Kumar A. 2006. Self-Help Groups, Women's Health and Empowerment: Global Thinking and Contextual Issues. Jharkhand J Dev ManagStud. 2006 Sep ;4(3):2061–79. Available from: https://papers.ssrn.com/abstract=1330911

Kumar N., Scott S., Menon P., Kannan S., Cunningham K., Tyagi P., et al., 2018. Pathways from women's group-based programs to nutrition change in South Asia: A conceptual framework and literature review. Global Food Security. Elsevier B.V. 17: 172–85.

Laborde D., Martin W., Swinnen J., Vos R. 2020. COVID-19 risks to global food security. Science (80) .;500:6503. Available from: https://ebrary.ifpri.org/

Lamontagne J.F., Engle P.L., Zeitlin M.F. 1998. Maternal employment, child-care, and nutritional status of 12-18-month-old children in Managua, Nicaragua. Soc Sci Med. Feb 1; 46(3): 403–14.

Lancet, 2020. Humanitarian crises in a global pandemic. [cited 2020 Nov 21].

Malapit H.J.L., Kadiyala S., Quisumbing A.R., Cunningham K., Tyagi P. 2015. Women's Empowerment Mitigates the Negative Effects of Low Production Diversity on Maternal and Child Nutrition in Nepal. J Dev Studies. Aug 3;51(8):1097–123. Available from: http://www.tandfonline.com/doi/full/10.1080/00220388.2015.1018904

Arokiasamy, Perianayagam, 2004. "Regional Patterns of Sex Bias and Excess Female Child Mortality in India." Population (English Edition, 2002-) 59 (6): 833–63. https://doi.org/10.2307/3654897.

Malapit, Hazel Jean L. and Agnes R. Quisumbing, 2015. "What Dimensions of Women's Empowerment in Agriculture Matter for Nutrition in Ghana?" Food Policy 52 (April): 54–63. https://doi.org/10.1016/j.foodpol.2015.02.003.

March C., Smyth I., Mukhopadhyay M. 1999. A Guide to Gender-Analysis Frameworks. Available from: www.oxfam.org.uk/publications

Menon P., Avula R., Pandey S., Scott S., Kumar A. 2019. Rethinking Effective Nutrition Convergence: An Analysis of Intervention Co-coverage Data | Economic and Political Weekly. Econ Polit Weekly. Jun 15 Moser C. 1993. Gender planning and development: theory, practice and training. Routledge;

Moseson H., Hamad R., Fernald L. 2014. Microcredit participation and child health: results from a cross-sectional study in Peru.J Epidemiol Community Health; 68: 1175–1181.

Nankinga O., Kwagala B., Walakira E.J. 2019. Maternal employment and child nutritional status in Uganda. PLoS One. Dec 1;14(12).

Nithya D.J., Bhavani R.V. 2018. Dietary diversity and its relationship with nutritional status among adolescents and adults in rural India. J Biosoc Science. May 1; 50(3): 397–413. Available from: https://doi.org/10.1017/S0021932017000463

Ojha S., Szatkowski L., Sinha R., Yaron G., Fogarty A., Allen S.J. et al., 2020. Rojiroti microfinance and child nutrition: a cluster randomised trial What is already known on this topic? Arch Dis Child. 105: 229–35.

Quisumbing Agnes, Kumar Neha, Meinzen-Dick Ruth, and Ringler Claudia, 2020. Why gender matters in COVID-19 responses — now and in the future. In: Swinnen Johan & McDermott John (Ed.) COVID-19 GLOBAL FOOD SECURITY. IFPRI. Washington, USA.

Ramachandran, V.K., Swaminathan, M., Rawal, V. 2010. Socio-economic surveys of three villages in Andhra Pradesh: A study of Agrarian relations. New Delhi: Tulika Books.

Ramalingaswami V., Jonsson U., Rohde J. The Asian enigma. 1996. In: UNICEF, editor. The progress of nations. New York: UNICEF New York; p. 11–7.

Rao N., and Raju S. 2020. Gendered Time, Seasonality, and Nutrition: Insights from Two Indian Districts. Fem Econ Apr 2; 26(2): 95–125. Available from: https://www.tandfonline.com/doi/full/10.1080/13545701.2019.1632470

Rao N. 2008. "Good Women Do Not Inherit Land": Politics of Land and Gender in India. Orient Black Swan. Delhi.

Rao N. 2017. Gender Differences in Adolescent Nutrition: Evidence from two Indian districts.

Sahoo, Subarnalataand Panda, Basumati, 2006. A Study of Nutritional Status of Pregnant Women of Some Villages in Balasore District, Orissa, Journal of Human Ecology, 20:3, 227-232, DOI: 10.1080/09709274.2006.11905932

Sinha D., Tiwari D.K., Bhattacharya R., Kattumuri R. 2016. Public services, social relations, politics, and gender: tales from a north Indian village. In: Himanshu, Jha P, Jerry Rodgers, editors. The Changing Village in India: Insights from Longitudinal Research. New Delhi: Oxford University Press. p. 401–436.

Smith L., Haddad L. 2014. Reducing Child Undernutrition: Past Drivers and Priorities for the Post-MDG Era. IDS Work Paper. Apr 1;2014(441):1–47. Available from: http://doi.wiley.com/10.1111/j.2040-0209.2014.00441.x

Smith L.C., Haddad L. 2015. Reducing Child Undernutrition: Past Drivers and Priorities for the Post-MDG Era. World Dev.68:180–204. Available from: http://www. sciencedirect. com/science/article/pii/S0305750X14003726

Sraboni E., Malapit H.J., Quisumbing A.R., Ahmed A.U. 2014. Women's empowerment in agriculture: What role for food security in Bangladesh? World Dev. 61: 11–52.

Summerton S.A. 2020. Implications of the COVID-19 Pandemic for Food Security and Social Protection in India. Indian J Hum Dev. Aug 10; 14(2): 333–9. Available from: http://journals.sagepub.com/doi/10.1177/0973703020944585

Swaminathan, Madhura, 2020. Reset rural job policies, recognise women's work. The Hindu. 4th July.

Tsiboe, F., Zereyesus, Y.A., Popp, J.S. et al., 2018. The Effect of Women's Empowerment in Agriculture on Household Nutrition and Food Poverty in Northern Ghana. Soc Indic Res 138, 89–108. https://doi.org/10.1007/s11205-017-1659-4.

UNFPA, 2020. COVID-19: A Gender Lens. Technical Brief. March 2020.

Villar J., Merialdi M., Gulmezoglu A.M., Abalos E., Carroli G., et al., Kulier R., et al., 2003. Nutritional Interventions during Pregnancy for the Prevention or Treatment of Maternal Morbidity and Preterm Delivery: An Overview of Randomized Controlled Trials. J Nutr. May 1;133(5):1606S-1625S. Available from: https://doi.org/10.1093/jn/133.5.1606S.

4

Design and Implementation of Gender Sensitive Extension Programmes

Nimisha Mittal* and Rasheed Sulaiman V.

1. Context

Women play a critical role in agriculture, and agriculture continues to remain the major livelihood strategy for women as well as millions of small, marginal and poor farming households in India. Despite this, farm women lack access to extension services as extension programmes rarely identify women as an integral client. Farm women and men have differential needs for knowledge, information, technology and skills as often they are involved in different activities spread over time and cropping seasons. Besides these, men and women farmers have differential access to assets, information, markets, credits, and other services that are necessary for using new knowledge, technologies and skills. Ignoring women while delivering services and technologies creates a gap as many of the agricultural operations are performed by women only.

It is estimated that if women had the same access to productive resources as men, they could increase yields on their farms by 20-30 per cent (FAO, 2011). Yields on plots managed by women are lower than those managed by men, as they do not have the same access to inputs. If they did, their yields would go up, they would produce more, and overall agricultural production would increase. It isn't merely about enhancing productivity but also about being fair and equitable. Women are a significant part of the agricultural workforce and they deserve to be consulted prior to formulation of agricultural development plans and policies. SDG 5 on Gender Equality "Achieve gender equality and empower all women and girls" is closely aligned to this aspect. Moreover, enhanced access to services can help women grow more food, have more say in the family, and could potentially lead to more income and food in women's hands-leading to better food security and nutrition for the entire family as they are the major caregivers in the family.

---

* *Author contact: nimisha61@gmail.com*

2. Gender and Extension in India

Efforts have been initiated in the recent past by both governments and non-governmental organizations to incorporate gender issues to ensure women's full and equitable participation in agricultural development programmes.Women's role in agriculture started to receive explicit attention in Indian policy circles during the Seventh Five Year Plan (1985-90). Since then, several programmes for women in agriculture were implemented in India that have included:

- Special donor assisted programmes on women in agriculture in select states
- The central sector women in agriculture programmes
- Women component plan
- Initiatives for gender mainstreaming

The most common approach used by the state Departments of Agriculture in India, in reaching rural women is by creating a group through the social organization of women and then implementing agricultural development initiatives through that group. Several government-anchored initiatives have adopted this approach.The capacities and skills of the groups are developed through several rounds of trainings and exposure. Through Agricultural Technology Management Agency (ATMA) scheme, several extension programmes are organized and these include training, demonstrations, farm schools, and exposure visits. According to ATMA guidelines, 30 per cent of the beneficiaries in the programmes have to be women (GoI, 2018). Krishi Vigyan Kendras (KVKs), which are funded by the Indian Council of Agricultural Research (ICAR), organize vocational training for youth, farmers, and rural women. Women do participate in many of these trainings but there is a wide variation across KVKs in this regard.

There are several provisions and package of assistance which women farmers can claim under various on-going Missions/Submissions/Schemes of DAC & FW, Ministry of Agriculture & Farmers Welfare. These include support to buy agricultural implements, form food security groups, establishing seed gardens, form women groups, promotion of farmer producer organizations, capacity building, skill development and other support services. However, several studies and reports indicate that the style and type of many of these projects and programmes implemented is micro-level and input driven and follows a general pattern: formation of SHGs; initiating thrift and credit and linking to credit from banks; organising training programmes mainly on production and post-harvest technologies; demonstrations; exposure visits; and distribution of implements or subsidies.

Some of the recommendations of the XII Five Year Plan Report (2012) of the Working Group on Women's Agency and Empowerment[3] were:

- Extension policy needs to **explicitly target women** in agriculture.
- Promote a **group approach to extension services for women farmers**. Women farmers' should be represented as major stakeholders in all decision making bodies of public and private extension services.
- All programmes in the field should be planned and implemented through farmers' groups which have atleast 33% women in both general body and executive committees.
- Agricultural policy and programmes should adopt an Integrated Farming Systems Approach with special focus to reach out to small and marginal women farmers.
- All programmes providing facilities such as distribution of agricultural inputs, subsidies on inputs, training and extension, should have 50% reservation for women beneficiaries, irrespective of whether they own land or not. -All productive assets provided under these schemes should be given in the name of woman.
- Steps need to be taken to **involve women in on-farm participatory research for agricultural technology and development of women friendly implements/ tools.**
- The **gender friendly tools** should be popularized through training and demonstrations in KVKs, Agri-clinics, gram sabhas, etc.
- Develop a **database of women friendly technologies/equipments** available for all stages in the agriculture value chain for bulk purchase with list of manufacturers.
- Institutional and funding support for the formation of women producers' associations and existing women's federations/cooperatives to process, store, transport and market farm produce, milk, fish, crops etc. should be provided.

However, most of these recommendations still need to be operationalized and capacities of the extension staff need to be built accordingly.Though enhancing capacities to adopt better production and management practices through training and demonstration are necessary, this is not sufficient to address gender inequalities in the face of newer challenges they face.These challenges included

---

[3] https://niti.gov.in/planningcommission.gov.in/docs/aboutus/committee/wrkgrp12/wcd/wgrep_women.pdf

deterioration of natural resources, fragmentation of farm holdings, rapid globalization, climate change, and introduction of new standards for production and marketing.

Women and men farmers require a wide range of knowledge from different sources as well as support in integrating these different bits of knowledge into their production contexts. While pursuing this new and expanded agenda, again one needs to be mindful of the fact that women and men have different needs, opportunities and challenges in these contexts. Women's access to markets is limited due to their limited social networks and engagement with markets and market actors due to social norms which restrict their mobility and social interactions. Their limited access to credit and other productive resources restrict their economic and entrepreneurial opportunities as well.

> **Box 1:** Extension and Advisory Services (EAS) EAS consist of all the different activities that provide the information and services needed and demanded by farmers and other actors in rural settings to assist them in developing their own technical, organisational, and management skills and practices so as to improve their livelihoods and well-being. Global Forum for Rural Advisory Services (GFRAS) recognises the diversity of actors in extension and advisory provision, much broadened support to rural communities (beyond information and knowledge) and embracing new functions such as facilitation, intermediation and brokering by EAS.
> Source: *GFRAS, 2012*

Women have the potential to become change agents in their households and communities-in transforming agricultural and food systems towards climate resilience-but their limited access to relevant knowledge and technologies hold them back. Extension and advisory services (EAS) needs to play a role in easing these barriers for women, working with other actors across various sectors and helping create an enabling institutional environment (Box 1).

Integrating gender within EAS programmes helps in better understanding of gender differences and better planning of interventions. As extension remains a significant source of information for resource-constrained farmers (the majority of whom are women), the inclusion of methodologies and approaches that address gender is critical for extension's success. Inadequate or wrong understanding of gender differences leads to inadequate planning and design of projects and the perpetuation of gender inequalities and diminished returns on investments.

3. Operationalising gender-sensitive extension programmes

Extension programmes are ideally placed to facilitate capacity development of farming communities, through training, strengthening the innovation process, brokering linkages and partnerships, etc., to support the bargaining position of farmers (Sulaiman and Hall, 2004). To be able to do that, extension needs to shift from the way it is structured at present. Operationalizing gender-sensitive extension would require a significant shift in the design and delivery of programmes throughout the programme cycle as elucidated in Table 1.

Table 1. Operationalising gender-sensitive agricultural extension: Key shifts

	From	To
Objective	Focus on Increasing production/productivity	Geared towards improved income, decent employment opportunities, and better well-being for both men and women farmers Recognize women's identity as farmers
Activities	Training on technologies	Training on value chains and enterprise management
	(mostly aligned towards transfer of technologies produced by research systems)	Helping women gain access to, and control over, resources like credit, markets, information, assets, etc.
	Forming SHGs	Forming commercially-oriented producer groups/producer companies Handholding and mentoring groups
	Distribution of inputs	Grooming women and youth agri-preneurs and producer companies that can engage in procurement and marketing of inputs and services provision
Selection of interventions	Selection of interventions based on participatory planning	Demand-led and based on analysis of farmer's data (social and gender analysis), matched with opportunities and availability of complementary support and services
	Centrally designed ideas	Farmer aspirations carefully analysed with local and external knowledge and support Be cognizant that social constructs might make some of the gender-responsive goals/ideas mismatch with farmer aspirations
Approaches	Fixed/uniform	Evolving/diverse
Working with	Women	Working in partnership with all actors who could support rural men and women, Engaging with women to influence social norms and gender relations
Monitoring & evaluation	Input and output targets	Behavioural and livelihood changes in clients & related organisations, including gender roles & norms
	Subjective evaluations	Objective evaluations using rigorous benchmark data Assessment of outcomes & impact, not merely outputs Capturing good practices for scaling up interventions and feed into planning cycle
Targeting the poor & women	Inclusion by accident	Inclusion by design

Treat all women (or men) as a homogeneous group experiencing similar challenges and opportunities and hence provide blanket solutions Understand intersectionality and develop effective strategies taking into account diversity within the groups.

It is important to incorporate gender perspectives into planning, implementation, monitoring and evaluation of extension programmes. EAS in most cases reflect traditional gender norms in the way they are structured and operate. Programmes often fail because they do not take into account the different roles, needs, and priorities of men and women. If we don't understand the underlying causes of the gender gap and the social dynamics that contribute to it when designing programmes and interventions, not only are they likely to fail, but they may actually fuel greater inequality and discrimination.

4. Capacities needed to implement gender-sensitive extension programmes

Gender analysis helps extension professionals understand the situation, context, opportunities and challenges women farmers face, and therefore, their needs and priorities. This can help them in designing programmes that not only respond to the needs of men and women farmers, but also contribute towards reducing the gender gap in agriculture and help lay a foundation for women's empowerment. However, staff (both male and female) having strong gender bias is a major barrier. Patriarchy is deep rooted in society. This has implications on the mindset and attitudes of staff, development organizations, media, government officials and lawmakers, as they too are affected by the norms and values prevalent in the broader social environment. Disrupting the status quo or approaching a very sensitive issue in a bold or unconventional manner are often not encouraged. EAS staff might sometimes be quite averse to triggering conflict arising from their interventions.

Changing people's behaviours and beliefs takes a long time, especially when it comes to addressing gender issues and gaps. In addition, people might have their own beliefs and biases about gender, especially when they come from different backgrounds and communities even within the same country. Often people are unaware that they carry this bias baggage within themselves (unconscious bias). If the staff are not sensitive, then they need to be sensitized.

Relevant competencies are needed to design and implement a good gender analysis and to use the findings to inform programmes and their implementation. Competencies (see box 2) are needed at both the individual (field, middle and senior level personnel) and organizational levels. While those at the field-level need competencies to facilitate a process of change at the community level, the middle and senior level staff need to support the field level functionaries in their endeavour through gender-sensitive planning, implementing and monitoring of activities. However, gender competence adds an additional dimension to the competency at all levels.

Gender competence is the ability of people to recognise gender perspectives in their work and policy fields and concentrate on them towards the goal of gender equality. Gender competence is a prerequisite for successful gender integration and vice versa, i.e., new gender competence is produced through the implementation of gender-responsive programmes and intervention. Gender competence consists of the elements of

- Intention (to work towards the goal of gender equality),
- Knowledge (on gender in all its complexity), and
- Ability (to apply gender perspectives in the work context). (http://www.genderkompetenz.info/eng/gender-competence-2003-2010/ Gender% 20 Competence.html)

> **Box 2:** A competency is a skill, attitude, or behaviour that enables an individual to do their job more effectively in order to contribute to the mission of the organization to which they belong. A competency is, therefore, a characteristic embodied by individuals within an organization that is clearly linked to the goals the organization seeks to achieve (Davis et al. 2017). Competency also refers to sufficiency of knowledge and skills that enable a person to act in a wide variety of situations. It is the ability to do something efficiently and effectively (i.e., successfully).

Gender-focused competencies (Table 1) for EAS describe both the characteristics of gender-sensitive extension and advisory systems as well as the abilities of a gender-sensitive extensionist who, in his or her work, is aware that differences in men's and women's needs, abilities, and endowments cannot be taken for granted, but require analysis and implementation to ensure that extension and advisory services will reach, benefit, and empower farmers (Davis *et al.*, 2017).

Table 1. Gender-focused competencies for EAS

Competency	Description
Gender-sensitive language	Speaking and writing in a way that does not discriminate against a particular sex, social gender or gender identity, and does not perpetuate gender stereotypes; Using gender-inclusive language is a powerful way to promote gender equality and eradicate gender bias. https://www.un.org/en/gender-inclusive-language/
Gender analysis	Understands basic gender analysis, and is able to apply principles for integrating gender analysis
Gender-responsive Agricultural Technologies and Practices	Understands gendered differences in production, access and control of resources participation in groups, etc., and apply this to the dissemination process and identify opportunities for improving the same; Channels information about women's needs to other actors involved in technology design, use, and dissemination; Promotes agricultural technologies that benefit men and women and is able to identify opportunities for improving

	how women and men can benefit from agricultural technologies.
Inclusive, market-oriented EAS	Understands the key issues related to gender, Extension and value chains development and operations; Is able to identify opportunities for men and women farmers and entrepreneurs to participate within different agricultural value chains; Understands men's and women's specific challenges to participating in & benefitting from value chain development based on their different roles and responsibilities in value chains; Identify and promote opportunities for men and women farmers and entrepreneurs to improve participation in agricultural value chain.
Women's empowerment and Gender Transformative Approaches (GTA)	Identify and challenge underlying gender norms that inhibit women's equitable participation and ability to benefit from agricultural activities.

Source: Adapted from Davis *et al.*, 2017

Field level extension workers often serves as the main interface between community members and influencing their farming decisions, use of technologies (type of seed used, way of seed bed preparation, level of mechanization in sowing, transplanting, weeding, harvesting, etc.). To ensure that these services are provided in a gender-sensitive manner, extension workers must recognize the gender norms, bias, and power relationships in the areas where they work. Not doing so can influence the quality and usability of the services or advisory provided. For example, does an extension worker provide the technique/training on storage of seed to men farmers without bothering to find out who in the family is actually storing seeds? Does another extension worker miss opportunities to provide the female farmers with farming-related information because he doesn't view them as farming clients? An extension worker might be a good communicator, a good trainer and a good subject expert, but if she/he doesn't apply the gender lens to his/her programme planning and implementation, s/he is ineffective and their best efforts will have only a very limited impact.

When extension and advisory services (EAS) providers are competent at addressing gender issues in agriculture, they can contribute to empowerment, achieve gender equality and enhance agricultural productivity and household income. Ideally, knowledge on gender and addressing gender inequality should have been part of the technical educational in agriculture and rural development as well as in-service training curricula of EAS personnel. But beyond personal motivation, commitment and conviction on gender by field personnel or top leadership, there should be a gender-sensitive organizational culture within which all these efforts should be anchored for a sustained impact on gender by EAS.

5. Assessing gender-sensitivity in an organization

Addressing gender is not merely about serving women farmers better or reducing inequality in rural areas, but it is also about fostering a gender-sensitive organisational culture and having adequate policies and strategies in place to do so.In order to gain a better understanding of how extension programmes can be improved to enhance their accessibility and relevance to rural women, they need to be assessed from time to time. The results of the assessment can be used to initiate efforts to build a gender-sensitive organizational culture with staff at all levels better able to keep their biases at bay while designing and delivering programmes.

6. The Gender and Rural Advisory Services Assessment Tool (GRAST)

GRAST is an easy-to-use tool and methodology that helps organizations carry out in-depth analysis of the gender sensitivity of their extension and advisory services (EAS) at policy, organizational and individual levels. It basically delves into the following seven questions (adapted from Petrics *et al.,* 2018):

1. Are rural women included as legitimate clients in EAS?
2. How are the time and mobility constraints of rural women addressed?
3. How are the literacy and educational constraints of rural women addressed?
4. Does the programme facilitate rural women's ability to represent their interests and voice their demands?
5. Are extension and advisory services designed and delivered in a way that allows rural women to effectively participate, benefit, and get empowered?
6. Does the organizational culture enable women to become and effectively function as EAS agents and managers?
7. Are there institutional mechanisms in place to ensure the effective implementation of a gender-sensitive EAS and hold staff accountable?

GRAST starts from the enabling environment level – where various policy support is scrutinized. Then it branches into the organizational dimension and moves to the individual level. For the purpose of this chapter, we have restricted ourselves to the organizational and individual dimension of the tool.

6.1. Organizational dimension

Organizational dimension refers to systems, procedures and institutional frame works that allowan organization to deliver gender-sensitive services to its clients. The organizational dimension has a major impact on how individual staff members developtheir competenciesand how they are abletousethem within the organization (FAO, 2010). Having gender-sensitive processes, practices, and policies in place is crucial forsetting expectations and shaping organizational culture.The organizational dimension can be assessed on broadly two parameters, that is, organizational culture and provision of services as described herewith.

1. The culture of an organizationis likely to influence employees' perception of gender roles and the importance of gender equality in their work; studies confirm that gender-blind organizations tend to deliver gender-biased services (Buchy and Basaznew, 2013). Therefore, at this level, the tool assesses the organization's stated commitments to deliver gender-sensitive extension and advisory services, its policies related to these commitments, and its implementation plans for putting such services into action. It also examines the degree to which the organizational culture supports gender-sensitive service provision (Petrics *et al.,* 2018). The EAS organizational culture refers to:
   - Gender parity in staffing is a stated goal, and there are policies in place to encourage there cruitment of women as EAS advisors and to retain women who are hired;
   - Women are represented at the management level of the organization;
   - Both women and men work as EAS advisors in all capacities (i.e., women are not only 'home economics' advisors);
   - The organization has a gender equality policy/strategy;
   - The organization has anti-discrimination and anti-harassment policies;
   - The organization allocates part of its budget to specific efforts to reach women farmers and to provide gender training for staff;
   - The organization provides gender training to staff at different levels (managers, field staff, sub-contracted staff from other organizations or lead farmers who provide farmer-to-farmer extension);
   - Women portrayed in training materials are shown undertaking productive activities on an equal footingwith men, rather than depicted only in home making or caregiving roles.

6.1.1. Provision of services

The EAS organization has a stated mission to provide advisory services to both women and men, and women are specifically included as clients. The organization deliberately provides advisory services that are inclusive and does not limit participation based on landholdings, position in the household,marital status production practices, etc. The EAS client selection process is written, transparent and does not directly exclude women.

Organizational policy makes specific mention of efforts to reach women, including:

- by considering women's time and mobility limitations (schedules and workloads, etc.);
- by considering women's literacy and educational levels;
- by considering women's ability to represent their interests and voice demands for EAS; and
- by prioritizing methods of delivery, topics and technologies that would not only reach and benefit women but empower them as well.

This tool was tested in India by CRISP with PRADAN and the results are shared in Box 3.

Box 3: Key insights from testing GRAST with PRADAN in Madhya Pradesh, India

For PRADAN, its stated mission is "to enable the most marginalized people, especially rural women, to earn a decent living and take charge of their own lives. We focus primarily on women because we believe that even if they are considered to be the most disadvantaged in society, they are capable of driving the change they need. Our aim is to stimulate and enhance the sense of agency of poor communities, especially women's collectives, who being at the bottom of the cross sections of class, caste and gender, are the most vulnerable."

(PRADAN 2017)

PRADAN selects and trains people who are motivated to work for women's empowerment and aspire to create an atmosphere of mutual support and learning around fostering social change and gender equality. PRADAN professionals are catalysts of change, who are groomed by the organization and are committed to its mission and values at all levels. One of the major institutional mechanisms that supports this commitment is PRADAN's Development Apprenticeship Programme, a year-long initiation for new staff. The apprenticeship programme seeks to match the individual's aspirations and motivations to the organisational mission and vision through regular self-reflection (Petrics *et al.*, 2018).

6.2. Individual dimension

Individual level analysis includes the perspectives of both programme staff and women and men clients as discussed here:

1. Perspectives of programme staff: At the individual level, GRAST explores the skills, behaviours, attitudes, motivation and values of programme staff members. To be able to tailor advisory services to gender specific demands, extension agents need to have the sensitivity and capacity to understand these demands and respond to them with adequate content and appropriate methods of delivery. At this level, the tool helps to assess EAS advisors' awareness and understanding of the different needs and priorities of rural women and men, as well as the advisors' capacities to respond to them. The tool also assesses to what extent EAS managers are implementing

gender-sensitive human resource policies and organizational culture, and explores their awareness of why these policies and culture are important. Interviews with EAS advisors provide a means to gain additional insight on the challenges and successes that staff face in working with rural women and men.

2. Perspectives of women and men clients: In the second part of the individual-level section, the tool considers clients' perspectives. Analysis at the level of EAS clients helps to validate the responses of the providers, as well as to identify what the organization does that works for rural women and what could be improved. This helps EAS organizations understand how policies and programmes are implemented on the ground, what areas need improvement, and how users perceive the impact of the programme on their livelihoods (Petrics *et al.,* 2018).

It is also important to account for clients' perspectives on service provision when assessing the manner in which the organization works. These are illustrated in Table 2.

Table 2. Perspectives of programme staff

EAS organization	Service provision
How managers are working to promote a gender-sensitive organizational culture;	The extent to which providers are implementing organizational policies on gender-sensitive service provision;
How staff experience gender sensitivity in the organizational culture;	The challenges and constraints of working with rural men and women; and
Staff insights on barriers to women's ability to work as EAS advisors and advance in the organization; and	Success stories.
The training staff receive on gender issues and women's empowerment.	

7. Gender budgeting

Budgets are universally accepted as powerful tools in achieving development objectives and act as an indicator of commitment to the stated policy of the organizations/governments, etc. Women stand apart as one segment of the population that warrants special attention due to their vulnerability and lack of access to state resources. Thus, gender-sensitive budget policies can contribute to achieving the objectives of gender equality, human development, and economic efficiency. The purpose of a gender budgeting exercise is to assess quantum and adequacy of allocation of resources for women and establish the extent to which gender commitments are translated into budgetary commitments.

Gender-Responsive Budgeting (GRB) or gender budgeting (Box 4) has emerged as an important tool for integrating gender issues as part of the ongoing struggle to make budgets and policies more gender-sensitive in several countries across the world in the last two decades. Beginning from the mid-1980s to date, over 90 countries have thus far endorsed GRB as a valuable tool for engendering budgets and policies all over the world (UN Women, 2017). Not only national governments, but other non-governmental organisations also have started earmarking budget to match activities that reflect their commitment towards gender-sensitive extension programmes.

Box 4: Gender Budgeting

Gender Budgeting is not an accounting exercise but an ongoing process of keeping a gender perspective in policy/ programme formulation, its implementation and review. It entails dissection of the budgets to establish its gender differential impacts and to ensure that gender commitments are translated into budgetary commitments (WCD, 2020). The Council of Europe defines gender budgeting as a 'gender-based assessment of budgets incorporating a gender perspective at all levels of the budgetary process and restructuring revenues and expenditures in order to promote gender equality'.

Source: https://eige.europa.eu/gender-mainstreaming/methods-tools/gender-budgeting#2

Gender Budgeting looks at the organizational budget from a gender perspective in order to assess how it addresses the needs of women. For example, some organizations create budgets for building toilets for female staff, provision for safer transport facility (hiring four-wheelers) for female staff if they need to travel to the field, budgetary provision for two women farmers to attend a training from one village if the training is away from the village, etc.Gender Budgeting does not seek to create a separate budget but seeks affirmative action to address the specific needs of women. For example, earmarking some budget within an organization to have a childcare facility when there are more staff members with young children in need of such a facility. However, this is only perceived to be beneficial to women staff. But this provision is beneficial to all staff irrespective of their gender.This exercise facilitates enhanced accountability, transparency and participation of women in the community. The macro policies of the government can have a significant impact on gender gaps in various macro indicators related to health, education, income, etc.

If gender budgeting has to be used as an effective tool for integrating gender in agriculture, capacity building of staff-at all levels-on gender budgeting is needed. This would entail, collection and analysis of gender disaggregated data, identifying differential priorities of women and men involved in agriculture across different castes and land holding, and then using the above for planning, budgeting and monitoring of schemes.

8. Conclusion

Gender-sensitive extension programmes can be formulated and implemented only if there is committed staff, who are motivated, has a supportive organizational culture, and resources to uphold the conviction that such programmes must be implemented. While several interventions have been made to address this 'gender' bias in extension delivery, there continues to be a shortfall between the kind of support that is provided and the needs and demands of rural women. This gap between supply and demand needs to be addressed in order to improve the lives and livelihoods of women in the rural farming sector. Addressing gender is not merely about serving women farmers better or reducing inequality in rural areas, but also about having gender-sensitive organisational policies and strategies in place. These capacities needs to be assessed first to make sure that EAS is gender-responsive. Tools, such as Gender and Rural Advisory Services Assessment Tool (GRAST) and Gender Budgeting, are essential pre-requisites in this context.

References

Buchy M. and Basaznew F. 2013. Gender-blind organizations deliver gender-biased services: The case of Awasa. Bureau of Agriculture in Southern Ethiopia. Gender Technology and Development 9(2): 235–251.

Davis Robb, Kuyper Edye, Bohn Andrea, Manfre Cristina, Russo Sandra and Rubin Deborah. 2017. Competency framework for integrating gender and nutrition within Agricultural Extension Services. INGENAES. (Available at https://ingenaes.illinois.edu/wp-content/uploads/INGENAES-2017_08-Nutrition-and-Gender-in-Extension-Competency-Framework.pdf)

FAO, 2011. Women in agriculture: closing the gender gap for development. The state of food and agriculture. Food and Agriculture Organization of the United Nations. Rome, 2011 (Available at http://www.fao.org/3/i2050e/i2050e.pdf)

GFRAS, 2012. The "New Extensionist": Roles, Strategies, and Capacities to Strengthen Extension and Advisory Services, Global Forum for Rural Advisory Services (GFRAS). (Available at https://www.g-fras.org/en/knowledge/gfras-publications.html?download=126:the-new-extensionist-position-paper)

GoI, 2012. XII Five Year Plan report of the Working Group on Women's Agency and Empowerment. Ministry of Women and Child Development. Government of India. (Available at https://niti.gov.in/planningcommission.gov.in/docs/aboutus/committee/wrkgrp12/wcd/wgrep_women.pdf)

GoI, 2018. Guidelines for support to state extension programmes for extension reforms (ATMA) scheme. 2018 ATMA Guidelines. Directorate of Extension, Department of Agriculture, Cooperation & Farmers Welfare, Ministry of Agriculture & Farmers Welfare, Government of India. Krishi Bhawan, New Delhi -110001. (Available at https://extensionreforms.dacnet.nic.in/PDF/atmaguid23814.pdf)

Petrics, H., Barale, K., Kaaria, S.K. and David, S. 2018. The Gender and Rural Advisory Services Assessment Tool. FAO. 92 pp.(Available at http://www.fao.org/3/CA2693EN/ca2693en.pdf)

PRADAN, 2017. Annual Report 2015-2016. PRADAN. New Delhi. (Available at http://www.pradan.net/wp-content/uploads/2017/01/AR-2017-PDFFull. pdf).

Sulaiman V.R. and Hall A.J. 2004. Towards extension plus: Opportunities and challenges for reform. NCAP Policy Brief No. 17. New Delhi: National Centre for Agricultural Economics and Policy Research. Pp. 4.

UN Women, 2017. Gender responsive budgeting: A focus on agriculture sector. (Available at https://ruralindiaonline.org/library/resource/gender-responsive-budgeting-a-focus-on-agriculture-sector/)

WCD, 2020. Budgeting for gender equity. (Available at https://wcd.nic.in/gender-budgeting).

5

Public Private Partnership for Gender Mainstreaming in Agriculture

*K. Ponnusamy**

1. Background

Agriculture sustains 58 per cent of population in India. Productivity and profitability are two major planks for food security and sustainable livelihood of farmers. About 86 per cent of farmers in India own less than two hectors of land. As a result, their marketable surplus is small which poses a major challenge in ensuring remunerative price. Limitations associated with quality of produce, storage, transportation, credit as well as vagaries of weather and market are other related problems. These limitations further aggravate if the producers happen to be women due to their socio-cultural and economic position in the society. Only a decentralised and income assuring approach alone can bring prosperity to the farmers apart from gender equality. This would ultimately propel the agricultural growth of the country.

The chapter discusses one of such approaches called the public private partnership (PPP). It has gained prominence in the wake of the demand for farm reforms, trade, commerce, climate change and gender equality in recent years.

1.1. Agricultural development in India

The average annual growth rate of agriculture and allied sector in India increased from less than 2 per cent during 1993-94 to 2003-04 to 4.1 per cent during 2004-05 to 2008-09 and 4.9 per cent during 2004-05 to 2013-14. Further, the average annual growth rate of the sector was 3.9 per cent during the period 2014-15 to 2018-19. This increase has been recorded despite the occurrence of several drought and floods in different parts of the country. Even during COVID-19 pandemic, while industry as well as service sectors registered deep decline agriculture and allied sector continued to grow at 3.4 per cent. These indicate that agriculture sector has developed strong resilience against odds and remains one of the strong forts in India's development. However, in order to sustain this growth and ensure the attainment of the set targets of sustainable development

**Author contact: ponnusamyk@hotmail.com*

goals by 2030, the agriculture growth rate should increase from the current 4 per cent to 7 per cent. This warrants structural changes in the existing pattern and system of agricultural production and marketing through innovative models of development.

1.2. Income distribution among farm households

All India Rural Financial Inclusion Survey by NABARD for the year 2016-17 (NABARD, 2018) depicted the average monthly income of agricultural households in India as Rs. 8931. It was found to range between Rs. 23133 in Punjab and Rs. 6668 in Uttar Pradesh. The same survey further revealed that 45 per cent agricultural households do not have any savings. Majority of the agricultural households had to depend on two or more sources of income. The distribution of farm household income is such that only 35 per cent of total income of farmers came from crop farming. Wage labour (34%), livestock rearing (8%) and other sources (23%) comprised the remaining 65 per cent (Figure 1). Though the trend could be explained in terms of the theory of part time farming, the farmers continued to be in low-income trap. Nearly 33.3 per cent of agricultural households are reported to be in poor income group (Haque and Joshi, 2019). The average income of non-agricultural households has been observed to be nearly 5 times higher than that of agricultural households (Haque and Joshi, 2018).

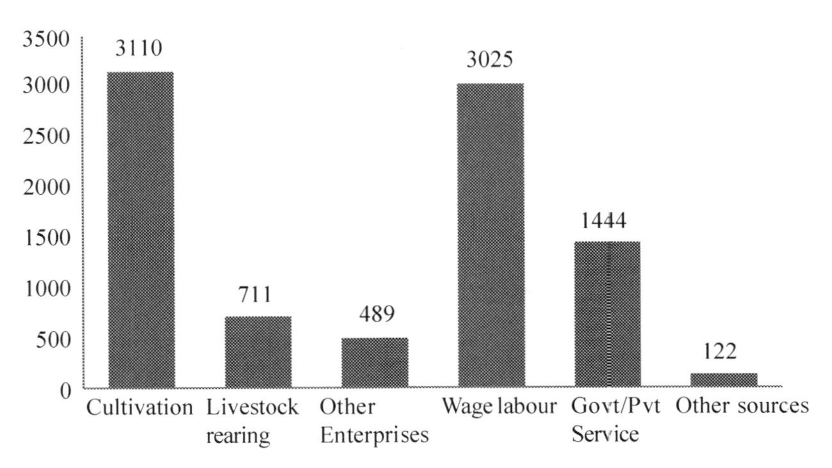

Fig. 1. Distribution of farm household income in India
Source: NABARD, 2018

2. Context of promoting gender mainstreaming in agriculture

Low income of farm households is due to multiple reasons primarily vagaries of climate, market, limited public and private investment in infrastructure as well as lack of institutional and policy innovations to involve key stakeholders especially women. The under-mentioned factors signify the importance of promoting gender mainstreaming in the field of agriculture and allied enterprises for higher and sustained agricultural growth.

- Persistent gender disparities in accessing resources and technologies impair the food and nutrition security. This has been linked to the failure of past farm innovations to proportionally benefit men and women which led to missed opportunities for women to capitalize the technologies and market trends.
- Weak property rights lead to sub-optimal decisions, severely affecting opportunities to enhance productivity
- Women do not take risk to adopt an unfamiliar farm technology
- Unequal income between men and women leading to limited ability of women to afford and invest in agriculture
- Rural women's access to finance happen to be mostly from micro-finance (smaller loans), with limited focus on livelihood from agriculture
- In areas where there are large numbers of small farmers, group work is more practicable than individual farm visits. Farmers often have a high awareness of group requirements and are controlled by powerful groups in the village. They prefer to initiate those values and technical competence similar to their own. All extension work should be based on group discussion, practical demonstration and participation. Extension should promote key stakeholders especially the women who play indomitable role for future food security of the nation.
- Lack of extension service provision exclusively for women. Worldwide women comprise only five per cent and in India 10 per cent of the Village level Workers (VLWs) engaged in extension work.
- Insufficient gender disaggregated data availability on the innovative extension models in the country prohibit planners to prepare gender specific programmes

2.1. Approaches for gender mainstreaming

As a concept, gender mainstreaming got recognition in development parlays after the World Conference on Women (1985), Nairobi. It was introduced as an approach of policy-making which takes into account the interests and concerns of both women and men. The following approaches has been observed to promote gender mainstreaming in the agri-sector.

1. Gender sensitive extension by recruiting more female extension officers

2. Facilitating involvement of women in technology transfer models such as custom hiring centres
3. Public private partnership models, women centric contract farming and farmer producer companies
4. Periodical gender auditing of development programmes
5. Women led entrepreneurships models in agriculture
6. Women centric policies in agriculture

3. Public Private Partnership (PPP) in agriculture

Public Private Partnership (PPP) emerged as an innovative approach to meet the challenges of agriculture development in the early years of the current millennium. PPP has ever since evolved as one the best experimented strategies in the areas of service, welfare, technology dissemination, manufacture and infrastructure development. It enabled the achievement of the actual targets within a framed span of time.

3.1. Concept of public private partnership

A main reason for evolution of PPP in various fields has been the constraints related to infrastructure facilities, human resources and time. In the current context, public authorities focus developmental and welfare schemes with PPP component to achieve the specified goals within the time frame and to modernize public services and infrastructure. This has established its dent in agriculture, health, science and technology, education, infrastructure development and extension. Through this approach, impossibilities are made possible with the contribution of both public and private partners resulting better economic conditions and livelihood of beneficiaries. PPP involves a contract between public and private sector entities wherein the private entity provides a public service or project and assumes substantial financial, technical and operational risk in the project with specified roles and responsibilities (Ponnusamy, 2013). The PPP approach supplements scarce public resources, creates a more competitive environment and helps to improve efficiencies and reduce costs. The rationale for public sector involvement differs between different kinds of services and influences the type of involvement required (Grout and Stevens, 2003). Risk allocation plays a vital role in PPP management. There is a need to delineate a framework for operationalizing suitable public-private partnerships (Ponnusamy *et al.*, 2014) based on past experiences and inferences derived. Pre-planned proposals with time frame, budget, methods and materials would be a prerequisite for the expected outcome of PPP.

3.2. Historical evolution of PPP in agriculture

PPP approach is adopted in various facets of agriculture such as research and development, quality enhancement, farm production, extension and marketing. Functional and operational factors of the PPP linkage tend to differ from enterprise to enterprise based on the capability of partners, budget and time frame. Although, the partnership between public and private players is not new in the history of agriculture, in PPP written agreements are to be honoured. This dictates the need of a definite mode or mechanism for proper implementation to achieve the intended targets. In fact, PPP models emerged in agriculture quite later as compared to sectors like infrastructure, health and education. Some significant milestones in the use of PPP in Indian agriculture are discussed.

Many of the studies on PPPs in developing countries suggest that it focused on agricultural biotechnology, biosafety regulation, intellectual property rights (IPR) and ways in technology transfer in support of pro-poor PPPs (Spielman *et al.*, 2007). Several research programmes in India actively sought increased links with private stakeholders as partners and research users (Harris *et al.*, 2005) which needed a variety of institutional innovations and incentives for better coordination under PPP (Byerlee *et al.*, 2005). Robust and sustained partnerships between public and private sector agencies in design and implementation of research projects led to greater ownership of outputs and their effective promotion (Lenne, 2008). Few early cases of PPP in agricultural research in India include the following.

1. MAHYCO Research Foundation (MRF), now Barwale Foundation (a non-profit organization), entered into an agreement with ICAR by contributing one crore of rupees each for three years for hybrid rice development work in the country.
2. MAHYCO also made partnership with ICRISAT for developing a CMS system in breeding a pigeonpea hybrid (Ayyappan *et al.*, 2007).
3. Crop Life India partnered with the NCIPM in the validation of IPM of certain crops.
4. Many a time, private pesticide companies conduct efficacy trials of their products in partnership with state agricultural universities.

PPP initiatives have been common in areas of biotechnology research and development. Some of the frontline areas where PPP could be successfully implemented are development of vaccines using recombinant technology, Enzyme Linked Immunosorbent Assay (ELISA) testing kits for disease detection, gene silencing, stem cell and gene therapy. Private sector participation in quality control of conventional and new generation vaccines has also been common (APCoAB, 2007). PPP for gender mainstreaming in agriculture was implemented in action research mode in six states of India benefiting farm women to access technology and market (Ponnusamy *et al.*, 2012).

- The World Bank funded NAIP project of ICAR established market oriented collaborative alliances comprising public and private partners resulting in 51 value chains covering marigold, cotton, agro-forestry, cobia, neutraceuticals, improvement in Trichogramma production etc (Kochubabu *et al.*, 2011).
- John Deere, a leading farm implements manufacturing company helped to promote mechanized farming in tribal region of Gujarat by establishing eight Agricultural Implements Resource Cnetres each covering 600 acres of cultivated land through PPP (Reddy and Rao, 2011).
- Agricultural Technology Management Agency (ATMA) in different states of India facilitated commodity based groups to partner with private agencies in production and marketing of basmati rice and medicinal plants in Bihar, maize in Andhra Pradesh and mango in Maharashtra (Ponnusamy, 2013).
- Public private partnerships for service delivery have revealed significant opportunities for women entrepreneurs and groups in delivering local services and creating conditions for empowerment. The PPP between Cadbury India, Kerala Agricultural University and DBT during past 23 years trained 250 women and established 28 cocoa chocolate units in different parts of Kerala. Thirumadhuram Pineapple project through PPP involving Kudumbhasree Project Mission, Department of Agriculture, women SHGs and Nadukkora Agro-processing centre could produce 25000 tonnes of pineapple in 500 ha and directly employed 12500 women (Rajendran *et al.*, 2010).
- The good impact of PPP in any field depends on involvement of institutions and industries in seeking collaboration and combining all available public and private skills (Peter, 2002). PPP has made positive changes in market linkage of farm produce, capacity building of farm families, reduction of risk and uncertainties, social mobilization and economic empowerment of farmers (Hisrich and Peters, 2002).
- The ICAR-Central Institute of Fisheries Technology (CIFT), Kochi, Kerala and M/s Bodina Naturals Private Limited (BNPL) entered into an agreement in 2020 for promoting three Seaweed Based Products developed by CIFT in PPP mode (www.icar.org) which have the anti-viral, nutritional and immune modulatory effects namely - ZAFORA Seaweed Hand Sanitizer, ZAFORA-360 Enriched Fucoidan Capsules and ZAFORA Gargle.
- The impact of PPP in agriculture should be realized by the local people who are living in a village, where the project was implemented. The key stakeholders in farm production especially women should be treated as partners while undertaking institutional innovations like PPP (Ponnusamy *et al.*, 2017).

A successful model of PPP in agriculture should be replicated all over the nation so that out a significant impact in production, technology adoption, crop management and marketing can be brought out.

3.3. PPP Models for gender mainstreaming in agriculture

In the domains of research, teaching and extension, women should be in the forefront due to their pre-eminent role in agriculture. In this regard, it is essential to demonstrate PPP models at field level in order to push forward the models among the key stakeholders. Such models developed and demonstrated by CIWA, Bhubaneswar in collaboration with various university partners are discussed as under:

Case I: Maize cultivation in Odisha

ICAR - Central Institute for Women in Agriculture (CIWA), Bhubaneshwar designed and implemented a PPP model for gender mainstreaming in maize cultivation during 2011-12. It involved both public and private agencies in Bantla village of Khurda district in Odisha targeting the poor tribal farm households in a fragile agro-eco system (Ponnusamy *et al.*, 2014).

Roles and responsibilities of public and private Agencies in the model: CIWA undertook the planning, implementing and monitoring of the project as well as capacity building of farm-women. Agricultural Promotion & Investment Corporation of Odisha Limited (APICOL), Agricultural Technology Management Agency (ATMA) and Department of Agriculture, Government of Odisha facilitated provision of funds under Integrated Scheme of Oilseeds, Pulses, Oil palm & Maize (ISOPAM) scheme and monitoring. Kamboj Seeds, Karnal in Haryana, served as the private partner and provided Quality Protein Maize (QPM) seeds. The maize producers in Bantla village group cultivated the maize as per the technical guidance and earned a remunerative price for their produce.

Partnership building: CIWA sensitised both men and women maize cultivators of Bantla village on the importance of group dynamics and economic benefits of QPM cultivation. A maize producers' group was formed through participatory process by electing a woman member as president and male member as secretary. During interaction, the villagers informed about the unsold maize of previous year. A private poultry entrepreneur was facilitated to purchase the entire quantity of 3000 kg and thereby could gain the confidence of villagers. A Memorandum of Understanding (MoU) was signed specifying terms and conditions and roles and responsibilities of each partner. The Department of Agriculture, Government of Odisha provided funding support of Rs. 1.2 lakhs for supplying quality input under ISOPAM scheme.

The QPM seeds (250 kg) were procured from Kamboj Seeds, Karnal, Haryana, @ Rs. 115/kg and distributed to 10 farm-women in Bantla village for sowing in 10 acres. Before sowing, soil tests were conducted & results were communicated to the farmers. A pre-season kharif training was organized for 50 farm men and women in the Bantla village. The project team helped the farm-women in carrying out maize cultivation scientifically. The farm implements such as seed-cum-fertilizer drill, improved sickle, maize sheller and chaff cutter were provided to reduce drudgery of farm-women in maize cultivation. A field day was organized on 2nd October, 2011 at the project site where the farm women could dispose off their maize @ Rs. 10.80/kg, which was more than the minimum support price (Rs. 9.80/kg) offered by the government in 2011-12. Impact evaluation revealed the excellent working of PPP model with enhanced knowledge, skill, confidence and decision-making power of women.

Case II: Vegetable Cultivation in Tamil Nadu

A PPP project on vegetable cultivation was implemented in the Ikkarai poluvampatti village of Thondamuthur Block in Coimbatore district involving Avinashilingam Institute for Home Science and Higher Education for Women and Coimbatore Marketing Committee as public partners. Ikkaraipoluvampatti Farmwomen Marketing Society (40 vegetable growing farmwomen as members) and Sree Annapoorna Sree Gowrishankar Group of Hotels (P) Ltd. were the private partners. A Farmwomen Marketing Society was registered under Tamil Nadu Registration Act, 1986. Based on the operational modalities in the PPP project, a MoU was signed by all the partners (Thangamani *et al.*, 2012). Technical knowledge on scientific vegetable cultivation and marketing skill were imparted to women group members to enhance their bargaining and negotiating skill. A trial supply of vegetables was started before signing of MoU by the project partners & it was only after three months of satisfactory supply that MoU was signed. The details of supplied vegetables during the project period are given in Figure 2.

The study revealed that participation of farm-women in social activities improved due to linkages with various developmental departments. Also the confidence level in managing the fund flow enhanced due to familiarity with management of accounts in the Farmwomen Marketing Society. Through this society, they could avoid intermediary charges and shared transportation charges and obtained good profit. The rates of vegetable in wholesale market were collected from the non-society members of Ikkaraipoluvampatti village. The rates given by Annapoorna Hotel were collected from the Farmwomen Marketing Society. A notable difference between the two rates was seen. For all vegetables, Annapoorna Hotel fixed a better price than the whole sale market.

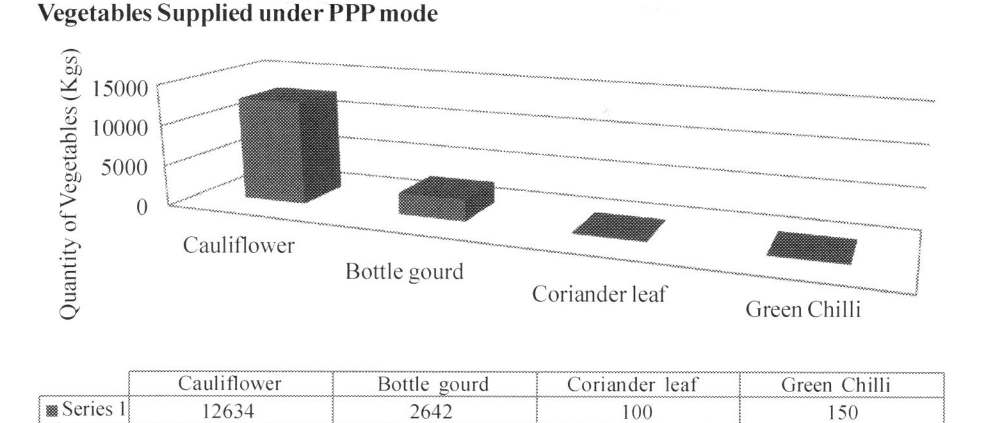

	Cauliflower	Bottle gourd	Coriander leaf	Green Chilli
Series 1	12634	2642	100	150

Figure 2. Supply of vegetables under PPP model to Annapoorna (First three months)

Profit Analysis using differential data on rates of vegetables in local wholesale market and received from Annapoorna was carried out. The rates fixed by Annapoorna Hotel and prevalent in local wholesale market on the same day were recorded for profit analysis. A significant difference between the two rates was observed, which determined the profit for farmers in the vegetable supply chain (Figure 3).

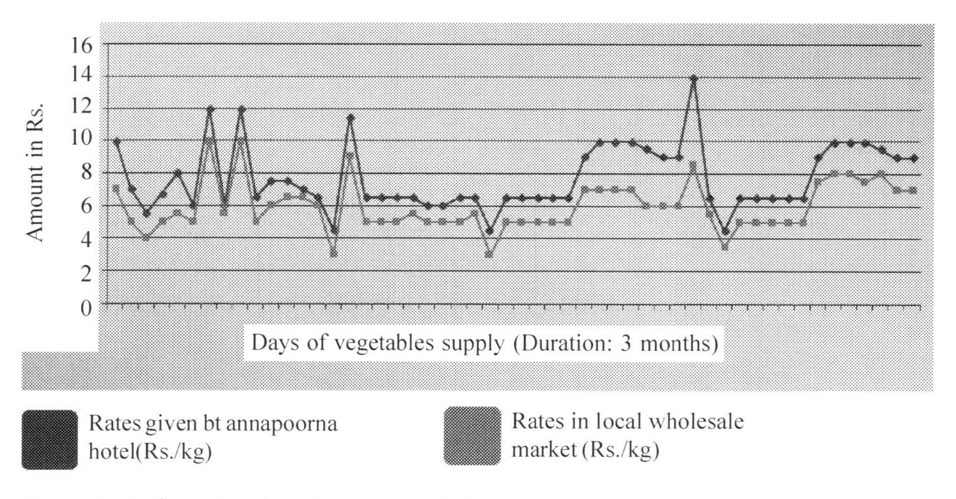

Figure 3. Profit analysis in PPP model project

Case III. PPP model for vermi-compost production in Assam

Vermi-composting has the potential to transform farm-women as environmental conservationists. The benefits of using vermin-compost in the fields were highlighted in Jorhat district of Assam through public-private-partnership (PPP) approach (Ponnusamy *et al.,* 2017). The farmers especially farm-women, were in forefront in practising and propagating the production and utilization of vermi-compost in their area. The impact of gender mainstreaming of the project was evaluated by undertaking a field study in the Jorhat district. Even though there was no hurdle in approaching the development agencies for acquiring the required knowledge, more efforts were required to bring forth the felt needs of the farming community. The PPP project seized the opportunity first to motivate the farmers, both men and women, about the technology and then mobilized them based on their involvement in SHGs, through the NGO partner. Most of the respondents viewed that they could visualize the benefits of applying vermi-compost in tea gardens and ridge gourd plots in terms of higher yield, reduction in pest and diseases, substantial reduction in cost towards chemical fertilizers and pesticides and disease-free healthier life. Although the PPP project was not directly responsible for the better access to credit from institutional sources, it had indirectly motivated the beneficiaries for thrift and savings. The project also facilitated the surplus produce selling through fairs and exhibitions. Each farm family learnt to pack vermi-compost in appropriate quantities and understood the enormous value of organic produce. There was no hurdle in accessing and utilizing the local resources like water and other common properties. The women felt enhanced access to services provided by the panchayat.

Case IV: Mushroom cultivation for developing women entrepreneurs from Kerala

Kerala Agricultural University (KAU), Thrissur implemented a PPP project to promote commercial mushroom production with better market tie ups, networking and collaboration among the private and public stakeholders for supply chain and marketing management. The project was implemented in Palakkad & Thrissur districts during 2011-12. The public partners include State Horticulture Mission for funding; Kerala Agricultural University for mobilization of SHG, hand holding & facilitation services and private partners include Meadow Mushrooms Pvt. Ltd. Elapully, Palakkad as well as Grow Rich Mushrooms, Kollazhy, Thrissur for marketing & input supply. Panchayat Raj Institutions of Palakkad & Thrissur served as legitimizing partners and Bank of India was the credit partner. State Department of Agriculture also partnered in the project as subsidy provider.

Two women groups consisting of 30 members in Palakkad and five members in Thrissur district were formed. After a series of sensitization and motivational meetings of the selected SHG members, hands-on trainings with help of both public and private partners were organized. The signed MOU by all partners formed the terms of reference for all negotiations under the project which include the quantity, quality, time and place of delivery and procurement price. Rates were fixed at Rs. 70/ kg for spawn & Rs 92.5/kg (including packing material) for mushroom bought back.

Impact assessment

The continuation of the project and refining of the terms and conditions of MOU was taken up to suit any emerging contingencies. Process documentation was followed to evolve PPP model for gender mainstreaming in agriculture. The model was validated in the women groups in the project through measurement of various women empowerment dimensions. An impact study was undertaken to assess the impact of three public private partnership (PPP) models in agriculture during 2017 for vegetable cultivation in Coimbatore district of Tamil Nadu, mushroom cultivation in Palakkad and Thrissur districts of Kerala and vermicompost production in Jorhat district of Assam (Ponnusamy *et al.*, 2017). The findings revealed that farm woman had better understanding about the concept of PPP and its utility. The relatives and neighbours although discouraged the PPP model in Tamil Nadu, whereas got whole hearted support in Kerala and Assam. Farm women realised 20 per cent more income and ten per cent additional employment generation in all the three states. PPP models enhanced the access of farm woman to training, market, extension and financial services. They also helped the farm women to build up their confidence and team spirit. The critical input from these field level demonstrated models would help to farm or strong extension strategy for promoting partnership based models in agriculture.

3.4. Lessons Learnt from PPPs for strategizing gender mainstreaming in agriculture

The results have indicated that implementation of PPP project with the help of local governance will give effective outcome. Long-term projects will make a significant change in adaptation level of farm technologies. Continuous motivational meetings and follow-up activities are very important to increase the participation of farmwomen in the project. Need-based activities during convenient times will increase the number of participants in the program. Lessons from the successful public private partnerships as well as focus of policy makers and planners give rise to the following strategies for gender mainstreaming:

1. Organizing farmers around commodities with focused attention provides good dividends
2. Linkage with market-oriented private players ensures remunerative prices for farm commodities

3. Proper implementation of PPP results in enhanced bargaining power of stakeholders
4. Higher productivity through adoption of new technologies is possible through PPP
5. Promotion of crop diversification in line with market demand is possible through PPP
6. Introduction of new crops (export oriented) and enterprises in the existing farming systems yields better returns
7. There is a possibility of quality production and value addition through PPP approach
8. Trust building and proper communication among stakeholders are vital for the success of PPP
9. Statutory support from government would reduce the number of failure cases of PPP
10. A well-defined plan of action and management pave the road map for ultimate success of PPP
11. Capacity building of farm women on technologies promoting diversification of cropping and farming systems would promote scientific farming and higher productivity
12. Legalizing land leasing could benefit large number of women farmers
13. Promoting situation and site-specific land-use plans for different agro-ecological regions
14. Dissemination of a long-term weather forecasting system through mobile based applications would strengthen the hands of all partners
15. Ensuring effective implementation of various social safety net programs with focus on women farmers can provide further momentum
16. Updating women farmers on price monitoring and forecasting of major food commodities frequently can help to take timely decisions
17. Strengthening price and marketing reforms and their periodic review of progress would enhance the confidence of all players.

4. Emerging areas for PPP

The following areas will have considerable scope for undertaking research, extension, education and other development through PPP mode.

Crop science	Animal science
Development of novel varieties	Development of new breeds or strains
Development and validation of fertilizers and pesticides	Vaccine development, field testing
Seed production	Feed and fodder production
Group dynamics-formation of farmer groups including FPC, SHG, CIG, CA, custom hiring centres, repair of farm machineries	Field extension for technology transfer including advisory services, AI, PD and distribution of inputs
Contract farming	Distribution of final products in market
Value addition and processing	Diagnostic services
Harvesting, labour engagement	Transport, cold storage

5. Conclusions

PPP have contributed to gender mainstreaming, food security, additional income and employment generation, poverty reduction, economic growth and agricultural production which have direct implications at the local level. The partnerships between private sector and public institutions are to be based on transparency ensured through written MoU. A favourable policy framework is important for creating coalitions that aim at local development, particularly women empowerment. The demonstrated models have suggested that while incentives and perceptions do differ between different models, sufficient common space can be facilitated through incentive structuring and mutual sharing of risks and benefits for strengthening partnerships. These suggest that PPP can play a constructive role in building strong and vibrant Indian agriculture and better livelihood opportunities for farm families.

Evidences from the field also indicate that there is no dearth of gender sensitive PPP models in technology transfer. But mostly, they are in the conceptualization stage or scattered testing in some places or else end up with completion of projects and fewer publications consequently. The stakeholders in the process of development have understood the role of women in building a strong and vibrant agriculture in the rural areas which can alleviate the poverty, malnutrition and unemployment. However, this requires appropriate mechanism, institutions, models and policies to facilitate the women to attain their full potential and thereby contributing to the development of all members of the society. Many of the models tested are not upscaled due to lack of support from policy and political angles. India will have to make concerted efforts in improving agricultural linkages, women empowerment, healthcare, sanitation, drinking water, nutrition awareness, and education by adopting a convergence approach so that PPP models can gain better prominence.

References

APCoAB, 2007. Brainstorming Session on Models of Public-Private Partnership in Agricultural Biotechnology - Highlights and Recommendations. p 24+viii. Asia-Pacific Consortium on Agricultural Biotechnology, New Delhi and Trust for Advancement of Agricultural Sciences, New Delhi.

Ayyappan, S., Chandra, Pitam and Tandon, S.K. 2007. Agricultural Transformation through Public-Private Partnership: An Interface. ICAR-Industry Meet organised by Indian Council of Agricultural Research, New Delhi. PP 1-152.

Byerlee, D., Diao, X. and Jackson, C. 2005. Agriculture, rural development and pro-poor growth: country experiences in the post-reform era. Agriculture and rural development discussion paper No.21. World Bank, Washington, DC.

Grout, Paul A. and Stevens, Margaret, 2003. Financing and Managing Public Services: An Assessment. Programme on Public Private Partnership in Social Sector Chapter 6. p1. Bella Vista Publication, Hyderabad.

Haque, T. and P.K. Joshi, 2018. 'Price Deficiency Payments and Minimum Support Prices: A Study of Selected Crops in India', Economic and Political Weekly, May 19, 2018, Vol LIII No. 20.

Haque, T. and P.K. Joshi, 2019. 'Agricultural Transformation in Aspirational Districts of India, Economic and Political Weekly, January 2, 2019, Vol. LIII No.51.

Harris, D., Richards, J.I., Silverside, P., Ward, A.F. and Witcombe, J.R. 2005. Pathways out of poverty. Aspects of Applied Biology, 75: 115-26.

Hisrich, R.D. and Peters, M.P. 2002. Entrepreneurship. Tata McGraw Hill Publishing Company Ltd. New Delhi-110095. Pp 1-663.

Kochubabu, Baboo. Bengali and Singh, Ashutosh, 2011. NAIP Publications Vol1, pp. 1-142. NAIP, ICAR, New Delhi-12.

Lenne, J.M. 2008. Research into Use: Managing Achievements for Impact. Outlook on Agriculture, 37(1): 23-30.

NABARD, 2018. All India Rural Financial Inclusion Survey, 2016-17, Mumbai.

Peter Scharle, 2002. Public-Private Partnership (PPP) as a Social Game Innovation. The European Journal of Social Sciences 15(3): 227.

Ponnusamy K., Mishra, Sabita; Prusty M and Dash, Jiban S S. 2012. Market Linkages for Women Farmers: Public-Private Partnership Shows Way Ahead. Indian Farming 61(12): 25-28.

Ponnusamy K. 2013. Impact of public private partnership in agriculture: A review. Indian Journal of Agricultural Sciences. 83(8): 803–808.

Ponnusamy, K., Bonny, B.P. and Das, M.D. 2017. Impact of public private partnership model on women empowerment in agriculture. Indian Journal of Agricultural Sciences, 87(5): 613–617.

Ponnusamy, K., Das, M.D., Bonny, B.P. and Mishra, S. 2014. PPP and gender mainstreaming in agriculture: Lessons from field studies. Agricultural Economics Research Review, 27: 147–55.

Rajendran P., Prasad R.M. and Bonny, Binoo P. 2010. Proceedings of National Workshop on Public Private Partnership for Gender Mainstreaming in Agri-entrepreneurship Development. pp 1-113. Nov 2011. Kerala Agriculture University, Vellanikara, Kerala.

Reddy, G.P. and Rao , K.H. 2011. Public Private Partenership in Agriculture- Challenges and Opportunities , Summary proceedings and recommendations of NAARM. National Workshop held on September 19 &20, 2011. National academy of Agricultural research Management, Hyderabad-500 407. Andhra Pradesh. India.

Spielman, D., F. Hartwich and K. von Grebmer, 2007. Sharing science, building bridges and enhancing impact: Public-private partnerships in the CGIAR. IFPRI Discussion paper 708. Washington, DC: International Food Policy Research Institute.

Thangamani K., Leelavathy K.C. and Meenakshi S. 2012. Mainstreaming of Farmwomen: An Experience of PPP Approach in Vegetable Marketing. Abstracts of Global Conference on Women in Agriculture. pp.159-160. 13-15th March 2012. New Delhi.

6

Gender in Agriculture Research: Trends Over the Years

Jayasree Krishnankutty, Shilpa Karat and Shinoji K.C.*

1. Introduction

Gender equality in agriculture is a long time demand to achieve the sustainable agricultural development goals. According to the World Bank (2009) differences in the socially constructed gender roles and gender relations affect agricultural development as it creates inequality in the distribution, access to and control over agricultural resources between men and women; that in turn affect the agricultural development outcomes. However, role of men and women in agriculture is not constant across the world due to the disparities existing in the allocation of productive farm resources to women and men among different societies and cultures. Most of the agrarian social systems recognize men as the key players of crop production due to their ownership over agricultural land and other resources. This often masks the significance of activities carried out by farm women in the crop production process. Currently agriculture provides employment to nearly 27.2 percent men and 25.4 percent women globally (World Bank, 2020). Irrespective of the unequal power relations between men and women, their relative representation in the agriculture workforce remained almost equal throughout the years. Hence, it became inevitable to address the concerns of both genders in agriculture rather than concentrating only on women centric issues while formulating programmes and policies aiming agricultural development. These trends in agriculture motivated social researchers like Safiliou (1990) to propose gender as a 'variable' similar to 'social class' to rationalize agriculture. Globally, gender based researches has covered various dimensions of contributions of men and women in agriculture as cultivators as well as agricultural labours in different time period. However, studies on topics related to women's rights, women empowerment, and gender equality in order to achieve food security and agricultural development has increasingly been carried out after the "fourth world conference on women: action for equity, development

---

**Author contact: jaysree.krishnan@kau.in*

and peace" held at Beijing (China) during 4-15 September, 1995. Though gender research in agriculture mostly revolves around farm women and their empowerment concepts like marketing extension, market-led extension etc., it has led to the empowerment of both women and men cultivators.

The chapter discusses the concept of gender in research and development, its evolution and its integration into research perspectives in agriculture. A brief description of gender analysis tools is also included.

2. Gender in research and development

The modern developing world practically opened its eyes to gender and development through the works of Danish economist Ester Boserup who highlighted the significance of women's contributions in African agriculture. Boserup believed women in Africa were excluded from development efforts even after working hard for long hours in farm as well as home. She described women as struggling using primitive techniques to ensure the food security of household while men as market working relatively short hours adopting modern technologies. As a response to the increasing bias skewed against women, an approach called Women in Development (WID) was formulated in the early 1970s. This was aimed at bringing women's issues in development projects. During the mid1970s, when the Farming Systems Research and Extension approaches gained ground, the technical agricultural research was allowed to be set up within a broad systems framework. Around this time, a global dialogue on gender equality also emerged and arguments for ensuring women's inclusion in development activities started gaining momentum.

In the 1980s the women centric WID concept was replaced by Gender and Development (GAD) concept as a way of restructuring development projects and programs based on gender relationships. According to Moser (1993) the GAD approach focuses mainly on the socially constructed differences between men and women. She highlighted the need for challenging the existing gender roles, relations, and the creation and effects of class differences on development. The approach was observed to be influenced by the academic arguments that social relationship between men and women lead to the subordination of women (Moser, 1993). Studies of economists Benería and Sen (1981) who assessed the impact of colonialism on development and gender inequality also influenced GAD perspectives. According to them, colonialism imposed more than a 'value system' upon developing nations by introducing an economic system designed to promote capital accumulation leading to class differentiation. In fact, the GAD approach uses two major frameworks viz., gender roles and social relations analysis to capture gendered inequalities in society. 'Gender roles' covers the social construction of identities within the household, expectations from men and women in their relative access to resources. But the 'social relations analysis' assess the social dimensions of hierarchical power relations embedded in social institutions, as well as its influence on determining 'the relative position of men and women in society (Razavi and Miller, 1995). This relative positioning tends to discriminate against women.

3. Analytical frameworks

In order to assess the differences in the economic participation of men and women the Harvard Analytical Framework also known as Gender Roles Framework was published in 1985 after the third World Conference on Women in Nairobi (Kenya).The framework was originally designed at Harvard University in 1980 in connection with the World Bank training on Women in Development and aim to assist planners in the effective allocation of resources to women and men so as to make development more efficient-a position named the "efficiency approach". This became a standard for gender analysis in agriculture.

The 1990s integrated more analysis tools like the Moser gender planning framework (1993), Social Relations Approach (1994) and The Women's Empowerment Framework (1995) to the gender analysis methodologies.

3.1. The Moser Gender Planning Framework (1993)

It was developed by Carol Moser was introduced as a tool for gender analysis in development planning. The framework considers three basic concepts such as the triple role of women in society (reproductive, productive, and community-managing), practical and strategic gender needs, and categories of WID/GAD policy approaches.

3.2. The Social Relations Approach (1994)

It was developed by Naila Kabeer, was introduced for gender analysis in socialist-feminist philosophy. The basic idea in the development of the approach is that it focuses on the interchange between patriarchy and social relationships. It aims to promote human well-being which consists of survival, security and autonomy. The approach looks at the relationships between the government, the market communities and the family.

3.3. The Women's Empowerment Framework (1995)

It was published by Sara Hlupekile Longweis considered as an effective tool for planners to measure the women empowerment in a more practical way as well as to evaluate whether a particular development initiative supports their empowerment. The basic premise in this framework is that the empowerment process can be viewed in terms of five levels of equality i.e., welfare, access "conscientization", participation, and control.

With the progress of the decade, there was a conceptual shift to empowerment of women, gender justice, and a socially-progressive agenda to address women's subordination. Also expressed was support for women's individual and collective agency- that women themselves become active participants in the change process.

In 2011, the Food and Agriculture Organization (FAO) stated that closing the gender gap in agriculture would generate significant gains for the agriculture sector and for society. It also stated that with the same level of access to productive resources women could produce 20–30 percent more yield from their farm lands compared to men. This could raise the total agricultural output of developing countries by 2.5–4 percent; that in turn could reduce the number of hungry people in the world by 12–17 percent (FAO, 2011). Subsequently, the First Global Conference on Women in Agriculture organized at Delhi (India) in 2012 focused on five areas for action to ensure that institutions promote women's ownership and control of resources. It strived for advocacy to raise awareness of women's issues, generating an evidence base to show the economic and social impacts of addressing women's issues. It also encouraged collective action and leadership among women to develop programmes that directly meet women's needs, addressing discrimination through appropriate policies, legislation and enforcement mechanisms (ICAR, 2012).

4. Gender research

Earlier research on gender shows evidence for the presence of division of labour in ancient societies. The earliest incidence of division of labour existed among the hunter gatherers (Kehoe, 1997). In these primitive societies, men went in search of food whereas women took care of children at home. This brought the concept of "man the hunter" which has been enrooted in our thinking from the early decades of gender research. While the hunting hypothesis continued to be endorsed in most scientific circles, its obvious gender biases were sharply attacked by some anthropologists and other scholars who are highly skeptical of this theory (Slocum, 1975). Anthropologist Eleanor Leacock strongly argued that male and female roles were not demarcated at an early stage of human evolution. But was later brought in through division of labour which paved way for such a delineation that further led to hierarchical relations of power and prestige. He also said that the hunting hypothesis did not consider the dynamics of social change or the social, historical, and cultural (rather than simply the biological) factors that may have shaped differences in status between males and females (Leacock, 1978). The postulate put forward by Tanner and Zihlman (1976), aimed to establish women in the central role of human evolution proposed gathering, rather than hunting, as the key economic innovation that led human species to their survival and success. According to them the most primary unit of kinship in early hominid societies was not the male-female pair bond but the mother-infant relationship. As scientists delved more into the life patterns of the early communities, the demarcation between hunting and gathering started to obliterate as there were many activities like preparing, identifying, planning and strategizing before the large animal hunting. However, historians for long had not taken into consideration these activities, and the processes and roles that contributed to their execution as such. In fact, the community members including children took part in many of these activities (Brumbach and Jarvenpa, 2006).

4.1. Gender roles in agriculture

The advent of agriculture accelerated the process of civilization globally, wherein women's role was more pronounced as food producers and this raised their social status to a small extent (Childe and Gordon, 1948). It is widely assumed that women being the gatherers and nurturers are the initiators of agriculture. This assumption is largely questioned by later researchers, who feel that even though women would have started the small time sowing of seeds and raising plants near the dwelling areas, cultivating in an extended area that provided food for the community would have been implemented by men who designed and mastered the early implements. Also men would have initiated domesticating and using animals to support these activities, which subjugate the role of women to a secondary position in early writings. But again, researchers argue that the participation of women and children lie intertwined in these activities. In early times, communities, according to their specific circumstances, geographical, and climatic conditions shared among their members these activities without a strict division of labour among the sexes (Bolger, 2010).

However, with the emergence of industrialization, men were actively involved in work that yielded productivity and monitory benefits. This confined women to the domestic sphere which undermined their share in nation building. Over the years, men took strong hold in modern agriculture and industries, limiting women into subsistence farming which made them to loose income and power (Razavi and Miller, 1995).

4.2. Gender discrimination in the access to resources

The challenge of 'gender discriminated access to resources' is an age old phenomenon with its roots from the patriarchal system. The disparity in access makes women farmers face the disadvantage of an increasing knowledge gap. Women's right to land plays a vital role in social status, economic wellbeing and empowerment (FAO, 2002). Gender inequality related to the right to land and other productive resources have an intimate relation with women's poverty and exclusion. The right to access and control over resources are linked to the belief that men, as the head of the household can only exercise power over them. Agricultural credit, for instance, is critical for farmers to manage the seasonality of agricultural income and expenditures, and to invest in technologies and long-term farm improvements (Ignaciuk and Tun, 2019). In fact, the multidimensional concept of gender inequity (Alkire *et al.*, 2013) has been identified to remain at the core of all these discriminations. This recognition has led to the emergence of a new paradigm governing agricultural research and outreach practices in developing countries. It encompasses different dimensions of inequity, such as decision making power over production and income, which vary independently across and within communities (Mason and Smith, 2003). In some communities, women enjoyed considerable decision making power over production and input. However, they remained disempowered with respect to asset ownership, control over income, or community leadership (Alkire *et al.*, 2013). Though there have been multiple legislations that stipulates ownership of land and other assets to women, the hard core patriarchal set up and handicapped implementation of laws forbids them from being effective. The gender prejudice exhibited by mainstream public institutions adds to this predicament (Ahmed, 2001).

There is agreement that gender inequalities and lack of attention to gender in agricultural development contribute to lower productivity, lost income, and higher levels of poverty as well as under nutrition. This has brought renewed interest in gender and agriculture which has produced several new initiatives, calls for action, and commitments from the international development community since 2005 Beijing summit (IFAD, 2003; IFPRI, 2007). In addition, guides, tool kits, and other resources on theory and practice of gender integration and promising programmatic approaches have been developed to streamline gender-specific agricultural development initiatives (Doss 1999; Mehra and Rojas 2008; Quisumbing and Pandolfelli 2010; UN-HABITAT 2006; World Bank *et al.*, 2009). However, there seem to be wide differences in the observations among researchers, about the access to agricultural inputs. Many researchers study mainly access to land as a resource and do not discern about other major and minor inputs, like technological, natural and human resources (Peterman *et al,* 2014). However, there has been agreement that the choice of crops and division of labor differ by gender within disparate regional and cultural contexts. For example, throughout Sub-Saharan Africa, lucrative cash crops are often perceived to be "male crops," and crops for home consumption are perceived to be "female crops" (Kasante *et al.,* 2001; World Bank/Malawi, 2007).

4.3. Women Empowerment

The 90s dawned with a dire need for feminism in all walks of life. The forward motion towards feminism accelerated by female pioneers in diverse fields spread the idea of women empowerment around the globe. Women empowerment was one of the most important progenies of feminism. Empowerment of women is considered a 'prerequisite' to achieving global food security. Gender systems, however, are diverse and complex. The nature and extent of gender inequity and the conditions necessary to empower women vary across countries, communities and regions (Akter and Rutsaert, 2017). In the last five decades, the concept of women empowerment has undergone a sea of change from welfare oriented approach to equity approach. Some of the key indicators pursued by researchers are women's household decision making power, financial autonomy, freedom of movement, women's acceptance of unequal gender roles, exposure to media, access to education, women's experience of domestic violence and their political participation (Nayak and Mahanta, 2012).

Social norms and family structures in developing countries like India manifest and perpetuate the subordinate status of women. The study carried out by Gupta *et al.* (2017) identified Indian farm women to be disempowered mainly due to lack of opportunities to express their leadership abilities, low level access and decision making power over productive resources. The authors found a strong relationship between market orientation and empowerment levels and hence linking women to markets could promote their empowerment in agricultural domains. Offering access to microfinance services to women can be another important way to increase women's empowerment (Huis *et al.*, 2017). However, in order to address the fundamental challenges to achieving women's empowerment and gender equality in developing countries, more innovative and expansive efforts are needed. In fact, understanding how innovation connects to women's empowerment provides powerful guidance for action (Malhotra *et al.*, 2009).

Women are marginalized in decision making by a variety of factors which begin as early as in childhood where they are brought up to have passive roles in family. They lack opportunities within the family as well as the society to exercise their right to decide. Higher levels of illiteracy and limited skill training among women adds to this predicament (Corner, 1997). Consequently, though women provide 40 per cent of the work force around the globe, only 15 per cent rise up to be administrators and managers. Despite all these challenges there has been considerable improvement with regard to decision making among women in the past decades. In India, women entrepreneurs constituted one tenth of the total in the past decade. Globally also women have more access to positions of authority than they had thirty years ago (O'Neil and Domingo, 2015). Moreover, the economic contribution of women has found to increase their role in household decision-making. This has also led to higher household-level outcomes with respect to health, education (especially education of the girl child), nutrition and family planning (Kumar, 2009). Further, women are found to exhibit a more democratic and participatory style of decision making than their male counterpart which yields higher efficiency. However, in case of rural households primarily engaged in farming and allied activities, decision making is still a man's autonomy. Women's involvement is negligent especially in economic matters (Raju and Rani, 1991). Women are still confined to labour and house hold chores where as men decide and implement. The role of women in purchase, sale and management of agricultural inputs is still warrant improvement (Pal and Haldar, 2016).

The wheat based agricultural societies in Bihar is an exemplary case where women are transitioning from workers to innovators and managers. Their decision making capacity has doubled over the years and this has strengthened them to be independent and dignified. The migration of men to town for non agricultural jobs has made it possible for women to decide for the family and farm. The emergence of women as decision makers is very crucial for their empowerment and for the social upliftment of the nation as a whole.

4.4. Gender and climate smart agriculture

Climate change and resilience are inexorably considered as a gender issue by the world development bodies. Inequality limits women's ability to adapt to the impacts of climate change. This vulnerability is exacerbated by viewing women as victims, rather than key actors who have critical knowledge of their society, economy, and environment, as well as practical skills. Effective climate change adaptation recognizes that women, men and children experience impact differently depending on where they live, how they sustain their livelihoods, and the roles they play in their families and communities. In 2015, the Paris Agreement under the United Nations Framework Convention on Climate Change acknowledged the importance of gender equality and women's empowerment, and called for climate action to be gender-responsive (Daze, 2019). Women commonly face higher risks and greater burdens from the impacts of climate change in situations of poverty, and the majority of the world's poor are women. Women farmers were less likely to adopt strategies to tackle climate change due to financial limitations and less control over land (Misha and Pede, 2017). As a result of natural disasters in India, girls were more likely stunted and underweight than boys. Women face an increasing workload to collect water and firewood for which they travel long distances and are prone to forest fires. Women also suffer psychosocial impacts of migration more than men as they have to take care of children and other family members. Women's unequal participation in decision-making processes and labour markets compound inequalities and often prevent women from fully contributing to climate-related planning, policy-making and implementation. United Nations climate change negotiations, void of gender-related texts and discussions until 2008, have more recently reflected an increased understanding of the links between gender equality and responding to climate change.

Women play an important role in food production, contributing 60 and 80 per cent of the food in most low-income countries. Agriculture interventions, considered 'climate smart', should therefore provide long-term benefits for women. In other words, when we look at the value and practicality of different approaches to climate smart agriculture, it is important to understand the gender aspects of their impact (Jost *et al.*, 2014). Climate smart agriculture has the potential to narrow the gender gap by bringing them to the mainstream agriculture. Ensuring equal access for women to productive resources, climate smart and labour saving technologies and practices is crucial for sustainable agriculture, food security and eradication of poverty (FAO, 2017).

Potential labor-saving CSA technologies for women farmers in areas facing high climate risks and CSA options for reducing the levels of labor drudgery (Chhetri *et al.*, 2020) are being explored. Gender and social differences are dynamic and nuanced within communities; a greater understanding of these differences is critical for climate-smart smallholder agriculture programming. (Bernier *et al.*, 2013). In order to translate these insights into action a comprehensive menu of practical tools for integrating gender in the planning, design, implementation, and evaluation of projects and investments in climate-smart agriculture has been developed by the World bank – FAO-IFAD coalition (World Bank, 2015).

Many other leading organizations have come forward with their programmes to empower women with regard to climate resilient agriculture. The Dimitra programme introduced by FAO is a participatory information and communication project. Dimitra clubs aid women in enhancing their ability to adapt to climate threats, improve food security with increased incomes. Farmer Field Schools also impart skill training focused on climate resilience particularly for women. The Women-led Climate Resilient Farming Model aims at repositioning women as the flag bearers for sustainable agricultural practices through economically viable and environment friendly cultivation. Women are imparted knowledge through campaigns, on site demonstrations and workshops along with a community resilient fund for financial support. The programme has empowered 41,000 women farmers in climate resilient agricultural practices and a 25 per cent increase in crop yield was seen due to mixed cropping and organic inputs.

In developing a broad evidence-based understanding of how gender and CSA technologies interact, studies are needed to provide detailed empirical knowledge. These will build up the body of evidence required to fully understand how gender relations influence adoption decisions. This evidence can then be used to improve extension service delivery and policy. Evidence-based comparative gender studies will help track short-to medium-term outcomes of innovation technologies, policies, and organizational interventions in agriculture (Farnworth and Badstue, 2017).

5. Measurement of gender inequities

Even equal access to resources need not be sufficient to ensure equal returns from productive resources. Specialized trainings on agriculture and child care should be provided so that women wouldn't have to compromise one for the other. This warrants design of targeted programs based on quantitative measures of gender inequities. This has led to the development of many tools for measuring gender inequity including the WEAI (Alkire *et al.*, 2013) which qualitatively bring out the factual picture in a developing country context. The other most prominent indices to assess the gender equality are UNDP's Gender-related Development Index (GDI) and the Gender Empowerment Measure (GEM), introduced in 1995. Also the Gender Equity Index (GEI) introduced by Social Watch in 2004, the Global Gender Gap Index (GGGI) developed by the World Economic Forum in 2006, and the Social Institutions and Gender Index of OECD Development Centre in 2007 are effective to bring out the existing imbalances.

6. Women in Indian agriculture research

Increased participation of women in Indian agriculture has been noticed by many researchers in the past decade. The Indian agricultural census 2015-16 reported an increasing trend of female cultivators in India. Researchers viewed the reasons behind this shift as casualisation of workforce, and unprofitable crop production. Also when distress migration weakened the male agricultural workforce in India, women started to take over the opportunities in agriculture even to the level of cultivators and entrepreneurs (Vepa, 2005).

Realizing the need for gender based research in Indian Agriculture, the Department of Agricultural research and Education (DARE), Government of India has established the National Research Centre for Women in Agriculture in the year 1996 under the Indian Council of Agriculture Research (ICAR). Later it got upgraded as ICAR-Central Institute for Women in Agriculture in 2014 with the mission of generating and disseminating knowledge to promote gender sensitive decision making for enhancing efficiency and effectiveness of women in agriculture.

7. Conclusion

The forgoing discussions suggest that the research on gender in agriculture has largely been centered on few mainstream ideas. The evidences indicate that the historical perspectives are mostly unexplored, and there can be wide variations in the roles, dynamics, equality patters and access across regions and communities. The modern era is said to go down on the role of women in agriculture, but looking at the larger picture, this might not be factual. A paradigm shift is required in agricultural research, development, and extension (R, D & E) systems in developing countries, from a focus on production toward a broader view of agriculture and food systems. The proposed paradigm should be inclusive of women's distinct role in ensuring the food security of their households (Meinzen-Dick *et al.,* 2014)

Thus there is a definite need to increase awareness that gender issues are not peripheral to agriculture but are fundamental to increasing productivity, incomes, nutrition, food security, sustainability, and ultimately in the contribution of agriculture to poverty reduction. Both research and firsthand experience play an important role in generating this awareness. Statistical and impact assessment agencies need to be involved to ensure that the data and methods are developed to capture gender differences in needs, contributions, and outcomes. Looking at the gender dimensions should become a mainstream component at least in social science research and not confined as the responsibility of a small group. This, in turn, requires strengthening the capacity of all involved, linking contextual knowledge about gender relations to broader patterns and even global lessons. Institutional mechanisms should endorse this and make financial, human, and time resources available as well as recognize and reward excellence in these endeavors.

References

Ahmad, 2001. Report on mechanization & technological adoption: scaling up micro enterprises to small scale enterprises. Monograph series on social science. Serdang: Penerbit Universities Putra Malaysia.

Akter, S., Rutsaert, P., Luis, J., Htwe, N.M., San, S.S., Raharjo, B. and Pustika, A., 2017. Women's empowerment and gender equity in agriculture: A different perspective from Southeast Asia. Food Policy, 69 : 270-279.

Alkire, S., Meinzen-Dick, R. S., Peterman, A., Quisumbing, A. R., Seymour, G., and Vaz, A. 2013. The women's empowerment in agriculture index. World Dev. 52: 71–91.

Beneria, L. Gender, development, and globalization: economics as if all people mattered. (2nd Ed.) Berik, Günseli,, Floro, Maria, NewYork. Online : ISBN 9780415537483. OCLC 903247621

Beneria, L. and Gita, S. 1981. Accumulation, Reproduction and Women's Role in Economic Development: Boserup Revisited in Signs, 7(2).

Bernier, Q., Franks, P., Kristjanson, P.M., Neufeldt, H., Otzelberger, A. and Foster, K. 2013. Addressing gender in climate-smart smallholder agriculture. ICRAF Policy Brief 14. Nairobi, Kenya, World Agroforestry Centre (ICRAF).

Bolger, D. 2010. The dynamics of gender in early agricultural societies of the Near East. Signs: Journal of Women in Culture and Society, 35(2):503-531.

Brumbach, H.J. and Jarvenpa, R. 2007. Gender dynamics in hunter-gatherer society: archaeological methods and perspectives. In S. M. Nelson (ed.) Identity and subsistence: gender strategies for archaeology, 169–201. Walnut Creek, CA: Altamira Press.

Chhetri, Khatri, A., Regmi, P.P., Chanana, N. and Aggarwal, P.K. 2020. Potential of climate-smart agriculture in reducing women farmers' drudgery in high climatic risk areas. Climatic Change, 158(1): 29-42.

Childe and Gordon, V. 1948. New Light on the Most Ancient East: The Oriental Prelude to European Prehistory. In: Kegan Paul and Trench, Trubner. (Eds.). Man Makes Himself. London: Watts.

Corner, L. 1997. Women's participation in decision-making and leadership: A global perspective. Proceedings of the conference on Women in Decision-making in Cooperative held by the Asian Women in Cooperative Development Forum (ACWF) and the International Cooperative Alliance Regional Office for Asia and the Pacific (ICAROAP): 7-9.

Daze, A. 2019 Why Gender Matters in Climate Change Adaptation. Online: https://www.iisd.org/articles/gender-climate-change.

Doss, C.R. 1999. Intra-household resource allocation in Ghana: the impact of the distribution of asset ownership within the household. In: Peters GH, von Braun J (eds) Food security, diversification, and resource management: refocusing the role of agriculture? Dartmouth Publishing Aldershot : 309–316.

FAO, 2002. Women and food security. Online: http://www.Fao.org.

FAO (Food and Agriculture Organization of the United Nations) (2011). The State of Food and Agriculture 2010–11. Women in agriculture: closing the gender gap for development. FAO, Rome.

FAO [Food and Agricultural Organisation]. 2017. FAO in India [on-line].

Available: http:// www.fao.org/Country_collector/FAO in India/Our office/India at a glance. [11 Dec 2020].

Farnworth, C.R. and Badstue, L. 2017. Embedding gender in conservation agriculture R4D in Sub-Saharan Africa: Relevant research questions. GENNOVATE resources for scientists and research teams. CDMX, Mexico.

Gupta, S., Pingali, P. L. and Andersen, P. 2017. Women empowerment in Indian Agriculture: does market orientation of farming system matter. Online: Food. Sec, 9: 1447-1463.

Huis, M.A., Hansen, N., Otten, S. and Lensink, R. 2017. A three-dimensional model of women's empowerment: Implications in the field of microfinance and future directions. Frontiers in psychology, 8: 1678p.

ICAR, 2012. Proceedings of the first global conference on women in agriculture, March 13-15, 2012, New Delhi, India: 82p.

IFAD (International Fund for Agricultural Development), 2003. Mainstreaming a gender perspective in IFAD's operations: plan of action 2003–2006. IFAD (International Fund for Agricultural Development), Rome.

IFPRI (International Food Policy Research Institute), 2007. Proceedings of the consultation on strengthening women's control of assets for better development outcomes. IFPRI (International Food Policy Research Institute), Washington, DC.

Ignaciuk, A. and Tun, N. 2019 Achieving agricultural sustainability depends on gender equality.Online:https://www.ifpri.org/blog/achieving-agricultural-sustainability-depends-genderequality#:~:text=Addressing%20gender%20inequality%20is % 20 essential, more % 20prone%20to%20food%20insecurity

Jost, C., Kristjanson, P. and Ferdous, N. 2014. Participatory approaches for gender-sensitive research design (CCSL Brief No 5). Copenhagen: CGIAR Research Program on Climate

Change, Agriculture and Food Security (CCAFS) Online: https:// cgspace. cgiar.org/ rest/ bitstreams/32200/retrieve

Kasante, D., Lockwood, M.,Vivian J. and Whitehead A. 2001. Gender and the expansion of nontraditional agricultural exports in Uganda. In: Razavi S (ed) Shifting Burdens: Gender and Agrarian Change under Neo-liberalism. Kumarian, Bloomfield: 35-66.

Kehoe, A.B. 1997. Gender in Archaeology: Analyzing Power and Prestige, by Sarah Milledge Nelson, AltiMira Press, Walnut Creek, CA. Bulletin of the History of Archaeology, 7(2), p.40. DOI: http://doi.org/10.5334/bha.07208

Kumar, A. 2009. Self-Help Groups, women's health and empowerment: Global thinking and contextual issues, Jharkhand J. Devpt. and Mgt. Studies. 4(3) (2009): 2061-2079.

Leacock, E. 1978. Women's Status in Egalitarian Society: Implications for Social Evolution. Current Anthropology 19(2): 247–75.

Leacock, E., Abernethy, V., Bardhan, A., Berndt, C.H., Brown, J.K., Chiñas, B.N., Cohen, R., De Leeuwe, J., Egli-Frey, R., Farrer, C. and Fennell, V.1978. Women's status in egalitarian society: Implications for social evolution [and comments and reply]. Current anthropology, 19(2): 247-275.

Malhotra, A., Schulte, J., Patel, P and Petesch, P. 2019 Innovation for Women's empowerment and Gender Equality International Center for Research on Women, Washington DC: 1-14 .

Mason, K.O. and Smith, H.L. 2003. Women's empowerment and social context: Results from five Asian countries. Gender and Development Group, World Bank, Washington, DC: 1-39.

Meinzen-Dick, R., Quisumbing, A.R. and Behrman, J.A. 2014. A system that delivers: Integrating gender into agricultural research, development, and extension. In Gender in Agriculture, Springer, Dordrecht: 373-391.

Mehra, R. and Rojas, M. 2008. A significant shift: women, food security and agriculture in a global marketplace. International Center for Research on Women (ICRW) citing FAO focus on women and food security. Food and Agriculture Organization of the United Nations, Rome. Online:http://www.icrw.org/publications/women-food-security-and-agriculture-global-marketplace.

Miller, C. and Razavi, S. 1995. From WID to GAD: Conceptual shifts in the women and development discourse (No. 1). UNRISD Occasional Paper.

Mishra, A.K. and Pede, V.O. 2017. Perception of climate change and adaptation strategies in Vietnam. International Journal of Climate Change Strategies and Management, 9(1): 112.

Moser, C .1993. Gender Planning and Development. Theory, Practice and Training. New York: Routledge. Online: ISBN 978-0-203-41194-0.

Nayak, P. and Mahanta, B. 2012. Women empowerment in India. Bulletin of Political Economy, 5(2): 155-183.

O'Neil, T. and Domingo, P. 2015. The power to decide: Women, decision-making and gender equality. Overseas Development Institute Report. Online: https://www. odi. org/sites/odi. org. uk/files/odi-assets/publications-opinionfiles/9848 on 13/08.

Pal, S. and Haldar, S. 2016. Participation and role of rural women in decision making related to farm activities: A study in Burdwan district of West Bengal. Economic Affairs, 61(1): 55-63.

Peterman, A., Behrman, J.A. and Quisumbing, A.R. 2014. A review of empirical evidence on gender differences in own land agricultural inputs, technology, and services in developing countries. In. Gender in Agriculture, Springer, Dordrecht: 145-186.

Quisumbing, A.R. and Pandolfelli, L. 2010. Promising approaches to address the needs of poor female farmers: Resources, constraints, and interventions. World Dev. 38: 581–592.

Raju, V.T. and Rani, S. 1991. Decision-making role of women in agriculture. Indian J. Home Science 20(1): 13-17.

Razavi S. and Miller C. 1995. Gender Mainstreaming. UNRISD, Geneva.

Safiliou, C. (1990). Agricultural education in gender issues: a necessity for rational agriculture. Online: https://edepot.wur.nl/240009 on 28.11.2020 .

Slocum and Sally, 1975. Woman the Gatherer: Male Bias in Anthropology. In Toward an Anthropology of Women. ed. Rayna R. Reiter, 36–50. New York: Monthly Review Press. Stordeur, Danielle, 2000.

Tanner, N. and Zihlman, A. 1976. Women in evolution. Part I: Innovation and selection in human origins. Signs: Journal of Women in Culture and Society, 3(1): 585-608.

UN-HABITAT (United Nations Human Settlements Programme), 2006. Mechanism for gendering land tools: a framework for delivery of women's security of tenure. Strategies and outline adopted at the High Status Round Table on Gendering Land Tools, 21 June, Nairobi, Kenya. Available: www.unhabitat.org/downloads/docs/4223_75221_gltn.doc.on 18 Nov 2020.

Vepa, S.S. 2005. Feminisation of agriculture and marginalization of their economic stake. Economic and Political Weekly 40(25): 2563-2568. Online: https:// www. jstor.org /stable/ 4416785

World Bank, 2007, World development report 2008: Agriculture for development. World Bank, Washington, DC.

World Bank, 2009. Gender in Agriculture Sourcebook. Online: http://www.fao.org/3/a-aj288e.pdf.

World Bank, 2015. Gender in Climate Smart Agriculture – Module 18 for the Gender in Agriculture Sourcebook. World Bank – FAO – IFAD: 96p.

World Bank, 2020. Employment in Agriculture (modelled ILO estimate). Online: https:// data.worldbank.org/indicator/SL.AGR.EMPL.FE.ZS on 28.11.2020

World Bank/FAO/IFAD, 2015. Gender in Climate-Smart Agriculture- Module 18 for the Gender in Agriculture Sourcebook: 84p.

World Bank/IFAD/FAO (International Fund for Agricultural Development/Food and Agriculture Organization for the United Nations).2009. Gender in agriculture sourcebook. World Bank, Washington, DC. Online: http://siteresources. worldbank.org/ INTGENAGRLI VSOUBOOK/Resources/ CompleteBook.pdf.

World Bank/Malawi, 2007. Malawi poverty and vulnerability assessment (PVA): investing in our future. Synthesis report: main findings and recommendations. Poverty Reduction and Economic Management 1, report 36546-MW. World Bank, Washington, DC.

7

Gender Dimensions in Natural Resource Management

Prema, A.*, Hema M. and Jyotsna C.

1. Introduction

"Earth provides enough to satisfy every man's needs, but not every man's greed"

Mahatma Gandhi

Natural resources form the core of ecology that is put to economic use. It is usually defined in terms of materials created in nature that are used to satisfy human needs. Soil, water and energy supplies (eg. coal, natural gas) that serve to satisfy human needs and wants comprise the major natural resources (Barsch and Burger, 1996). Based on the impact on an environment, cost of extraction and rate of exhaustion, natural resources are broadly categorized into renewable and non-renewable resources. Major renewable sources comprise of wind, solar energy, tidal energy, geothermal energy, forest etc. which have the potential to be renewed or replaced through natural processes over time, once consumed. Unlike this, nonrenewable resources such as coal, petroleum, natural gas and similar types have finite reserves that cannot be regained once consumed. However, environmental degradation, aggravated by the growing population and climate change vagaries, has intensified the competition for these scarce resources in recent years. This has necessitated sustainable natural resource management strategies and advanced experimentations to convert nonrenewable energy sources to renewable forms as they form the basis of all present day development and economic growth. Increasing demands on natural resources have warranted a more prudent and diligent efforts in ensuring sustainable natural resource management. The dichotomy of unlimited human wants and limited resource availability is at the heart of all natural resource management principles that guide economics of resource allocation and optimum utilization.

Globally it is well documented that, economic reforms that favour both domestic and global market expansion, pressures associated with population growth,

**Author contact:prema.a@kau.in*

migration, urbanization and commoditization are changing the patterns of natural resource use. Along with these political, economic and social factors, depletion of renewable resources has also destabilized livelihoods, adversely affecting ecosystems and challenging development. Hence the conservation and management of natural resources, especially renewable resources has gained added significance in such situations and has led to the evolution of the concept of natural resource management. The term Natural Resource Management (NRM) was first coined in the United States in the early 1960s, and it focused on the management of land, water, forest etc. in a futuristic perspective. It is based on the philosophy of conserving natural resources for both present and future generations without affecting the quality. NRM perspectives also valued the resources for the ecosystem services such as provisioning, regulating, supporting and cultural, which are essential for the wellbeing of the human kind.

As natural resources form the basis for human livelihoods, their management assumes critical significance in sustainable development. The cardinal questions that need to be addressed include whether the environment and for that matter, natural resources affect men and women differently, in a society. Observations of environmental activist Shiva (1986) that all pre-colonial societies were based on an ontology of the feminine as the living principle and rural, indigenous women as the original givers of life and rightful caretakers of nature has been contested by WED (Women, Environment and Development) as a spiritualist and cultural premise. WED's logic was that women were adversely affected by environmental degradation due to an *a priori* gender division of labour. There is sufficient literature available which highlights the strategic role played by women in managing and conserving natural resources including water, fisheries, forestry, natural farming etc (Resurreccion and Elmhirst, 2008).

2. Gender in natural resource management

Gender is defined as the socio-cultural roles and characteristics played primarily by men and women in society (Parker, 1993; IFAD, 1999). The gender perspective in NRM emerges from the inequality and the needs of men and women within a specific social context rather than the sex differences *per se*. In any society, the decisions related to access, control and management of resources are found to be influenced by the roles and responsibilities men and women exercise in the society. Even though environmental interactions have an effect on both men and women, the impact is often found to be gender sensitive. The interactions, men and women foster with the environment, also may be different, which leads to varying opportunities to protect and manage resources in a sustainable manner. As rightly argued by Agarwal (1992), women in rural settings often face the duality as victims of environment degradation, and as active agents when it comes to its protection. In the majority of the cases, men held the key role in decision making and income generation, while women served more or less as the care-takers of the resources. The socio cultural barriers in many communities and societies prevent women from participating community activities including natural resource use. Limited participation of women in natural resource management programmes challenges the effective utilization of the vast indigenous knowledge otherwise could be used for the betterment of society.

3. Major gender issues related to NRM

Gender and development formed an integral component of all NRM projects implemented across the globe. However, many of these reports documented uneven representation of gender concerns in NRM projects. There was also conceptual difference on what constituted gender concerns between projects and even between people working in the same project. Many of them included gender issues as an initial commitment which was lost in the project implementation cycle. Ecofeminism and its essential stereotypes need to be addressed under the following subheads. Pioneering works of Jackson (1993a, 1993b) that proposed a gender analysis framework has been useful in delineating power relations between women and men. This indicated that the factors influencing disaggregated gender roles in NRM are socially and historically constructed which are discussed below in detail.

3.1. Lack of formal tenure rights

Women's property rights are important, as these are fundamental to women's economic security, social and legal status, and sometimes even their survival. Without property rights, women have limited say in household decision-making and no recourse to the assets during crises (be it divorce or death of a husband or any other difficult situation). The lack of property rights may also result in domestic violence. Women's control over land formed a major constraint in many societies that limited the participation of women in NRM projects to the role of users rather than decision makers. The trend is prevalent among different communities in the form of legislation that bars women from ownership, sometimes through socially indoctrinated subordination of women's ownership rights by the men, or at times as a total dependency in terms of land access.

The social milieu and norms in many countries also restrain women from securing land use rights or towards a total denial after marriage or divorce. Lack of ownership on land has dissuaded women from making farm investments, and also from obtaining credit. Picardo (1996) had reported that in Uganda, due to the unfriendly land rights to women, more and more women are choosing to remain out of wedlock to avoid losing land upon the death of the spouse or divorce. Many research findings have proved that a country's development cannot be sustained only if women are granted property rights. Women property rights form the corner stone for gender equality, which can eventually lead to development. According to UN Habitat (1999), one in every four developing countries have laws that impede women from owning property which needs to be corrected in favour of women to have a more inclusive and sustainable NRM regime.

3.2. Unaccounted family chores related to NRM

In many communities social and cultural barriers are found to inhibit or even prohibit women in deciding about natural resource use; despite being critical actors in natural resources use and management. May it be in the management of water or energy use; it is women who are daily involved. But they seldom have the role of decision-makers. The workload for women is found to increase either if NRM programme results in a negative environmental impact, or if the programme is a success (due to the influx of male migrants limiting the benefit for women). Though the success of any programme warrants the participation of all stake-holders right from the planning phase, often women are excluded citing social, cultural and logistic barriers such as timing and length of meetings. When women are denied participation in the planning process, then the decision taken may be less favorable for them. This becomes all the more significant in the recent times of climate change and the increased frequency of climate extreme events. The societal norms or cultural barriers are often seen to leave women more vulnerable to disasters than men. It has been reported that in cases of floods and other natural disasters, societal norms and cultural barriers leave poor and disadvantaged women more vulnerable than men to severe disaster impacts (Azad *et al.,* 2013). Women being the prime users and managers of natural resources, refusal to involve them in planning for NRM programmes may also lead to negligence of their indigenous knowledge about the use and management of natural resources. This assumes all the importance in the context of climate change vagaries and its impact on natural resources.

4. Gender dimensions associated with major natural resources

In order to have a more sensible analysis of the gender dimensions which are clearly noticeable with regard to important renewable natural resource management are outlined below.

4.1. Water resources

Water is one of the important natural resources which is getting scarce day by day and, its sustainable and equitable management is of paramount importance.

According to the reports of the World Bank (2018), about 40 per cent of the world population live in water scarce areas. It also reports that by 2025, about 1.8 billion people will be living in regions of countries that have absolute water scarcity. Recognizing the importance of conserving this natural resource, the United Nations had observed the International Decade for Action from 2005-2015 with the hash tag *Water for Life*, wherein women's participation and involvement in water-related development efforts was the focal theme. The close interlinkages between gender equality and women's empowerment and access to water and sanitation has also been laid down in the Millennium Development Goals (MDG) of the UN. This formed a testimony to the significance of gender dimensions in water use across the world. Moreover, there has been an increase in the number of women appointed as water and environment ministers. On a positive note, the exciting trend would provide an impetus to gender and water programmes worldwide (UN water, 2019).

UN identifies equitable access to water supply, equitable access to land rights and water for productive use as the major factors that need to be addressed while implementing a gender approach to water resources management. Even though rural women contribute substantially towards food production in developing countries (60-80 per cent) they are often constrained by very little or no access to irrigation water for agricultural purposes (UN water, 2019). Despite all these efforts, many communities are still reluctant to accord ownership rights to women or recognize them as land holders who contribute to the development process. Therefore, there is an urgent need to assess and account the role and knowledge of women as users and managers of the water. The documents of prehistoric times depicted that women always played an active role in water conservation programmes. In the household chores also, the water collection for domestic use mostly rested with the home makers and in majority cases, it also has been the women folk. The drudgery associated with long distance travel and the laborious nature of water collection works necessitated them to devise methods and practices for the conservation of water resources in wells, ponds, lakes or rivers especially, in South Asian countries. The active involvement and participation of women in water users' groups are also important.

Reports from Kerala showcase the active involvement of members of women self-help groups (SHG), southern most State of India under *Kudumbasree* (poverty alleviation programme through self-help concept implemented in Kerala from late 1990s) and other organisations in restoring, cleaning, desilting and conservation of various water bodies. These activities helped in reviving many abandoned open wells, channels, ditches, ponds and even digging of new ponds as a water harvesting measure is worth mentioning. The initiatives undertaken by the Government of Kerala since 2016 through its flagship programme, *Harithakeralam* deserves a special mention in the rejuvenation of existing water bodies, ground water recharge and sanitation in many parts of the State. The mission has renovated 390 km of rivers and 41529 km of streams and recharged 55000 wells in Kerala. These activities were led by Mahatma Gandhi National Rural Employment Guarantee Scheme (MGNREGS) through the support of local self-governments that mostly had women as labourers. A campaign organized under this for ensuring the free flow of rivers could clean 7291 km of water stream during 2016-2019 in Kerala (GoK, 2020).

4.2 Land resources

The land is a key economic resource intimately linked to access to, use of and control over other economic and productive resources. Thus it formed a critical resource for poverty reduction, food security and rural development (FAO, 2003). Globally, more than 400 million women are engaged in farm work, but they lack equal rights in land ownership in more than 90 countries. Around 80 per cent of farm work is done by women in India, but they own only 13 per cent of the land (Oxfam, 2013). The recent statistics show that women constitute over 42 per cent of the agricultural labour force in India (NCAER, 2018), but own less than two per cent of farmland. The proportion of female agricultural land owner to total land owners in India is around 12.8 per cent which imply that nearly 88 per cent of the land ownership rests with men. As a key factor of production, the ownership of and/or control over land has been a facilitator for poverty reduction, food security, and inclusiveness. According to the India Human Development Survey (2018), 83 per cent of agricultural land in the country is inherited by male members of the family and less than two per cent by their female counterparts. As the key input for agricultural production, lack of ownership over land denounces women's access to financial resources. The majority of women access land to cultivate it, yet they do not control production, access to property and inheritance rights over the land. A study by UNDP, India found that a savings of Rs. 35,000 per household by consuming own farm grown food utilizing natural farm inputs which minimize the use of natural resources and dependence on external inputs. The annual savings helped the women to get the additional reserve for meeting other family expenditures which enhanced the family economic status. This also implied conservation of natural resources will be possible only through the active participation of people who depended on it for their livelihood which again depended on their land entitlement.

4.3. Forest resources

Forest resource served as the basis for various livelihood options such as food, timber, fuelwood, minor products and other non-timber products. World Bank (2001) statistics show that at least 1.2 billion people around the world depend on forest resources for livelihood. Traditionally, men were involved in activities like felling and logging of timber while women engaged in tasks like fuelwood collection and NTFP (NonTimber Forest Products) procurement. Women were often considered as the primary users of forest resources like fodder, fuelwood, honey, medicinal plants, lac and other products (Aguilar *et al.,* 2011). The sale of these products provide income during the lean seasons of the year. Primarily, smaller quantities of these products are used for household consumption, thus neglecting its economic value altogether and are not considered as an inflow into the family income. The gathering of forest produces involved traversing long distances which amounted to drudgery and also there was the risk involved in accessing forest (Mehta, 2018). However, Women's greater focus on food and health is likely a key driver of differences in conservation priorities between men and women (Wan *et al* , 2011).

The role of women as conservationists of forest resources especially trees can be seen in eco movements like Chipko (India), Ban Thung Yao (Vietnam), Green Belt Movement (Kenya) etc. However, these movements could not translate women rights into ownership of land rights or monetary rights over sold Non Timber Forest Products (NTFP). The Environment and Gender Index (EGI) generated by IUCN indicates that women are underrepresented in forest resource governance and natural resource governance (EGI, 2021). This has a detrimental effect on the success of conservation strategies and programs. A study on Joint Forest Management (JFM) in Madhya Pradesh and Karnataka revealed that gender and caste intersections lead to exclusions in JFM implementation (Marlène *et al.,* 2020). In another study on women's involvement in forestry and their role in JFM in West Bengal, Sarker and Das (2002) observed that despite women's greater involvement in forestry, their participation in community institutions was meagre. However, the success story of integrated forestry from Nepal highlighted the importance of involving rural women in various development programmes. The active participation of women led to improved contribution of forestry in poverty reduction, income generation, health, and nutrition. It also contributed to broader outcomes like a creation of an atmosphere for protecting the environment at the household level, change in the thinking of the local community and enhanced self-confidence of women (Gurung *et al.,* 2000; Aguilar *et al.,* 2011).

4.4. Mangroves

Mangrove lands often described as *wastelands* are being confronted by the opposing forces of conservation and development. Often they are exploited by the external forces whose life was not directly dependent on it. The study conducted by Hema and Devi (2014) in Kerala reported that mangroves conservation efforts were undertaken mainly by the residents and inland fishermen. Women played an active role in planting mangroves especially *Bruguiera gymnorrhiza* along the boundary of their households in the coastal areas of Kerala. This bio-shield protected the area from soil erosion, strengthened the coastal boundary line and significantly offered protection against storm surges in areas that are in proximity to the sea. The mangrove boundary walls gained more prominence close after the Asian Tsunami of 2004 when large tracts of mangroves protected the life and assets in places like Pichavaram in Tamil Nadu. The women involved in mangrove conservation were mostly from economically weaker sections and were doing it as part of sustaining their livelihoods.

4.5. Fisheries resources

Fishery resource is one of the major natural resources that includes both marine fisheries and inland fisheries. An estimated 59.51 million people are engaged in the primary sector of fisheries and aquaculture, of which 14 percent are women (FAO, 2020). India is the second largest fish producing country in the world accounting for 6.56% of global production and contributing about 1% to the country's Gross Value Addition (GVA). The sector provides livelihood support to about 160 lakh people at the primary level and almost twice the number along the value chain in the country. This is facilitated by India's long coastline of 8118 km and equally large areas under estuaries, backwater, and lagoons conducive for developing capture as well as culture fisheries. The inland fishery resources include 3.8 lakh kms stretch of rivers and canals 27.03 lakh hectare reservoirs, 24.7 lakh hectare pond and tanks, 4.3 lakh hectare of beels, 3.48 lakh hectare derelict water bodies /oxbow lakes and 9.65 lakh hectare brackish water areas (GoI, 2019).

Women make up nearly half of the overall workforce in the fisheries sector and the presence of women is visible throughout the fish value chain in harvesting, processing, marketing and trading. The FAO estimates that women comprise 15% of the workforce involved in harvesting and fill 90% of the jobs in fish processing. However, gender inequalities persist in the sector prevent women from fully participating in economic opportunities and decision making, eventually limiting the potential of the sector. Traditional fishing practices like gleaning fishes or catching fishes with hands or small gears are employed by fisherwomen especially in mangrove areas. In other natural resources, fishing is considered as male oriented activity owing to its risks and cumbersome nature. In these regions, women's role and involvement in fishing activities are often undervalued. However higher in the value chain, women play key role in fish processing and retail sales.

5. Gender sensitive indicators in NRM

Addressing gender and socio-economics concerns is considered key to promoting sustainable agricultural development and natural resource management (NRM). Developing suitable gender sensitive indicators (GSI) and baseline standards against which change can be monitored is the step towards fulfilling the objective. According to FAO, Gender sensitive indicators are those that are developed to create awareness of the different impacts of development interventions on men and women taking into consideration their socio-economic and cultural differences. Two types of GSIs have been observed in NRM. The first type formed a gender sensitive impact indicator which describes the actual gender related changes due to a NRM project and the second one is gender sensitive output indicators which describe the actual NRM project in a gender sensitive way. (eg: number of men and women trained in roof water harvesting techniques). Another classification of GSI is based on the qualitative or quantitative nature of assessment. Quantitative GSIs use numbers/ ratios/ percentage to measure change and qualitative GISs are sociological information derived from a more qualitative process of investigation like focus group discussions which gives the views or perception of those experiencing change. GS indicator for natural resource management is usually situation specific even though a general framework could be observed in identifying the indicators. The FAO developed GSI for NRM is based on the PSR (Pressure- State- Response) indicator framework designed by (FAO, 2003) has been included in Figure 1.

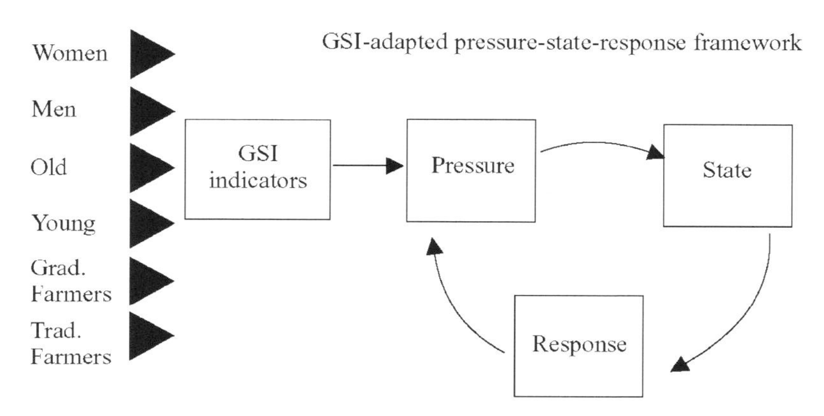

Fig. 1. PSR (Pressure-State-Response) indicator framework.
Source: FAO (2003)

Major steps in developing gender sensitive indicators for NRM proposed by FAO, (2003) is outlined below:

1. *Identifying* the natural resource management issues in the area

2. *Conducting* a gender analysis at the field level using appropriate tools and methods to find out what men and women do with the various natural resources.
3. *Developing* gender-sensitive indicators that are verifiable objectively (quantitative) and subjectively (qualitative), and that fulfil the *SMART* criteria (Specific, Measurable, Achievable, Reliable and Time bound).
4. *Access* whether the data exists, if not seek ways to collect it.
5. *Monitoring* the indicator over time to ascertain whether it suffices? or needs changes.

6. Gender considerations in NRM programme implementation

Pieces of evidence world over on NRM programme implementation points towards a positive correlation between sustainable NRM and the contextual role women take in the resources management decisions. Picardo (1996) had highlighted three components of gender analysis, which applies to natural resource management also. They include gender relations, gender roles and responsibilities and socio-spatial dynamics. In gender relations, one explores gender-based disparities in access to and control over resources, within the social unit of analysis (eg. household, community, livelihood system). Varying interests and motivations of men and women to conserve, protect or manage their resource are ascertained while defining gender roles and responsibilities. Economic consideration is one of the biggest concern while discharging the responsibility concerning conserving & managing the natural resource. Socio-spatial dynamics take care of mobility, displacement patterns, and the spatial location of human activity and how they are commonly differentiated by gender. The study conducted by Picard (1996) [WU1] [A2] in Africa highlighted women's invisibility as farmers and resource managers in the design and implementation of agricultural programs and research. Hence while implementing NRM programmes, a regular gathering of baseline information on the differential interaction of men and women in using resources and updating for monitoring and evaluation purpose need to be taken care of. Sufficient caution may also be exercised so that technologies and procedures are gender sensitive to ensure full integration into NRM programmes. Picard (1996) [WU3] also suggested that five different components for addressing gender issues which include (1) decision making, (2) female participation, (3) access to resources, (4) control over resources, and (5) benefits.

He also identified that projects which are based on the needs and priorities of community members (men and women) will have a greater chance of success. Ever since the inclusion of people in programmes as important stakeholders have been adopted which assured a sense of responsibility in natural resource management. It is also important that the stakeholder groups should comprise people of all age both young and old. Participation of both genders in both technical and institutional aspects of the programme including planning and implementation has also been considered a prerequisite. Regular evaluation and monitoring should be done to assess the gender-wise implication of the programme. The active participation and success of the programme are observed to have a multiplier effect on the community for future course of action. A case of successful integration of women in NRM by North East Network (NEN) has been included as Box-1.

Box-1: North East Network (NEN), women and NRM

North-East region (NER) of India, one of the world's biodiversity hotspot, is bountifully endowed with rich minerals, forest wealth, wildlife, water resources, agro-biodiversity and rich traditional knowledge systems. NEN works for the conservation of natural resources and believes that any work in isolation without the community's support is impractical. They have identified the critical role and participation of women in the processes of Natural Resource Management and its governance. Women in rural communities have in-depth knowledge, practices and understanding of the natural environment which remains unaccounted for and invisible. NEN, therefore, works towards strengthening women's role and amplify their voices in influencing policies at the community, state and global platforms on issues of environment, climate change, biodiversity, sustainable livelihoods, food and farming. These strategies are used to advocate for women's participation in all decision making processes. *Source:* North East Network

7. Conclusion

The gendered nature of natural resource management reiterates direct linkages with poverty and power positions in society. This has created gender gaps that act as barriers to sustainable development and livelihoods. But it is significant to note that it has further reinforced the already skewed power imbalances and gender inequality in many societies. This suggests that the sustainable solutions lie in building on the complementarity between the roles of men and women with recognition to the diverse roles played by women in different contexts. Women's practices and indigenous knowledge base will have a direct bearing on the processes. Also, context-specific understanding of the relationship of women groups with environmental resources can help in evolving better NRM policies and programs. This warrants more mediated dialogues to decipher the complex relations of women with men, kin and other social actors in the realm of NRM. This calls for continually reformulating relational and power dimensions for better gender structuring that favour woman who have remained the custodians of natural resources.

References

Agarwal, B. 2010. Gender and Green Governance: The Political Economy of Women's Presence Within and Beyond Community Forestry. Oxford University Press, United Kingdom.

Aguilar, L. Quesada-Aguilar, A. and Daniel Shaw, 2011. Wrapping up: the status of international forest policy and gender. In: Aguilar, L., Quesada-Aguilar, A. and Shaw, D. M. P. (Eds.). Forests and Gender. IUCN and New York. 123p.

Azad, A. K., Hossain, K. M. and Mahbuba Nasreen, M. 2013. Flood-Induced Vulnerabilities and Problems Encountered by Women in Northern Bangladesh. Int. J. Disaster Risk Sci., 4 (4): 190–199.

Barsch, H. and Bürger, K. 1996. Naturressourcen der Erde und ihre Nutzung. Klett-Perthes, Germany. 264p.

Sarker, D. and Das, N. 2002. Women's Participation in Forestry: Some Theoretical Issues. Econ. Pol. Wkly., 37(43), 4407-4412. Retrieved April 7, 2021, from http://www.jstor.org/stable/4412774

EGI (Environment and Gender Index), 2021. Environment and Gender Index Datasets. Available: https://www.wocan.org/resources/environment-and-gender-index-egi-datasets. [06th Apr. 2021].

FAO (Food and Agriculture Organization), 2003. Gender-Sensitive Indicators for Natural Resources Management (online). Available: http://www.fao.org/tempref/docrep/fao/010/a0521e/a0521e00.pdf [06th Apr. 2021].

FAO (Food and Agriculture Organization), 2020. The State of World Fisheries and Aquaculture 2020. Sustainability in action. Food and Agriculture Organization, Rome. 224p.

GoI (Government of India), 2019. Handbook on Fisheries Statistics 2018. Publishers: Department of Fisheries Ministry of Fisheries, Animal Husbandry & Dairying Govt. of India, New Delhi. 190p.

GoK (Government of Kerala), 2020. "Now, Let me flow". Report on campaign "Ini Njan Ozhukatte" to rejuvenate streams and its tributaries. Haritha Keralam Mission. Government of Kerala, Thiruvananthapuram. 76p.

Gurung, B., Thapa, M.T. and Gurung, C. 2000. Guidelines on gender and natural resource management. The International Centre for Integrated Mountain Development, Nepal. 36p.

Hema, M. and Devi, P.I. 2014. Factors of mangrove destruction and management of mangrove ecosystem of Kerala, India. J. Aquatic Biol. Fisheries, 2: 184-196.

IFAD 1999. Gender and water (online).Available: https://www.ifad.org/documents/38714170/39135645/Gender+and+Water+-+Security [06th Apr. 2021].

Jackson, C. 1993a. Doing what comes naturally? Women and environment in development. World Development. 21(12): 1947–1963.

Jackson, C. 1993b. Environmentalists and gender interests in the Third World. Development and Change 24: 649–677.

Marlène, E., Grosse, A. and Campbell, N. 2020. Unpacking 'gender' in joint forest management: Lessons from two Indian states. Geoforum, 111: 218-228.

Mehta, A. 2018. Gender gap in land ownership (online). National Council of Applied Economic Research (NCAER). Available: https://www.ncaer.org/news_details.php?nID=252&nID=252. [06th Apr. 2021].

North East Network for Women and NRM (online). Available: https://northeastnetwork.org/. [06th Apr. 2021]

OXFAM (Oxford Committee for Famine Relief). 2013. When Women Farm India's Land: How to Increase Ownership? Policy Brief. Oxfam India, New Delhi. 4p.

Parker, A. R. 1993. Another Point of View: A Manual on Gender Analysis – Training for Grassroots Workers. United Nations Development Fund for Women (UNIFEM), New York. 116p.

Picard, M. 1996. A Guide to the Gender Dimension of Environment and Natural Resources Management Based on a Sample Review of USAID NRM Projects in Africa (online). SD Publication Series, Office of Sustainable Development Bureau for Africa, 30p. Accessed on 27-02-2021.

Resurreccion, B.P. and Elmhirst, R. 2008. Gender, Environment and Natural Resource Management: New Dimensions, New Debates. In: Resurreccion B. P. and Elmhirst, R. (Eds.) Gender and Natural Resource Management. Earthscan, London, pp. 3-22.

Shiva, V. 1986. Ecology Movements in India. Alternatives XI. 11(2): 255-273.

UNCHS (Habitat), 1999. Women's Rights to Land, Housing and Property in Post conflict Situations and During Reconstruction: A Global Overview (online). Available: https:// mirror.unhabitat.org › downloads › docs [06th Apr. 2021]

UN Water (United Nations), 2019. Climate Change and Water UN-Water Policy Brief. Technical Advisory Unit, UN-Water. Genève 2 – Switzerland. 28p.

Wan, M., Colfer, C.J.P and Powell, B. 2011 Forests, women and health: opportunities and challenges for conservation . International Forestry Review, 13(3): 369-386.

World Bank, 2001. A Revised Forrest Strategy for the World Bank Group. Draft 30 July 2001, WB, Washington, DC. 99p.

World Bank, 2018. Water Scarce Cities: Thriving in a Finite World. World Bank, Washington, DC. 65p.

8

Climate Change and Gender Vulnerabilities in Agriculture

Chitra Parayil, Aiswarya.T.P. and Binoo P. Bonny*

1. Introduction

The world's temperature has already warmed up to almost 1.2°C since the pre-industrial levels and this warming impact is visible in the form of extreme weather events, sea level rise and diminishing Arctic sea ice. Heat waves and drought in Europe and China, forest fires in the U.S., dust storms and extreme rainfall in India (including the Kerala floods 2018) and high precipitation in Japan and other island nations are all examples of the disasters which have occurred within a single year of 2018. With a further 0.5°C warming, these effects would be even more pronounced than the scientists' previous predictions. A 1.5°C warmer world will see higher temperatures, increase in frequency and intensity of precipitation, higher sea levels, and floods, droughts and heat waves (Venkatesh, 2018). The climate change induced economic damage has been increasing in the past few decades and is likely to continue growing because of population growth, urban development and changing land use pattern (IPCC, 2012). Kerala encountered the most disastrous floods in its history since 1924, between June 1st and August 19th of 2018. As the torrential rainfall and associated storm thrashed the state, the entire state got buried under water with only few areas remaining above water. The combined precipitation received by the state during this period was 42 per cent in excess of the typical normal. The exceptional spell of rainfall inflicted heavy damage on the life and properties of thousands of people in the state.

Since 2012, the United Nations Framework Convention on Climate Change (UNFCCC) has considered gender and climate change as a stand-alone agenda item under the Conference of the Parties. This is because it has been already understood that climate change has a greater impact on those sections of the population which are more reliant on natural resources for their livelihood. People who have the least capacity to respond to natural hazards are also affected

**Author contact: chitra.parayil@kau.in*

more. Majority of women, who are considered among the poorest of the poor are at a greater disadvantage because their income is mostly derived from informal natural resources dependent livelihoods. Globally, more than 400 million women engage in farm work in more than 90 countries. Agriculture being a climate sensitive sector, climate change takes a huge toll on this area. Women are usually engaged in subsistence agriculture and labor-intensive works which worsens their susceptibility to climatic change (Lambrou and Piana, 2006). Hence, during extreme weather events, women experience greater impacts and vulnerability than men and become economically insecure following disasters.

2. Gender vulnerability to climate change

Roxy *et al.* (2017) have reported that each year, floods in India from extreme rainfall results in a loss of around three billion dollars which constitutes about 10 per cent of global economic losses. A large number of agriculture dependent rural households, most of which are involved in subsistence agriculture, were found to have borne the brunt of the unprecedented deluge as it vandalised the agricultural fields. Even though these extreme weather events have become a new normal, studies relating to gender and climate change have been limited in India. Based on a study undertaken in the flood affected regions of Kerala to assess the gender vulnerability of agricultural households to extreme weather events, it was found that female-headed households (FHH) had greater livelihood as well as economic vulnerability compared to male-headed households (MHH). The study was carried out in the BPL agricultural households of Mala and Vellangallur blocks in Thrissur district. The study was mostly based on primary data which was collected from male and female respondents through personal interview method. The vulnerability of female-headed households were studied in comparison to male-headed households. Their livelihood vulnerability was found using the approaches of LVI (Livelihood Vulnerability Index) and LVI-IPCC (Livelihood Vulnerability Index-Intergovernmental Panel on Climate Change) designed by Hahn *et al.* (2009). LVI is a composite index consisting of seven major components viz. socio-demographic profile, economic status, social networks, health, food, water and natural disasters and climate variability. These major components were grouped to form the three contributing factors of vulnerability according to IPCC viz. adaptive capacity, exposure and sensitivity. They were quantified into the composite index using weighted method as discussed here.

2.1. Livelihood Vulnerability Index (LVI)

The LVI values for the female headed households (FHH) estimated at 0.39 indicated a higher LVI than male headed households (MHH) with scores at 0.32 (Fig. 1). This depicted relatively greater vulnerability of women to extreme weather events compared to male-headed households. The component wise estimate of livelihood index presented as the spider diagram (Fig. 1) indicated that the women were more vulnerable with respect to all the seven components. The scale of the diagram ranged from 0 (least vulnerable) at the centre of the web, increasing to 0.6 (most vulnerable) at the outer edge of the web with an increment of 0.1 unit.

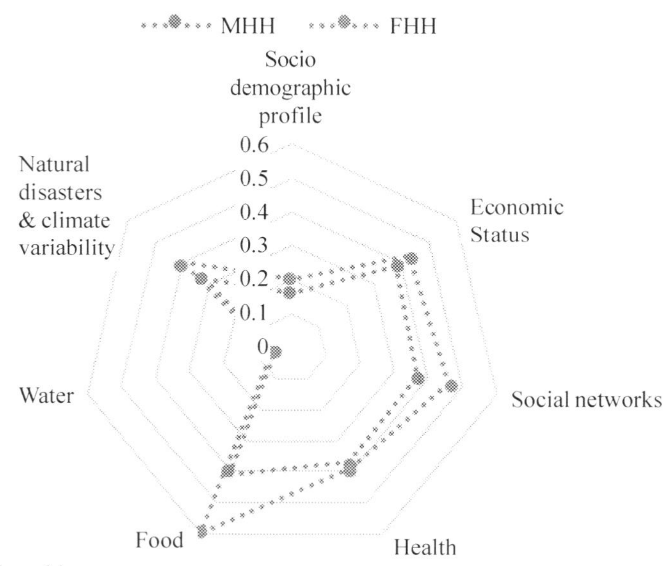

0.6 - Most Vulnerable

0 - Least Vulnerable

Fig. 1. Spider diagram for major components of LVI

A gender wise comparison on selected socio-economic variables among FHH and MHH is presented as Table 1.

Table 1. Socio-economic profile of the respondents (N=150)

Socio-economic variables	MHH	FHH	Total*
Age (years)			
Below 40	5 (4.5)	3 (7.5)	8 (5.3)
40-49	31 (28.2)	9 (22.5)	40 (26.67)
50-59	41 (37.3)	17 (42.5)	58 (38.6)
60 and above	33 (30)	11 (27.5)	44 (29.3)
Average age			
Family size			
2-4	84 (76.3)	32 (80)	116 (77.3)
5-10	26 (23.6)	8 (20)	34 (22.6)

Average family size	3.8	3.35	3.575
Dependency ratio			
0-0.49	49 (44.5)	18 (45)	67 (44.6)
0.5-0.99	20 (18.1)	5 (12.5)	25 (16.6)
1-1.99	34 (30.9)	14 (35)	48 (32)
2-3	7 (6.3)	3 (7.5)	10 (6.6)
Average dependency ratio	0.628	0.62	0.624
Educational status			
Illiterate	2 (1.8)	2 (5)	4 (2.6)
Primary	3 (2.7)	2 (5)	5 (3.3)
Upper primary	8 (7.2)	4 (10)	12 (8)
Up to 10th std	57 (51.8)	19 (47.5)	76 (50.67)
Higher secondary	28 (25.4)	8 (20)	36 (24)
Graduate	12 (10.9)	5 (12.5)	17 (11.3)
Farming experience (yrs.)			
<10	29 (26.3)	14 (35)	43 (28.6)
10-25	62 (56.3)	26 (65)	88 (58.6)
>25	19 (17.2)	0	19 (12.6)
Land holdings (ha)			
Marginal farmer	84 (76.3)	35 (87.5)	119 (79.3)
Small farmer	23 (20.9)	5 (12.5)	28 (18.67)
Large farmer	3 (2.7)	0	3 (2)
Main occupation			
Agriculture only	37 (33.6)	6 (15)	43 (28.6)
Agriculture as main occupation	27 (24.5)	7 (17.5)	34 (22.6)
Agriculture as subsidiary occupation			
Govt. service	4 (3.6)	0	4 (2.6)
Private service	8 (7.2)	11 (27.5)	19 (12.6)
Self employed	18 (16.3)	8 (20)	26 (17.3)
Agricultural labourers	8 (7.2)	5 (12.5)	13 (8.6)
Non-agricultural labourers	8 (9.2)	3 (7.5)	11 (7.3)
Sub total	46 (41.8)	27 (67.5)	73 (48.7)
Annual Income (Rs)			
Total household income (farm income + non-farm income)	2,37,949.36	1,93,806.25	4,31,755.6
Average agricultural income	1,49,085.27	1,06,206.25	2,55,291.52
% share of agricultural income to total income	62.65	54.8	58.72

* Figure in bracket shows the percentage to the total, MHH-Male-headed households, FHH-Female-headed households, Total number of MHH=110, Total number of FHH=40.

Socio-economic profile of the study area showed that more number of female-heads were deprived of high school education (20% compared to 11.8% in men) although a greater percentage of women had graduate level education (12.5% compared to 10.9% in men). Those deprived of high school education were mostly the eldest among the respondents and had either lower or upper primary level of school education. Poverty might be the main reason for deprivation of education and the variations in educational attainment simultaneously affected population growth, economic growth, greenhouse gas emissions, and vulnerability to climate change impacts.

In terms of land holdings, women in the study area were mostly marginal farmers (holding size <1 ha) who practised agriculture as a subsidiary source of income (67.5%). They had lower agricultural diversification (2.08) compared to male-headed households (2.17) which makes them more vulnerable in a changing climatic scenario. In most cases, women were involved in two major activities viz. agriculture and small scale livestock or poultry rearing, whereas in MHH, many of them carried out multiple allied sector activities along with agriculture. Fish rearing and growing chicken in large numbers on contract basis were found in some of the MHH.

However, quite a large number of MHH were completely dependent on the income from agriculture for their livelihood. The proportion of agricultural income was also higher among them. Since agricultural sector is sensitive to climate change, their income level may vary according to changes in monsoon pattern or occurrences of extreme weather events. That apart, male farmers had greater productive assets and income from agriculture. Availability of liquid assets and savings which indicates the household's capacity to cope with shocks were lower or almost absent among the FHH. They also had lower productive assets like land and diversification within agriculture. Due to lower land entitlements, women found it difficult to receive agricultural loans which also contributed to their greater vulnerability. Altogether the greater economic vulnerability of FHH compared to MHH had emanated from these factors. While natural disasters and dependency ratio of the family positively influenced the economic vulnerability of FHH, high school education, assistance from local government, family size and agricultural diversification index were found to negatively influence their vulnerability. Greater vulnerability to shocks like extreme weather events exacerbate their poverty and in turn their vulnerability to future shocks.

The household income of almost all the respondents were severely disrupted in the months immediately after the floods of 2019 for an average range-3 to 5 months. Most of the households did not have any income during this time as their farms were inundated by the floods. Most of crops which were ready to be harvested were damaged in the flood waters and those that remained were given to relief camps. Other employment opportunities also did not resume soon after the floods. Altogether, many of the BPL families barely survived with the food kits and clothes provided by the government flood relief schemes. Some of the MHH reported that men started going for non-farm employment after two to three months of the disaster which helped them to come out of the grim situation, reducing their vulnerability, compared to that of FHH.

Most of the FHH perceived that if they lost their primary livelihood source, they had no other option than relying on family or relatives for support. This is mainly because of the social obligations that the concept of gender has placed on women, which in turn has brought down their roles into house-keeping eventually making them less self-reliant. These indicate that women had more difficulty in realizing their financial requirements after a disaster compared to men. This was further substantiated from the fact that most of the MHH reported that they would rely more on other existing income-generating activities or find a new informal job. This found evidence from the reported cases where the male farmer had switched-on to supplying snacks to a bakery as the income from paddy farms had significantly reduced after the floods. Even the number of helps received through social networks were higher among FHH indicating their more vulnerable situation. These social network helps were either in the form of money or in the form of medicines, clothes or food during times of disasters and that too only in cases of emergency. Most of the respondents were prudent enough not to borrow money even in worst situations as they perceived it to increase their financial burden. Also, for some respondents, their relatives or friends were also poor or in even worse situations which prevented them from receiving helps. However, most of the women were members of at least one social group like Kudumbashree or Joint Liability Group (JLG) which helped them financially during times of stress.

Even with less physical capabilities and many reported issues of nutritional deficiencies, women had no choice than to bear the burden of the household chores and health care of the children, elderly and sick at home. Out of the respondents, FHH had a greater percentage of chronically bedridden people at home than MHH (20% compared to 13.2% in MHH). Many women revealed that they were not spared from household chores even when they themselves fell sick. One woman reported that increase in back pain due to ageing has deprived her from carrying out most of the farm activities that she herself used to do earlier such as turning the soil and providing props for banana. Now such tasks have to be carried out by hiring labour which reduces the profit from the farm. Among FHH, it was found that many of them were taking medicines for hypertension. These have put them at disadvantage during times of flood and crop loss. Women were really worried not just about the farm alone, but also about everything at home. None of the FHH had a health insurance whereas

10.9 per cent of MHH reported to have health insurance. Most of them were insured under the Rashtriya Swasthya Bhima Yojana Scheme (RSBY) under Government of India which aims to provide financial security for hospitalization-related expenses to BPL families.

From the study area, it was found that FHH were more vulnerable (0.40) to natural disasters and climate variability compared to MHH (0.32). When asked about whether they believed that they live in a disaster-prone area, 100 per cent of the FHH responded affirmative and were even more worried thinking about shifting things to safer places, disruption of children's education and cleaning the house after the floods. However, the percentage of households that did not receive a warning about the pending disaster were higher among FHH (0.13 versus 0.20 in MHH). This was evident from field reports where a woman, whose husband worked abroad, responded that she had been attentively noting the alerts and warnings given by the government even on the day before which flood waters entered their home. But she had no clue about the floods entering her house when it did happened. She opined that getting to know the rainfall alerts will be helpful only if they were more region-specific. Having had to be evacuated from homes suddenly with the children was an unmanageable task for her being the only elderly person at home. Some of the male respondents also told that they did not care the rainfall alerts and claimed that they knew their place better.

When asked about whether the 2018 and 2019 flood experiences helped them to stay prepared for the future, most of the respondents were desperate and told that nothing could prevent damage when floods of this magnitude occurred and that all they could do was to save their lives. However, some of the FHH reported that during the month of August in the following year (2019), most of them kept emergency kits containing clothes, medicines, torch lights and important documents ready beforehand. Most of the households also shifted their household objects to safer places in the house.

2.2. Livelihood Vulnerability Index (LVI-IPCC)-Intergovernmental Panel on Climate Change

LVI-IPCC also yielded a similar result as depicted in Fig.2 which showed that FHH were more vulnerable in terms of all the three contributing factors ie., adaptive capacity, exposure and sensitivity. Female-headed households had a lower adaptive capacity and higher exposure as well as sensitivity.

The triangle illustrated that FHH had a higher sensitivity with score at 0.39 against the MHH score of 0.28 and exposure scores at 0.40 and 0.32 respectively to extreme weather events. They also showed a lower adaptive capacity (0.512) than MHH (0.58).

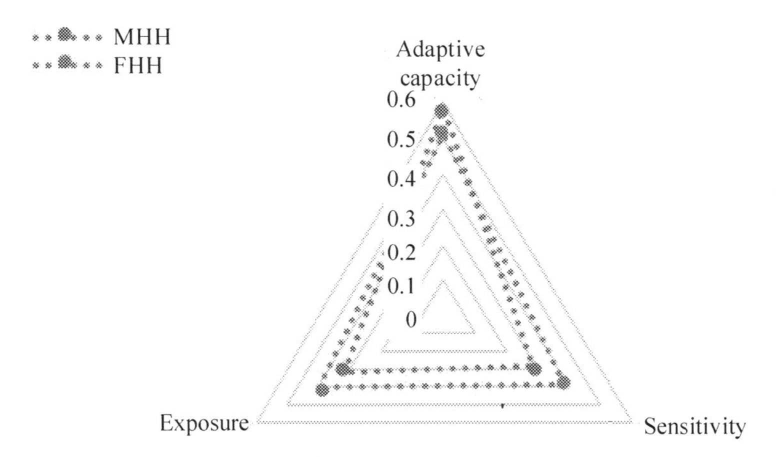

0 - Low Contributing Factor

0.6 - High Contributing Factor

Fig. 2. Triangle diagram for contributing factors of LVI-IPCC

3. Conclusion

It could be observed from the studies discussed that female-headed households were more vulnerable than male-headed households in terms of livelihood vulnerability as well as economic vulnerability to extreme weather events. They had a lower adaptive capacity and higher sensitivity and exposure to natural disasters. Natural disasters and dependency ratio positively influenced the economic vulnerability of female-headed households. Livelihood diversification was the most preferred adaptation strategy among them.

However, various initiatives under Kudumbashree have really helped the female respondents in terms of social empowerment as well as in gaining income. Initiatives like collective farming, Mahila Kisan Sashakthikaran Pariyojana (MKSP) and Polivu were examples that proved to be successful in enabling women with the needed support. Collective farming aimed at encouraging crop cultivation among neighborhood groups. This initiative could bring significant changes in the lives of the poor and also helped in increasing agricultural production by converting fallow and cultivable waste land into agricultural use. It also has significance as a food security measure in the local setting. Women participation in this programme was as cultivators and not just as agricultural wage labourers. They also helped them with a control over the means of production and access to formal credit which aided in increasing the returns from farming. There were some women from the study area who were part of collective farming through Joint Liability groups (JLGs). These JLGs are structured according to NABARD (National Bank for Agriculture and Rural

Development) guidelines. It also provided an open bank account in the name of the JLGs which was also brought under the purview of Interest Subsidy Scheme (ISS) of Kudumbashree. All such initiatives supported women to achieve the much needed financial security, a prerequisite to all forms of empowerment and pro-activeness, especially in times of distress and disasters.

References

Hahn, M.B., Riederer, A.M., and Foster, S.O. 2009. The Livelihood Vulnerability Index: A pragmatic approach to assessing risks from climate variability and change — A case study in Mozambique. Global Environ. Chang. 19(1): pp.74-88.

IPCC (Intergovernmental Panel on Climate Change), 2012. Managing the Risks of Extreme Events and Disasters to Advance Climate Change Adaptation (Special Report of the Intergovernmental Panel on Climate Change). Cambridge University Press, Cambridge, 594p.

Lambrou, Y. and Piana, G. 2006. Gender: The missing component of the response to climate change, Rome, pp. 1-58.

Roxy, M.K., Ghosh, S., Pathak, A., Athulya, R., Mujumdar, M., Murtugudde, R., Terray, P. and Rajeevan, M. 2017. A threefold rise in widespread extreme rain events over central India. Nat. Commun. 8(1): 1-11.

Venkatesh, S. 2018. Warming up to catastrophe?. Down to Earth 27(11): 36-43.

9

Gender Sensitization A Step Towards Gender Inclusive Development

Sreeram Vishnu and Archana Bhatt*

1. Introduction

The intricate relationship between agriculture and women has been widely featured in the development discourse for long. Globally, more than 400 million women are engaged in farm work directly or indirectly. In other words, 42 per cent of economically active women are engaged in agriculture and they comprise about 43 per cent of the total workforce in agriculture (Dash and Srinath, 2013). In India, agriculture and allied sectors continue to be the most immediate avenues of employment and income for about 160 million rural women who work as farmers, co-farmers, farm labourers and farm entrepreneurs (Oxfam 2017; Sadangi *et al.,* 2009). It is estimated that about 65 per cent of rural women workers are engaged in agriculture as cultivators and agricultural labourers in the country as opposed to 49.8 per cent of male workers. In these capacities, women perform multiple tasks such as land preparation, sowing, transplantation, harvesting and rearing of animals. Most of these operations are labour intensive and not mechanized (Pachauri, 2019). Besides, they are engaged in off-farm domestic activities such as cooking, cleaning, childcare, water and fuel collection and community activities. Shrinking remunerative opportunities in the farming sector has triggered migration of rural male folk to urban areas and also there is a shift in their job priorities to pursue employment in the rural non-farm sector in recent years. Since women did not feature predominantly in either of these trends, most of them are still engaged in agriculture, mostly as labourers and cultivators (Mehrotra, 2020). Vindicated by the fact that the number of female agricultural labourers in India increased by 24 per cent between 2001 and 2011, even though 7.7 million farmers left farming during the corresponding period. During the period when up to 34 per cent of men in rural areas migrated in search of employment and better economic opportunities, the corresponding figure for rural women is merely 3.6 per cent. A recent study of National Sample Survey Office (NSSO) found that 55 per cent of employed women in India are

**Author contact: sreeram.vishnu@kau.in*

working in the farm sector (MoSPI- PLFS, 2020) and their concentration is highest (28%) among small and marginal farmers. Active participation of women in agriculture is reported from most of the Indian states, except Kerala, Punjab and West Bengal where women are actively participating in non-agricultural activities (Ghosh and Ghosh, 2014). As per the estimates of the International Labour Organization, an average employed Indian woman works 44.4 hours per week as against the developing country average of 35-36 hours (ILO 2018). Rural women continue to face strong socio-cultural barriers and direct resistance from male members towards social and economic liberalization and independent decision making (DRWA 2014). The percentage distribution of female workers by broad industry division in India is depicted as Figure 1.

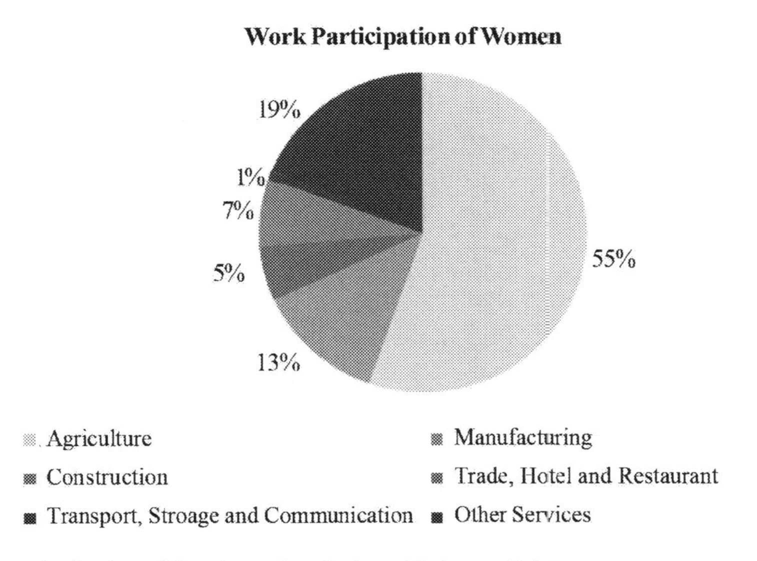

Fig. 1. Distribution of female workers by broad industry division
(*Source:* NSSO PLFS Report, 2018-19)

1.1. Women in agriculture

Various studies report that the role of women in agriculture is not well acknowledged. Their increased work participation has hardly translated into equal employment opportunities or remuneration. Women employment in agriculture sector is mostly seasonal and provisional adding up to all sorts of uncertainties in their income cycle (Nirmala and Venkateswarlu, 2012; Mehorta, 2020). Discrimination of wages and working status still prevails for women agricultural labour (Ghosh and Ghosh, 2014). This is despite the role of women in providing family labour input in agriculture and allied sectors like livestock keeping where their contribution exceeds that of men (Tipilda and Kristjanson. 2008). These suggest that their involvement has not been given due place and their share in income from livestock is considered negligible (Dudi *et al.*, 2019). These indicate that gap in access to productive resources and services persists and the gender-specific constraints are undermining women's potential to contribute to sustainable agricultural growth. Moreover, it tend to perpetuate disparity in earnings for men and women (Dash and Srinath, 2013). This finds support in the study report of Oxfam (2019) which state that the gender wage gap is highest in Asia with women wages recorded at 34 per cent below men, even in conditions of equal qualification and work. Further, according to prevailing social norms, women are supposed to perform domestic work like child-rearing and collection of fuel and water collection. Prevalence of illiteracy, lack of knowledge, improper training and less opportunities for skill development worsen their subdued existence (Majumder and Shah, 2017). These constraints would make them deprived of the land rights, access to systems and leave them incompetent concerning decisions related to agricultural transactions. This further deepens their precarious socioeconomic status and exclusion from paid labour. The low rank of India in the Global Gender Gap report 2020 of World Economic Forum, also reflect the low levels of economic participation and opportunity for women in the country. A study report by CIAE (2013) indicates declining agricultural workforce in the coming years and predicts continuance of this trend till the wages offered in the farm sector are comparable to other sectors. Such a scenario would lead to mechanization of most of agricultural operations and brings the need to upskill and reskill the farm labourers. With the present trend of increasing share of women labourers in agriculture, it may be a daunting challenge to equip the female labour force to fit into the employment opportunities in the farm sector. The change in the agricultural labour force dynamics in the country is presented in the Figure 2.

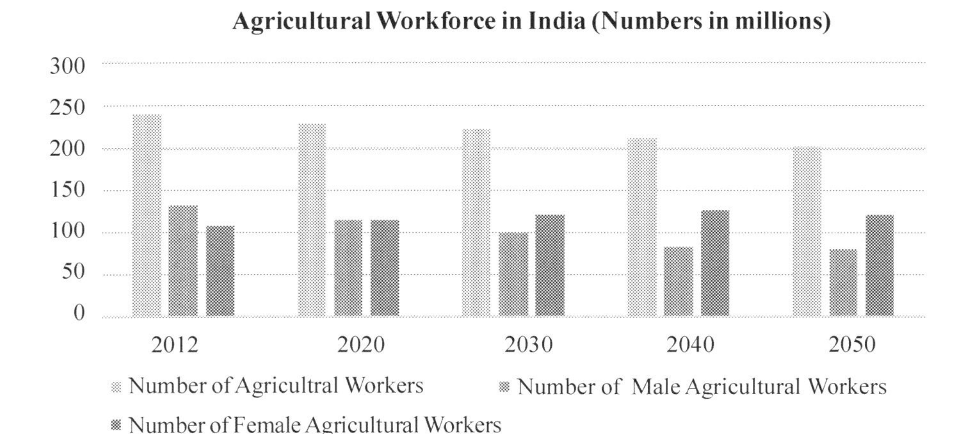

Fig. 2. Population dynamics of Indian Agricultural workforce over the years

Access of farm women to agricultural extension and advisory services (EAS) is another serious concern (CIWA 2018; Beevi *et al.*, 2018). EAS and information on new technologies are almost exclusively directed to men, even though women are increasingly contributing to the farm work (Kelkar 2011). Though multitude of factors contribute to the current scenario, one of the prime causes is attributed to the lack of required degree of gender sensitivity of our extension system. Extension mostly fails in adoptinggender-focused extension approaches and models for dissemination (CIWA 2018). Another significant cause emerges from the barriers created by the prevalent socio-cultural values and taboos that exist between extension agents & the farm women (Sadangi *et al.* 2009). There are global reports from Ethiopia that reveal social norms restricting the male extension officials from interacting with women farmers (Meinzen-Dick *et al.*, 2011). Also the gendered perspectives cause differences in the extension needs of women and men as they are often involved in different activities (CRISP n.d).

Finally, there also arises the issue of whether women are recognized as farmers per se for availing the benefits of the extension services (Meinzen-Dick *et al.*, 2011). When extension officers do contact women, it is often to provide information and advice that pertains to women's household, rather than farming (Berger *et al.* 1984). Thus lack of awareness among male extension workers about specific gender-based realities of agricultural information access (Lamontagne-Godwin *et al.*, 2019) deepens the marginalization of women. Moreover, reports suggest that women tend to be excluded further from the ambit of EAS with the service delivery platforms.

These conditions of inequality may persist, if gender is not well integrated into the agricultural development projects. There are studies which argue that women are capable of bringing positive contributions to development outcomes when they are given the opportunity and resources (Meinzen-Dick *et al.*, 2011). It is in these pretexts, the gender sensitization issues in agriculture are explored in this chapter. The focus is on transformative approaches which question the long-held view of women's subordinate position in various socio-economic spheres and propounds for a gender-equitable society. Measures which advocate or reinforce better integration of women into agriculture are also discussed with the support of evidence from literature and review of case studies.

2. Gender sensitization

Gender sensitization is the process of making a person aware of the differential ways in which men and women will be affected by policies, programs, and its outcome (Meinzen-Dick *et al.*, 2011). It intends to change the stereotypic mindset of men and women, which considers both as unequal entities and prompts them to function in separate socio-economic space (Dash *et al.*, 2008). The process intends to increase the sensitivity of people at large towards women and their issues and is considered as the primary step in recognizing gender equity and equality. It puts light into the role of women farmers and their differential needs in agriculture. As a result of sensitization, men is expected to share more responsibilities with women and actively back them in the pursuit of equality.

2.1. Gender sensitization in agriculture

Among various social identities, gender is one of the critical determinant in deciding the individual's relation with the society (Shrestha *et al.*, 2020). It refers to the socially constructed roles and responsibilities of women and men (Holmes and Jones, 2010). It has assumed greater significance in agriculture with the sweeping changes that is happening in terms of roles and technology use in recent years. Despite the spread of education and awareness, gender bias is still a glaring reality in our society, more particularly in rural areas. In agriculture it is manifested in myriads forms in almost all aspects of farming such as labour, control of resources in production to marketing causing marginalization of women (Meinzen-Dick *et al.*, 2011). These indicate that agriculture sector tend to reinforce a patriarchal ideology and a socio-cultural value system that keeps women bound to villages to perform domestic chores and agricultural work as argued by Mehrotra (2020). Though with the migration of men, women became de facto heads of many rural households, it has not resulted in their enhanced participation in decision making (Nelson *et al.*, 2002). Further, access of men as well as their control over the productive resources is much higher than women in most of the farming systems (ICAR, 2007). According to India Spend (2019), only 13 per cent of the women tillers owned their land. This lack of control on land resources has created serious impediment in EAS delivery as it is primarily targeted to farmers who are landowners and as such women continue to be secluded from the purview of extension agencies. This inequality has even pervaded into the competencies of women to deal with uncertainties related to climate change. Though women and men farmers are vulnerable to adverse impacts of climate change, women are less likely to act to reduce vulnerability by adopting climate-smart farming practices (Kristjanson *et al.*, 2017). Women perceive greater gender inequality than men do and encourage the implementation of measures to increase awareness and address the problem (García-González *et al.*, 2019). Without a deliberate challenge to gender relations, women farmers are not likely to benefit from EAS on par with their male counterparts (Ogawa 2004). Therefore, specific interventions are required to improve women's integration in agricultural activities by examining their social and economic condition. At the same time, these efforts must raise awareness about the contribution of women to farming as well as their other protective and household management roles among male members. Interventions for gender sensitization have been resorted as a means in various sectors including agriculture to highlight the role of women and stimulating actions for transformative changes. This include sensitization training, campaigns, workshops etc.

2.2. Process of gender sensitization

Gender sensitization aims to bring change in the thinking of individuals about the practices and approach towards gender issues. According to Dash *et al.* (2008), in sensitization process individuals undergo the following definite stages of sequential changes:

- Perceptional change
- Recognition
- Accommodation and
- Action

Perceptional change: The first stage of the gender sensitization process involves changing the inherent mindset of men and women about their roles and gender division of labour in the society. Traditionally women are seen in the subordinate role and considered as a weak and unequal entity. This very assumption is questioned and scrutinized at this stage. As perceptional change sets in men start to realize that women are equally important in the society and have the capacity to make independent decisions at various levels.

Recognition: In the recognition phase the multifarious roles played by women in productive, reproductive and community management functions of family and society are brought to the fore. Generally, these roles are less recognized and appreciated by men. At this stage, men start to look into the positively endowed qualities of women. Interestingly, women will also start to realize their vital roles and contributions at different levels. Thus the role of women and the significance of their actions gains more visibility at this stage. Such an orientation would prompt a realignment in the gender relations.

Accommodation: This is an adjustment phase in which the barriers between men and women start to disintegrate. Men tend to understand the difficulties and issues faced by women with great empathy. Rather than overlooking or aggravating the differences with their counterparts, men at this stage would try to discuss and resolve the issues. In other words, men would try to rationalize their behaviour by shedding their ego to accommodate concerns of women. In this way gender relations would be improved by narrowing the difference between men and women and both will try to work in harmony.

Action: In this final stage, the changes will be materialized into action and would become more apparent. Men would become the main proponents in promoting the welfare of women and their efforts would be directed to improve gender equations at all levels. Sensitized persons would play the lead role in nurturing women's talents and helping them to check discrimination and to claim their status. Women at this stage would tend to act collectively to address their issues which are necessary for their empowerment.

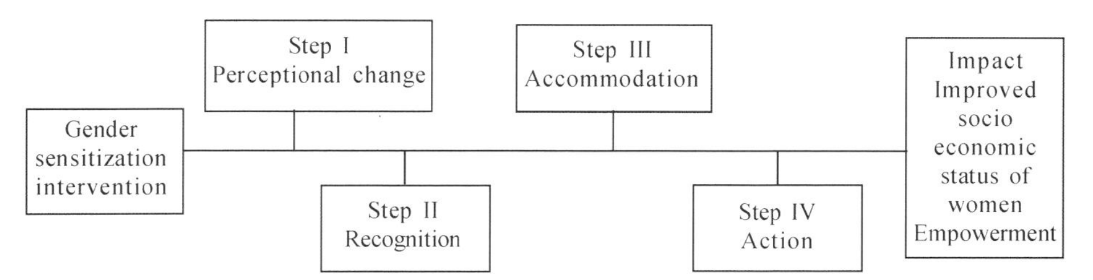

Fig. 3. Gender sensitization process

3. Promoting gender sensitization-Overview of some cases

Efforts to integrate the idea of gender sensitivity into the agricultural development interventions has been active from the last decade of twentieth century. The aim has been to redesign the service delivery mechanisms so as to improve the reach to women and increase the visibility of their multifarious roles. Acknowledging the growing role women play in India's key agricultural sector, policymakers has started paying attention to the gender aspects by incorporating gender component in most of its development programmes, increasingly. Economic Survey (2019) had highlighted increasing feminization of farm sector, while calling for inclusive transformative agricultural policy that aimed at gender-specific interventions to support the women farmers. These find reflected in the pro-women steps initiated by the Department of Agriculture, Cooperation & Farmers' Welfare (DAC&FW). It has earmarked 30 per cent of its annual funds for women under its various major schemes /programmes. These schemes are also aimed at promoting collectivization of women to improve their access to resources and mandating measures to ensure women staff in various programmes.

Many development organizations have intensified their efforts to build gender sensitiveness into their programmes in recent years. A few successful cases reported from different parts of India that structured extension interventions in a gender-sensitive manner has been included.

Case I: The *Pashu Sakshi* (meaning friends of livestock) initiative under the National Rural Livelihood Mission to support the women farmers engaged in rearing of small livestock such as sheep, goat and pig. Through this community-led livestock extension and advisory service approach, rural women are equipped through technical training and support to take lead in generating demand for inputs and provide services to livestock farmers. By building their capacities, over 4712 *Pashu Sakhis* have been promoted in 16 Indian states reaching to over 2.5 lakh small livestock farmers daily (Kumar n.d). The intervention is an excellent example for providing the EAS in a gender-sensitive manner.

Case II: Digital Green, an NGO in India works with women-led self-help groups to produce and screen locally adapted agricultural extension videos. The chief intention here is to increase women's participation in agricultural extension provision in contrast to the conventional extension approaches. Study reports suggest that the intervention is 10 times more effective for the money spent than the traditional approach on a cost per adoption basis (Gandhi *et al.*, 2009).

Case III: The interventions of State Rural Livelihood Mission (Kudumbashree), the network of 4.3 million women in Kerala, focuss on social and economic empowerment of women through collectives is another case worth mentioning. Kudumbashree organizes gender sensitization programmes to bring definite and conscious thinking and intervention of the staff and community members. The multifaceted strategic interventions include trainings for staff and community members, campaigns and thematic workshops. A gender sensitization module was also developed towards this purpose. The main objective is to integrate a gender perspective into its programmes to achieve gender equality (Thomas n.d).

Box-1: Meena Raju Manch- A unique initiative to promote gender sensitization among school children

The NGO, CORO was formed initially to propagate adult literacy in the slums of Mumbai. Later, it started working to promote the leadership of women in their community and empower them economically through entrepreneurship and self-help group-based activities. Gender sensitization is one of the areas in which the NGO works to make social change. Meena Raju Manch (MRM), a unique initiative of CORO aims to raise awareness about gender sensitization among students and teachers. The initiative launched in 2012, seeks to ensure that 'gender equality' is not just a book concept, but also a change in the behaviour. MRM sessions are based on different themes – including fair distribution of food and the need for equal nutrition amongst girls and boys. Gender sensitization campaigns are conducted in 25000 schools in Maharashtra in collaboration with UNICEF. The central idea is that, if correct messages about equity and equality are given to adolescents, we can surely expect the next generation to be more gender-sensitive and fair. Though the programme had a previous version, Meena Manch, it was restructured later to accommodate boys as well in the campaigns. When 'Meena-Raju Manch' is established, students take an oath of equality and swearing campaign. Further, children are made aware of the typified gender roles in their households through variety of PRA exercises. Through these activities, boys will be sensitized about the onerous tasks performed by their female counterparts and would be more accommodative in their behaviour.

4. Towards gender-sensitive extension advisory services (EAS)

A more gender-responsive agricultural system calls for a comprehensive look at the system. Actors of different stages, users of the technology, and needs that are to be addressed at each stage need to be delineated. The whole process from priority setting and implementation to evaluation and impact assessment need to be addressed. The most common approaches adopted to bring about better gender sensitization in extension organizations include organizing training programmes for gender staff, formation of women farmer collectives with micro-credit assistance and conducting targeted interventions for them and designing women friendly tools and equipments (CRISP n.d). EAS should be aware of various social and cultural norms which restricts women's access to advisory services and redesign and deploy services to benefit them (Williams and Taron 2020). Creation of institutional structures to accommodate the voice of farm women in policy and decision making is imperative for making the sector more gender-sensitive.

Box-2: Developing a Gender-Sensitive Extension Model-CIWA Experience

Central Institute of Women in Agriculture (CIWA), Bhubaneswar had conducted a study on developing a gender-sensitive extension model by deploying Village Level Para Extension workers (VPEWs). A male and female VPEWs each were recruited for a village and were given capacity building trainings. Besides they were equipped to provide farm advisory services and conduct discussions, meetings, method and result demonstrations and organize Women Self Help Groups (WSHG). Their roles and responsibilities were aligned to meet the needs and issues of farm women. Institutional mechanisms were in place to monitor their performance regularly and to give necessary incentives. VPEWs could also mediate knowledge transfer from agricultural experts to these women including the farmers from marginalized communities. Outcomes suggested increased contact of these extension staff with women farmers, better knowledge exchange and technology dissemination and improved reach of advisory services. Similar experiences were also reported by the *Jeevika* intervention of the Bihar Rural Livelihoods Program. These field experiences reinforces the argument that such interventions are crucial in making the extension service delivery mechanisms more gender-sensitive.

In order to ensure that farm women get a fair deal at the hands of change agents, one of the remedial measures that needs to be undertaken is the induction of women staff. A sizeable number of well-trained women personnel in training and extension programmes of agricultural development agencies at all levels and more so at the grass-root level can ensure it (UN Women, 2013). Extension interventions can make a better impact when they can leverage the network of female farmers than males, as demonstrated by many studies (Magnan *et al.*, 2015). Moreover, design and development of ergonomic tools may help to lessen the burden of women farmers. Presently, barriers to adoption of technology range from its design to access to credit, land, and information to purchase, access, or use the technology (Jones, 2019). The limited institutional capacity of the women should be strengthened to improve their access to production resources and technologies. Trainings and capacity building programmes may be specifically designed for farm women after considering their specific roles in each phase of the production process. Any training programme targeting rural women workers should be gender-friendly, given the patriarchial social norms governing women's mobility..

5. Emerging Trends

An overview of the discourse on gender-sensitive agriculture would suggest that the topic has assumed greater relevance among the development practitioners. Recently, many methodologies and tools are developed to promote gender responsiveness of the technologies and EAS delivery mechanisms. For instance, Integrating Gender and Nutrition within Agricultural Extension Services (INGENAES) Technology Assessment Toolkit developed by United States Agency for International Development (USAID) under the Feed the Future Programme helps in assessing gender sensitivity of agricultural technologies in terms of design, use and dissemination (Heinz, 2018). Similarly, Food and Agriculture Organizations' (FAO) developed Gender in Rural Advisory Services Assessment Tool (GRAST), to improve the gender responsiveness of rural advisory services (FAO, 2016). The information generated by the tool can be used to design and deliver gender-sensitive rural advisory services. Similarly, FAO (2017) has developed a manual to develop capacities to address gender issues in rural advisory services. The manual intends to impart knowledge and tools to improve the understanding of how gender issues and provide participants with the skills to design and deliver participatory and gender-sensitive training themselves. In order to enable gender-sensitive policy and programme design and implementation to maximize the effectiveness of social protection, the UK Department for International Development (DFID) has developed a toolkit. Gendered vulnerability analysis is used to inform the design of gender-sensitive social protection (Holmes and Johns, 2010). Further, women farmers are subjected to the vagaries of climate change to a greater extent. The tool, Climate Resilient Agricultural Module (CRAM) comprises a group of participatory research tools to design inclusive and gender-sensitive programs in climate-resilient agriculture. CRAM can also be used to identify opportunities for enhancing climate change adaptation for women and vulnerable groups (Douxchamps *et al.*, 2017). Jafry and Sulaiman (2013) proposed the 'New Consultative Design Process' (NCDP) for designing of extension services to women farmers. It enables, rural women, to represent their aspirations and desires for improving their livelihoods and tries to incorporate varied support from different stakeholders. Such tools are increasingly been used by development organizations to integrate gender concerns in their interventions to make it more inclusive.

6. Conclusion

The process of gender sensitization aims to modify the behaviour and attitude of both women and men so that there is greater awareness and empathy to create gender equality. By increasing awareness and sensitizing men on the discrimination women face in their communities, they can become supporters instead of barriers in a women's life (Vyas *et al.,* 2019). At the same time, it prompts women to proactively look into their versatile roles and contributions to the society at large. Also, gender sensitization programs can play an important role in forming and changing gender attitudes during adolescence and have the potential to alter their short and long-term beliefs. Lack of sensitization at different levels, may lead to poor implementation and outcome of development interventions. In the face of increased feminization of agriculture and participation of women in the farm labour force, gender aspects need to be better integrated in design and delivery of services. Development practitioners should be made aware about the variety of tools and methodologies at their disposal for gender-sensitive planning and implementation of interventions. Moreover, specific extension interventions to build the capacity of farm women and institutional structures to enhance their access to production technologies and resources are very much required. Sensitization efforts are required at grass root level to advocate for women's equal rights in decision making a social norm, by changing the attitude of male members. Finally, it is imperative to track the dynamic gender issues and priorities in different spheres for better designing and targeting of interventions (DRWA, 2014).

References

Beevi, C.A., Wason, M. Padaria, R. N & Singh, P. 2018. Gender sensitivity in agricultural extension. Current Science. 115(6): 1035.

Berger, M., Virginia, D. & Amy M. 1984. Bridging the gender gap in agricultural extension. U.S. Agency for International Development. Retrieved from https://www.icrw.org/wp-content/uploads/2016/10/Bridging-the-Gender-Gap-in-Agricultural-Extension.pdf

CRISP (Center for Research on Innovation and Science Policy). n.d. Reaching Rural women.Retrieved from http://reachingruralwomen.org/gendersens. htm#:~: text= This%20came%20 about%2C%20as%2 0there,source%20of% 20livelihood%2 0for%20women% 3B&text=So%2C %20there%20is% 20a%20need, %2Dsensitive %20agricultural%20planning%E2%80%9D%20methods.

CIWA (Center Institute for Women in Agriculture), 2018. Compendium Training Programme On Gender Sensitization for Strengthening Women Perspective in Agriculture. Retrieved fromhttp://icarciwa.org.in/gks/Downloads/Technical% 20 Bulletins/GenderSensitisationTraining.pdf. Accessed on 31.10.2020

CIAE (Central Institute of Agricultural Engineering), 2013. Vision 2050. Bhubaneswar, India: Directorate of Research on Women in Agriculture. Retrieved from http://www.ciae.nic.in/WriteReadData/UserFiles/file/ciae_vision_2050_31-12-2014.pdf

Dash, H., K., Srinath, K. and Sangi, B., N. 2008. Gender Sensitization: Role in reforming the society. Gender Notes. National Research Center for Women in Agriculture. Retrieved from http://icar-ciwa.org.in/gks/Downloads/Gender%20Notes/Gender%20 Notes(1).pdf

Dash, H.K. and Srinath, K. 2013. Promoting agricultural education among rural women: A critical intervention for sustaining farm and home. Current Science. 105(12): 1664-1665.

Douxchamps, S., Debevec, L., Giordano, M., & Barron, J. 2017. Monitoring and evaluation of climate resilience for agricultural development–A review of currently available tools. World Development Perspectives, 5: 10-23.

DRWA (Directorate of Research on Women in Agriculture), 2014. A model for gender mainstreaming in agriculture for village development. Retrieved from http://icar-ciwa.org.in/pdf/TB/ICAR-CIWA-TB(22).pdf

Dudi, K., Devi, I and Kumar, R. 2019. Contribution and Issues of Women in Livestock Sector of India- A Review. International Journal of Livestock Research. 9(8), 37-48. DOI: 10.5455/ijlr.20190421064818

Economic Survey, 2019. Ministry of Finance. Retrieved from https://www. indiabudget.gov.in/budget2019-20/economicsurvey/index.php

Food and Agriculture Organization (FAO), 2016. The Gender and Rural Advisory Services Assessment Tool (GRAST). Retrieved from http://www.fao.org/3/i6194en/I6194EN.pdf

Food and Agriculture Organization (FAO), 2017. Developing capacities in gender-sensitive rural advisory services. A training of trainers manual. Retrieved from http://www.fao.org/3/a-i7507e.pdf

Gandhi, R., Veeraraghavan, R. & Toyama, V.R. 2009. Digital Green: Participatory Video and Mediated Instruction for Agricultural Extension.Information Technologies and International Development. 5(1): 1-15.

García-González, J., Forcén, P & Jimenez-Sanchez M. 2019. Men and women differ in their perception of gender bias in research institutions. PLoS ONE. 14(12): e0225763. https://doi.org/10.1371/journal.pone.0225763

Ghosh, M. & Ghosh, A. 2014. Analysis of women participation in Indianagriculture. International Journal of Gender and Women studies. 19(5): 1-6.

Heinz, K. 2018. Stories of success: Integrating gender and nutrition within agricultural Extension services, Retrieved from https://www.agrilinks.org/post/stories-success-integrating-gender-and-nutrition-within-agricultural-extension-services

Holmes, R. & Jones, N. 2010. How to design and implement gender-sensitive social protection programmes. A toolkit. ODI, London UK

ICAR (Indian Council of Agricultural Research), 2007. Annual report 2006-07. https://icar.org.in/content/dareicar-annual-report-2006-2007.

ILO (International Labour Organization), 2018. India Wage Report: Wage policies for decent work and inclusive growth. Retrieved fromhttps://www.ilo.org/wcmsp5/groups/public/-asia/-ro-bangkok/-sro-new_delhi/documents/publication/wcms_638305.pdf

India Spend. 2019. 73.2% of rural women workers are farmers, but own 12.8% land holdings. Retrieved from https://www.indiaspend.com/73-2-of-rural-women-workers-are-farmers-but-own-12-8-land-holdings/

Jafry, T. & Sulaiman, R., V. 2013. Gender-Sensitive Approaches toExtension Programme Design, The Journal of Agricultural Education and Extension, 19(5): 469-485.

Jones, M. 2019. Innovative approaches to including gender within agricultural mechanizationRetrieved from https://www.agrilinks.org/post/innovative-approaches-including-gender-within-agricultural-mechanization

Kelkar, G. 2011. Gender and productive assets: implications for women's economic security and productivity. Economic and Political Weekly. 46(23): 59-68.

Kristjanson, P., Elizabeth, B. Quinn, B. Jennifer, T. Ruth Meinzen, D. Caitlin, K. Claudia R. Christine, J. & Cheryl, D. 2017. Addressing gender in agricultural research for development in the face of a changing climate: where are we and where should we be going? International Journal of Agricultural Sustainability. 15(5): 482-500, DOI:10.1080/14735903.2017.1336411

Kumar, S. n.d. Pashu Sakhi - An alternative livestock extension approach. Vikaspedia. Retrieved from https://vikaspedia.in/agriculture/best-practices/extension-practices/pashu-sakhi-an-alternative-livestock-extension-approach

Lamontagne-Godwin, J., Cardey, S. Williams, F. Dorward, P. Aslam, N. & Almas, M. 2019. Identifying gender-responsive approaches in rural advisory services that contribute to the institutionalisation of gender in Pakistan.The Journal of Agricultural Education and Extension. 25(3): 267-288, DOI: 10.1080/1389224X.2019.1604392

Magnan, N., Spielman, D.J. Lybbert, T.J. & Gulati, K. 2015. Levelling with friends: Social networks and Indian farmers' demand for a technology with heterogeneous benefits. Journal of Development Economics. 116: 223-251.

Majumder, J. & Shah, P. 2017. Mapping the role of women in Indian Agriculture. Annals of Anthropological Practice. 41(2): 46-54.

Meinzen-Dick, R., Quisumbing, A. Behrman, J. Biermayr-Jenzano, P. Wilde, V. Noordeloos, M. Ragasa, C & Beintema, N. 2011. Engendering agricultural research, development and extension. International Food Policy Research Institute.

Mehrotra, I. 2020. Is having more women in agriculture a good thing? Retrieved from https://timesofindia.indiatimes.com/blogs/developing-contemporary-india/is-having-more-women-in-agriculture-a-good-thing/

MoSPI (Ministry of Statistics and Programme Implementation), 2020. Annual Report. Periodic Labour Force Survey July 2018 to June 2019. Retrieved from http://www.mospi.gov.in/sites/default/files/publication_reports/Annual_Report_PLFS_ 2018_19_HL.pdf

Nelson, V., Meadows, K. Cannon, T. Morton, J. & Martin, A. 2002. Uncertain predictions, invisible impacts, and the need to mainstream gender in climate change adaptations. Gender & Development. 10(2): 51-59.

Nirmala, G. & Venkateswarlu, B. 2012. Gender and climate-resilient agriculture: an overview of issues. Current Science. 103(9): 987.

Ogawa Y. 2004. Are Agricultural Extension Programs Gender Sensitive? Cases from Cambodia. Gender, Technology and Development. 8(3): 359-380. doi:10.1177/097185240400800303

OXFAM, 2019. India Inequality report 2018. Widening Gaps. Retrieved fromhttps://www.oxfamindia.org/sites/default/files/Widening Gaps_IndiaInequalityReport 2018.pdf

Pachauri, 2019. The invisibility of gender in Indian agriculture. Retrieved from https://www.downtoearth.org.in/blog/agriculture/the-invisibility-of-gender-in-indian-agriculture-63290

Sadangi, B.N., Dash, H.K. & Mishra, S. 2009. Strategy for Gender Sensitive Extension in Agriculture and Allied Fields.Technical Bulletin 11. DRWA Bhubaneswar pp:1-35.

Shrestha, G., Freund, D. & Clement, F. 2020. Unravelling gendered practices in Nepal's public water sector. Retrieved from https://wle.cgiar.org/thrive/2020/09/24/unravelling-gendered-practices-nepals-public-water-sector

Thomas, S. n.d. Gender sensitisation for transformation. Kudumbashree Writeshop. Retrieved from https://www.kudumbashree.org/storage/files/amlyt_soya.pdf

Tipilda, A. & Kristjanson, P. 2008. Women and Livestock: A Review of the Literature. ILRI Innovation Works Discussion Paper 01-08. Nairobi: International Livestock Research Institute. Retrieved from. www.ilri.org/Innovation Works.

UN Women, 2013. Annual report 2012-13. Retrieved fromhttps://www.unwomen.org/-/media/headquarters/attachments/sections/library/publications/2013/6/unwomen-annualreport2012-2013-en%20pdf.pdf?la=en&vs=1457

Vyas, A.N., Malhotra, G. Nagaraj, N.C. & Landry, M. 2020. Gender attitudes in adolescence: evaluating the Girl Rising gender-sensitization program in India. International Journal of Adolescence and Youth. 25(1): 126-139.

Williams, F.E. & Taron, A. 2020. Demand-led extension: a gender analysis of attendance and key crops. The Journal of Agricultural Education and Extension, 1-18.

10

Gender Sensitive Monitoring and Evaluation: Prospects and Challenges

Archana Raghavan Sathyan* and Anu Susan Sam

1. Introduction

Gender remains a critically important and largely ignored lens to view development issues across the world (Jayaraman, 2017). Gender inequality is the greatest human rights challenge of our time and also a critical economic, moral and social challenge (UN, 2015). There exists a significant gender gap in various sectors such as health, labour market opportunities, education and political representation all over the world. This mostly cause the restriction of women to stay at home for household chores. However, the associated disempowerment resulting from the adverse effects of restricting their way around outside world are seldom addressed in the development evaluation and monitoring processes (UNICEF, 2019). Therefore, gender equality and women empowerment are made explicit throughout the Sustainable Development Goals (SDGs). Drawing urgency to these efforts, SDG5 aims on full gender equality and women empowerment as a cross-cutting theme along with more than 30 related targets across other SDGs.

According to FAO (2011), rural women constitute about 43 percent of world's agricultural labour force and are the agents of change and resilience builders. Even then they undergo greater constraints compared to the men counterparts in accessing technological interventions, productive resources, market information, services and financial assets. The contribution of women to food security often remains undervalued and invisible (FAO, 2011). This leads to inadequate reflection of women's role in policy, legal and institutional frameworks. These have created difficult situations for them to address the multiple tasks of natural resource management, child rearing and family well-being, without more rural women inclusive development interventions. There would be a drastic reduction in the number of hungry population and malnourished children in the world if women had the same access as men to productive resources which in

**Author contact: archana.rs@kau.in*

turn helps to increase yields on their farms significantly (FAO, 2018). Furthermore, they are also under-represented in grassroot level institutions and administrative mechanisms and assumed to have less decision-making ability. In addition to these constraints, women face an imprudent work burden, at the same time much of their labour remains unpaid and unrecognized under the prevailing gender norms and discrimination. It is high time to include them in decision making process within the households, communities and institutions and thus to enhance women's participation in governance (Bayeh, 2016).

Most of the global, national and sectoral policies, projects, schemes, legislation and investment plans for food security and nutrition do not always encapsulate women's role, effort and contribution effectively (FAO, 2015). Therefore, these interventions absolutely fail to address and react to their specific needs and challenges. To evaluate the impact of such initiatives, the assumed assessment mechanisms especially the Monitoring and Evaluation (M&E) must be gender-responsive (Espinosa, 2013). The assessment mechanisms should be formulated in such a way that they identify the gender inequalities in attitudes, opportunities, perceptions and access to resources and decision-making. It should also identify the impacts of these interventions and policies on social understandings i.e. what these efforts mean/indicate to a woman or a man especially on gender relations in the community as a whole (Birchall, 2016).

Gender-sensitive M&E systems are as important as a gender-responsive project design. In order to assess how far development programme has achieved its gender goals and equality, the indicators must be gender-sensitive. Many of the projects are gender-blind and these can be revised by including gender-responsive indicators and assuring gender-sensitive evaluations. Thereafter, it is crucial to monitor various contributions of project to the goal of gender concerns and equality throughout the execution of the project (Moser and Moser, 2005). The information produced out of these evaluations can be used to advance the agendas of women's empowerment as well as to advocate for gender equality (Wagner *et al.*, 2005). With this background, this chapter discusses the major concepts of gender, need for gender-sensitive M&E, various methods for gender-inclusive M&E and recommendations for future.

2. History & concepts

Gender is defined by FAO (1997) as 'the relations between men and women, both perceptual and material'. Gender is constructed socially and not biologically determined. 'It is a central organizing principle of societies, and often governs the processes of production and reproduction, consumption and distribution' (FAO, 1997). In nutshell, the term gender is being used as a shorthand terminology which encodes a very crucial point that our basic social identities as men and women are socially constructed rather than based on fixed biological characteristics. Gender equality exists when the rights, responsibilities and opportunities of individuals will not vary irrespective of whether they are born male or female (Warth and Koparanova, 2012).

There have been several attempts to mainstream gender since the Fourth International Women's Conference conducted at Beijing in 1995. In gender mainstreaming, the integration of gender perspective into policies and strategies can happen at different levels i.e. at field level, in various programmes and projects; at institutional level and at government level within development institutions (European Institute for Gender Equality, 2020). There are two elements for gender mainstreaming viz. analytical element and normative element. The analytical element indicates the picture of power relations between men and women while the normative element emphasis on ensuring enhanced gender equality so that both genders have equitable access and control of productive resources such as land, capital and associated benefits (UNDP, 2007). Thus, adequate care must be taken to include gender representation and gender-responsive dimensions in all phases of the policy-making process (UNW, 2014). The gender mainstreaming process can be illustrated as four stage cycle that consists of defining, planning, implementing and checking stages (Figure 1), and the gender-sensitive M &E falls under the 'checking tools' stage.

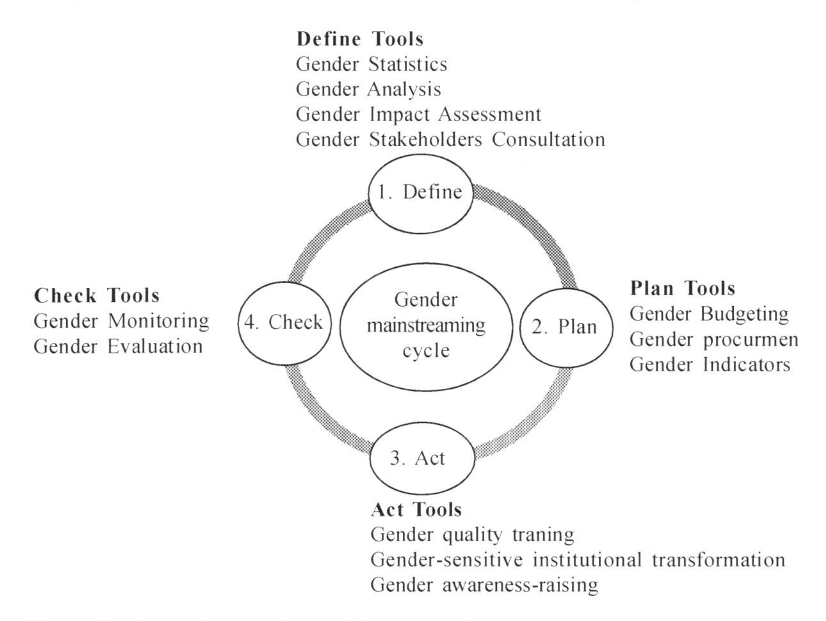

Fig. 1. Gender mainstreaming multi-stage cycle
(*Source:* Modified from European Institute for Gender Equality, 2020)

M&E term is generally used to outline and assess projects/programme against its objectives, nevertheless both are often used interchangeably. However, monitoring is defined as the period and continuous collection, analysis and distribution of information/data on the progress of the interventions and project activities implemented, whereas evaluation work at a broader level (Walters *et al.*, 1995). Another similar term, impact assessment looks at the positive and negative effect of the project (Goyder *et al.*, 1998) while evaluation reports whether the positive outcomes planned by the project have been achieved or not (Walters *et al.*, 1995). Gender-responsive M&E supports aforementioned analytical and normative components of gender mainstreaming by systematically analysing the effects of an intervention in achieving the goal of gender equality as well as power relations between men and women. It also recommends corrective measures to improve the effectiveness of a project activity and thus to address the varying needs of women and men in the social, political or economic realm.

3. Need for gender-sensitive M&E

Gender, being a socio-cultural variable, refers to the comparative, relational or differential roles, responsibilities, and activities with emphasis on culturally-based expectations about men's and women's identity and behaviour (FAO, 1997). It is important to recognise that these gender roles may vary among and within societies over time. However, the tools used to monitor and assess the development interventions and policies have mostly neglected the gendered responses. In fact, the differential impacts of these initiatives on both men and women can only be identified if M&E mechanisms are responsive to gender dimensions. This enables inevitable adjustment of projects and policies to fit and respond to gender issues in a more comprehensive way and eventually to corroborate that the desired intentions in planning and policy forefront are met (Brambilla *et al.*, 2001). Thus gender-sensitive M&E is used to disclose whether a project/policy addresses the different priorities and needs of women and men, to assess if it has an impact on gender relations, and thus to ensure the crucial gender aspects that need to be inculcated into M&E systems.

A gender-sensitive M&E can help the policy makers and project implementing agencies to improve project performance and accountability during implementation. It also facilitates learning and making adjustments in project activities to achieve gender equality objectives and to derive lessons for future projects. The M& E provides feedback on how the activities affect the various groups of beneficiaries both women and men, disaggregated by age, ethnicity, caste, education, employment and geographical location.

Even though gender is considered to be a cross-cutting issue within international and national organisation's development interventions, more often the gender mainstreaming commitments vanish even before implementation of the intervention and thus only remains as a commitment on paper. The main reason behind this is the lack of precise, clear and simple guidelines and checklists to monitor gender-sensitive impacts during the project implementation phase (Stephens *et al.*, 2018). Another critical fact is that the project staff lacks specific skills as well as commitment to address the various gender constraints and benefits, and thus to draft gender-sensitive outcomes during the project planning. Yet another important reason for gender-sensitive M&E is to foster human rights. Women and children are more affected by hunger and poverty than men because of traditional cultural beliefs and social structures which restricts them from accessing productive resources in the same way as men do (ILO, 2020). Gender-sensitive M&E can help to delineate the critical factors which restrict men and women from enjoying equal rights and thus to rethink on aiming future interventions for improving the well-being of all.

4. Methods for gender-sensitive M&E

There are a number of M&E frameworks that can be used and adapted to capture the complexity of change in women's rights and gender equality work. These M&E frameworks are categorised according to their underlying assumptions in tracking and understanding the nature of change (Batliwala, 2010). Accordingly, the major M&E frameworks in use are as follows:

- Causal Framework
- Contribution Framework
- Gender Analysis Framework
- Advocacy and Network Assessment Framework

The *Causal Frameworks* aims to demonstrate the causal and logical chains that lead to programme impact and the *Contribution Frameworks* attempts to track the multiple and variable forces involved in producing the change and highlight the contribution of change agents to the social change process and intended outcomes. The *Gender Analysis Frameworks* and *Advocacy and Network Assessment Frameworks* draw from causal and contribution frameworks. *Advocacy and Network Assessment Frameworks* aim to assess the way that change happens through an advocacy lens and accounting for complicated network structures (Batliwala, 2010).

According to Sudarshan *et al.* (2015), there are three types of gender-responsive evaluation methods. They are:

i) Gender-instrumental
ii) Gender-specific and
iii) Gender-transformative

The gender-instrumental evaluation only considers gender by disaggregating data by sex whereas the gender-specific method examines an intervention's impact on sex or gender specific needs (Sudarshan *et al.*, 2015). The gender-transformative type of evaluation is concerned with learning the influence of an intervention in respect to changing oppressive power dynamics (Sudarshan *et al.*, 2015). In order to holistically view the gendered effects of policies and programmes, the different starting points of women, men, boys, girls and transgendered people, are considered under the gender-transformative evaluation method. The major ways for gender-sensitive M&E are given in Figure 2.

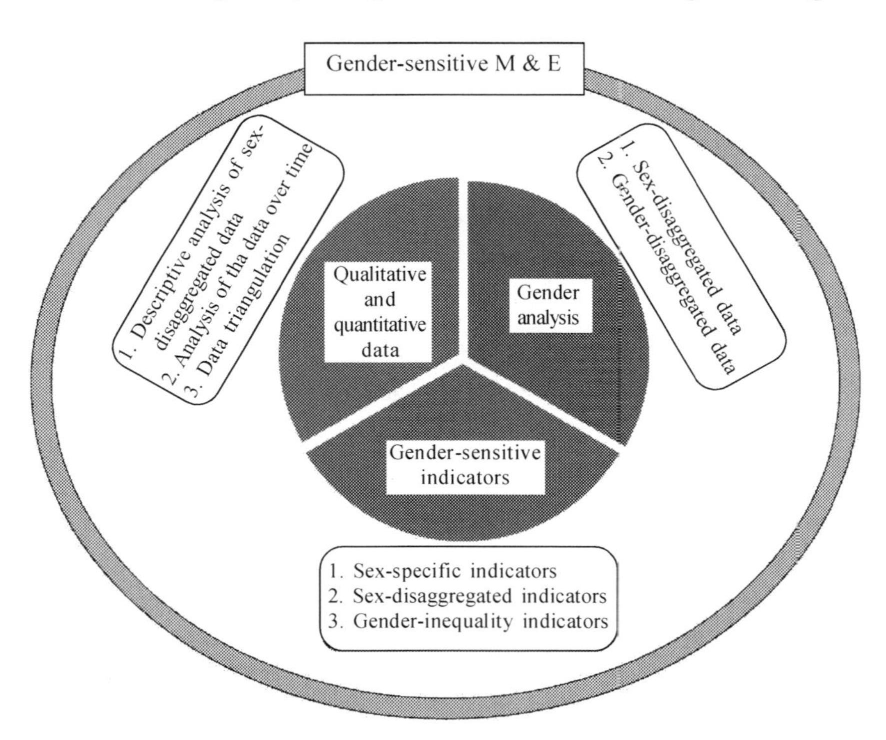

Fig. 2. Different methods of gender-sensitive M&E (Source: Author compilation)

4.1.1 Gender-responsive situation analysis/gender analysis

Gender analysis is an essential component of M&E and it works with information that brings into focus the similarities, differences and inequalities that exist between women's and men's experiences. Otherwise, gender analysis is a systematic analytical process based on sex-disaggregated and gender information that describes gender differences in a particular context (UNDP, 2007). Three basic steps to a comprehensive gender analysis as reported by Connelly and Barriteau (2000) includes: descriptive analysis of sex-disaggregated data analysis of these data over time, in comparison with other groups and in relation to other norms; and data triangulation that examines the broader social, cultural, economic legal and health systems factors.

Such an analysis helps to address the policy and normative frameworks of the programme or project and identifies the structural causes of gender discrimination and inequalities in employment and occupation.

Gender analysis is closely linked to the quality control of an intervention by ensuring that it adequately considers gender concerns throughout its planning and implementation, regardless of whether the intervention explicitly targets the empowerment of women or gender equality (Connelly and Barriteau, 2000). The relevance, effectiveness and sustainability of an intervention will be drastically reduced by inadequate addressing of gender concerns. In fact, this reinforces the existing and unequal power relations between the sexes and diminish women's status. Hence, it is necessary to identify and understand the relevance of gender roles and power dynamics in order to monitor and assess how an intervention affects the gender relations and gender equality (Stephens *et al.*, 2018). It helps to ensure equitable participation of women and men in various development processes and projects and plays a determining role in outcomes. The analysis should identify the division of labour between women and men and their access to and control over productive resources and benefits. In addition to that the practical and strategic needs of women and men, their challenges and opportunities and capacity to promoting gender equality in respect of the capacities of government, workers and ability to mainstream gender and promote gender equality must be identified.

There are 85 gender relevant and 54 gender specific priorities across various SDGs. Gender-differentiated data and information must be available for policy makers so that they may be able to assess the situation and develop appropriate, evidence-based responses and policies (Valero, 2019). Without this data 50% of the world's population will be missed out in development. It captures the specific realities in the lives of men and women. Thus gender-disaggregated data is needed in all types of M&E, auditing or impact assessment process. Data will often need to be also disaggregated along other lines such as age, urban/rural, ethnic group, disability, etc.

4.1.2 Gender-sensitive indicators

Gender-sensitive indicators help to explain manifestations of gender inequality that are often invisible in conventional indicators (Espinosa, 2013). There are three following types of gender-sensitive indicators as reported by WHO (2016).

1. Sex-specific indicators which pertain to only women or only men, or subgroups among them.
2. Sex-disaggregated indicators which measure differences between women and men in relation to a particular metric. The data should be disaggregated by relevant stratifiers such as age, socioeconomic status, ethnicity or place of residence to enable a deeper analysis of the contributing factors and dynamics in society to the specific situations of women and men.
3. Gender-inequality indicators which directly measure gender related inequalities.

Good indicators should be reliable enough ensuring the detection of trends over time and should make explicit comparison to a norm. For instance, if an indicator assesses the status of women, the comparison group could either be men or another group of women in the same country. The development of indicators should be evolved through a participatory process as far as possible with execution in a participatory way (Estrella and Gaventa, 1998; World Bank, 2005). However, a *technocratic one-size-fits-all* indicator development is inappropriate and impossible as gender-sensitive indicators are very specific for each programme or activity level (Hochfeld & Bassadien, 2007). For example, impact assessment of Action Aid programme in Bangladesh revealed that men gave higher priority to changes related to women's development while women gave higher priority to access to credit, increased decision-making and greater mobility for women (Goyder *et al.*, 1998). It was interesting to note that both women and men expressed their preference to participate in income generating activities under the programme.

Adequate indicators for gender equality should focus on the output and the quality of outcomes. One of such indicators is participation rate of an underrepresented or disadvantaged sex in a given specific activity such as a training programme or even discussion of a new policy. When there is a higher participation rate from the disadvantaged sex, nonetheless we may not be able to conclude that a project is gender-responsive or contributing to gender equality just because a high rate of women has taken part in its activities. A 'gender-responsive' or 'gender-sensitive' indicator should also measure the gender-related changes of the specific intervention over time.

4.1.3 Qualitative and quantitative data

Effective gender-responsive M&E needs to include both qualitative and quantitative data that measure the impact on gender relations (Stephens *et al.*, 2018). Collect both quantitative and qualitative information about different population groups through a variety of methods including focus group discussions, interviews, direct observation, and community consultations. Such data can reveal gender-based distinctions, for example, in analysing disparity in access to services or of protection needs among different population groups and thus providing a clear insight into what is driving those differences. Without sufficient data, a meaningful analysis of the impact on gender equality is very difficult. This also implies that, as a minimum/at least, all data should be collected, presented and analysed in a sex-disaggregated manner.

Sex-disaggregated data simply make a distinction between men and women while gender-disaggregated data help us to analyse these differences, allowing a more accurate understanding of the situation (FAO, 2011). Gender-disaggregated data may show differences between groups of women and men but fail to show the gendered power relations between these groups and hence qualitative analysis will be needed.

Gender issues are inextricably linked to cultural values, beliefs, social attitudes and perceptions. Hence, a variety of indicators must be needed for engendering quantitative as well as qualitative information. Moreover, qualitative analysis helps to understand various social processes, why and how a particular situation can be measured using indicators and how such situations could be changed in the future (CIDA, 1997).

4.1.4 Who measures gender sensitivity and who responds

It is crucial to have ample trained people to collect and analyse high-quality relevant, gender-sensitive data to plan and budget ahead to support gender-responsive M&E. It is preferred to have both women and men in the data collection teams. It is also important to ensure that the composition of data collection teams accords with the following:

i) Nature of the programme or project

ii) Socio-cultural factors, and

iii) What is most comfortable for respondents

In some cases, it may be necessary to have both female and male evaluators working together, while in other circumstances, it may be critical that female evaluators collect data among women and girls, and male evaluators among men and boys. Beginning with the project implementation phase, build a data collection team who are gender-aware i.e. those who are capable of perceiving the influences and manifestations of gender roles in a crisis-affected population.

Most of the times, monitoring is directed by the specific interests of the stakeholders who implement the project. Then there are higher chances for a biased monitoring process and this may lead to the neglect of relevant information on changes to gender relations. Eventually the assessment is likely to be focused on specific issues that could be instrumental in achieving the results desired by these stakeholders (Brambilla, 2001). In order to overcome such issues, it is encouraged to carry out consultations with women and men separately and together and in age-separated groups in participatory ways. In participatory M&E process, the target groups contribute genuine input in developing indicators and this allows the M&E process to be 'owned' by the target group rather than imposed on them by cosmopolite channels.

4.1.5 Challenges in gender responsive M&E

Gender responsive M&E data should be collected in a disaggregated manner e.g., by sex, ethnicity etc. Inclusive evaluations need to take into account all affected groups, stakeholders and rights bearers and at the same time responsive to differences among the stakeholders (Brambilla, 2001). Such M&E disaggregates groups by relevant criteria and pay attention to which groups benefit from, and which groups contribute to the intervention under review. In addition to this, participatory and reflective evaluations engage stakeholders in meaningful ways to ensure that their voice on what will be evaluated and how the evaluation will be undertaken are considered. According to ILO (2020), the assessment process analyses the stakeholder's participation level in the design, implementation and monitoring of the intervention and report and reflect upon their engagement both in the intervention and evaluation phase.

Another important aspect is conducting respectful participatory evaluations where all stakeholders, particularly vulnerable groups that are marginalised and impoverished are treated with due respect for their culture, language, sex, location and abilities (ILO, 2020). At the same time, the design and conduct of the evaluation should be transparent and responsive (ILO, 2020). It should also be taken care that the results of the M&E are publicly accessible with feedback to stakeholders about the process, results and use of the evaluation.

However, in ensuring integration of gender-equality into M&E we need to overcome a number of challenges we are confronted with in the process of design and implementation. Gender blind conventional M&E systems do not capture gender differences in accessing the resources and impacts of the intervention. When a household survey is conducted, the prevailing norm is to define, the household head as a male member and is often referred as the only source of information. This is typical example of the assumed "gender neutrality" of conventional M&E methods and processes which are addressed in gender responsive M&E methods (Stephens *et al.*, 2018).

Another important challenge is the lack of awareness of the staff on gender equality issues especially those who are engaged in preparing monitoring plans, evaluation terms of reference for conducting M&E. Concerns on the time and cost spend on collecting data from both gender groups also needs to be addressed. This has to be given emphasis at the initial phase of project formulation itself as it is often neglected during the planning and budgeting of the M&E exercise.

5. Conclusions and recommendations

The impact of development programmes and policies on target groups is assessed through M &E processes and are essential to track if targets are being met. Gender-sensitive M&E is necessary to track changes on the situation of women and men over time and on their relations at household and community level. This chapter discussed about various ways for gender-sensitive M&E mechanisms and the challenges in conducting such assessments. Based on the discussions and interpretations from the chapter the following recommendations are made which can be taken into consideration while assessing the gender sensitivity of the project activities.

1. Gender-disaggregated indicators complemented by qualitative and baseline data analysis help to avoid gender bias due to lack of gender awareness and cultural attitudes of the people involved in the process.
2. M&E systems must be a part of a gender-sensitive planning cycle with precise objectives against which the results should be measured.
3. Execute participatory M&E and they must be designed in consultation with the target group.
4. Impart gender-responsive training to NGO/government officials who are directly involved in the M&E process.
5. Ensure accountability and transparency of project activities among the stakeholders about the expected outcomes and missed opportunities.

Thus, it can be concluded that it is imperative to include both qualitative and quantitative gender-sensitive indicators with emphasis on empowerment issues in the M&E frameworks. Nevertheless, the indicators included should be representative to provide information on gender relations and amenable to comparisons at different levels of region, language or cultural-specificity.

References

Batliwala, S. 2010. Capturing Change in Women's Realities. A critical overview of current monitoring and evaluation frameworks and approaches, AWID, Canada.

Bayeh, E. 2016. The role of empowering women and achieving gender equality to the sustainable development of Ethiopia. Pacific Science Review B: Humanities and Social Sciences, 2(1): 37–42. https://doi.org/10.1016/j.psrb.2016.09.013

Birchall, J. 2016. Gender, Age and Migration- an extended briefing, Institute of Development Studies (IDS), ISBN: 978-1-78118-301-4

Brambilla, P. 2001. Gender and monitoring: A review of practical experiences. Direktion für Entwicklungszusammenarbeit und Humanitäre Hilfe. BRIDGE, Switzerland.

CIDA.,1997. A Project Level Handbook. The Why and How of Gender-Sensitive Indicators. Quebec: Canadian International Development Agency.

Connelly, J.L. and Barriteau, P. 2000. Theoretical Perspectives on Gender and Development. IDRC.

Estrella, M. and J. Gaventa, 1998. 'Who Counts Reality? Participatory Monitoring and Evaluation: A Literature Review', IDS Working Paper 70, Brighton: Institute of Development Studies.

Espinosa, J. 2013. Moving towards gender-sensitive evaluation? Practices and challenges in international-development evaluation. Evaluation, 19(2): 171–182. https://doi.org/10.1177/1356389013485195

European Institute for Gender Equality, 2020. Gender-sensitive monitoring and evaluation. European Institute for Gender Equality. https://eige.europa.eu/thesaurus/terms/1217

FAO, 1997. Gender: the key to sustainability and food security. SD Dimensions, May 1997 (available at www.fao.org/sd).

FAO, 2011. FAO policy on gender equality: Attaining food security goals in agriculture and rural development. FAO.

FAO, 2015. Enhancing the Nutritional Impact of Agriculture Investment Programmes: A Checklist and Guidance for Programme Formulation, pp 6-12, Rome.

FAO, 2018. Empowering Rural Women, Powering Agriculture: FAO's work on Gender. http://www.fao.org/3/CA2678EN/ca2678en.PDF

Goyder, H., Davies R. and Williamson, W. 1998, Participatory Impact Assessment. A Report on a DFID Funded ActionAid Research Project on Methods and Indicators for Measuring the Impact of Poverty Reduction, ActionAid, London.

Hochfeld, T. and Bassadien, S.R. 2007. Participation, values, and implementation: Three research challenges in developing gender-sensitive indicators. Gender & Development, 15(2): 217–230. https://doi.org/10.1080/13552070701391516

ILO, 2020. Integrating gender equality in monitoring and evaluation. http://www.ilo.org/wcmsp5/groups/public/@ed_mas/@eval/documents/publication/wcms_165986.pdf

Jayaraman, S.R. and N. 2017. Gender issues in India: An amalgamation of research. Brookings.https://www.brookings.edu/research/gender-issues-in-india-an-amalgamation-of-research/

Moser, C. and A. Moser, 2005. 'Gender mainstreaming since Beijing: a review of success and limitations in international institutions', Gender and Development 13(2): 11/22

Stephens, A., Lewis, E.D. and Reddy, S. 2018. Towards an Inclusive Systemic Evaluation for the SDGs: Gender equality, Environments and Marginalized voices (GEMs). Evaluation, 24(2): 220–236. https://doi.org/10.1177/1356389018766093

Sudarshan M.R., Murthy K.R. and Chigateri S. 2015. Engendering Meta-Evaluations: Towards Women's Empowerment. New Delhi: Institute of Social Studies Trust.

UN, 2015. Gender Equality. https://www.un.org/en/sections/issues-depth/gender-equality/

UNDP, 2007. Gender Mainstreaming in Practice: A Toolkit. Part II & I. Regional Bureau for Europe and the Commonwealth of Independent States (RBEC), Bratislava.

UNW (United Nations Women), 2014. Gender mainstreaming concepts and definitions. Available at: http:// www.un.org/womenwatch/osagi/conceptsandefinitions.htm

UNICEF, 2019. Difficult Dialogues: A Compendium of Contemporary Essays on Gender Inequality In India. UNICEF Global Development Commons. https://gdc.unicef.org/resource/difficult-dialogues-compendium-contemporary-essays-gender-inequality-india

Valero, S.D. 2019. Learning from good practices: The Caribbean RSDS, https://www.unescap.org/sites/default/files/Caribbean%20RSDS.pdf

Wagner, D.A., Day, B., James, T., Kozma, R.B., Miller, J. and Unwin, T. (n.d.). 1995 Monitoring and Evaluation of ICT in Education Projects. 154. Walters, H., Monitoring and Evaluation from a Gender Perspective. A Guideline. Netherlands Development Organisation (SNV), Holland.

Walters, H., Hermans, A. and Van der Hel, M. 1995. Monitoring and Evaluation from a Gender Perspective: A Guideline, The Hague: SNV

Warth, L. and Koparanova, M. 2012. Empowering Women for Sustainable Development, Discussion Paper Series, No. 2012.1, United Nations Economic Commission for Europe, Geneva, Switzerland.

WHO, 2016. A tool for strengthening gender-sensitive national HIV and sexual and reproductive health (SRH) monitoring and evaluation systems, UNAIDS, Switzerland.

World Bank, 2005. Gender Issues in Monitoring and Evaluation in Rural Development. A Tool Kit. Washington, DC: World Bank.

Part-II

Gender in Allied Sectors of Agriculture

11

Gender Responsive Agribusiness Development: An Indian Perspective

K.P. Sudheer, Sreelakshmi K. Unni and Ann Annie Shaju*

1. Introduction

Agriculture and its allied sector is the largest economic sector in India and Indian economy depends heavily on the performance of this sector. Along with production, changes are occurring in supply, machinery, processing, distribution and marketing in the agriculture sector. All over the world, agriculture is passing through a phase of transition, and elements such as discrimination free trade, fair trade, transparency and its resulting predictability, scope for fair competition are being facilitated in to the agriculture sector through different norms and agreements. Due to the concerns about the environment, requirement of clean fuel which can be derived from vegetable sources are also getting increased. All these developments have created new opportunities and scopes for agriculture and agribusiness. Agripreneurship is considered as an employment in the strategy in the country especially for rural people. It promotes our economy by improving productivity and integrating the products into different markets including international markets. It helps in the reduction of cost and avoiding the uncertainties related to markets. To promote micro to medium scale industries, agripreneurship has a major role. Agripreneur faces constraints in related to identification of scope, proper time management and allocating budget etc. and proper project management skills are required to counter these constraints. Agripreneur should also consider sustainability, social and economic factors also (Gupta *et al.*, 2017).

The male-centered business models were considered as the norm till the advent of the present century. But now we have entered an era where there is consensus that women entrepreneurs' have huge role in the national economic development. Moreover, a significant number of studies show that even though there is a gap between the number of men and women entrepreneurs, the contribution from

---

**Author contact: kp.sudheer@kau.in*

women entrepreneurs has been constantly on the rise. This is despite the challenges faced by women entrepreneurs with respect to availability of credit, training, networks and information access and policy constraints. In order to ensure sustainable economic growth, it is necessary to embrace women's entrepreneurship as it gives new market opportunities and greater development impact. In this chapter, emphasis is given to women entrepreneurship and its contribution to national economy, social empowerment and national integration. Also the challenges and opportunities faced by women in agri-entrepreneurship and changes in the perceptions on women entrepreneurship are discussed.

2. Emergence of women in entrepreneurship development

A country's economic growth and prosperity depends on the performance of its enterprises. In India, small business and entrepreneurship play huge role in creating new job opportunities and distribution of income. Social and cultural variables determine the behavior of entrepreneurs and it varies from country to country. As per the sociological theories, the entrepreneurship development depends on factors such as social sanctions, cultural values, role expectations, religious faith etc. The advent of Sustainable Development goals (SDG) in the current millennium has brought a transformational shift from traditional patterns to versatile global practices in entrepreneurship. This has also brought solutions to the problems of low literacy rate, less participation in labor, and the inadequate presence in political sector faced by women, mostly of the third world countries. In addressing these issues, many countries took bold policy initiatives such as the National Policy for Empowerment of Women by the Government of India in 2001. One of the major objective under this was to ensure women participation as agents of socio-economic change and development (Mokta, 2014). This has also brought forth many indicators for the measurement of Women's empowerment. Some of the indicators are based on education, control over income, gender of household head, and control over assets at the time of marriage (Gupta *et al.,* 2019). One of the widely used multidimensional indicator is the Global Gender Gap Index (GGGI) established by the World Economic Forum and this index measures several aspects such as health, education, economic participation, and political empowerment. In 2012, Women's Empowerment in Agriculture Index (WEAI) was introduced to measure Women's Empowerment in the agriculture sector. This index considers factors such as women's access to resources and ability to make decisions in different domains in agribusiness such as production, resource use, control over income and leadership (Alkire *et al.,* 2013). But the use of WEAI has been restricted to household survey and it face challenges with respect to the adaptation of questionnaires with respect to local contexts and sensitivity analysis.

Women entrepreneurs are considered as the new engines for the development of economies in the developing countries. Female entrepreneurship in developing countries has recorded a significant increase due to increased interest in their role in the industrial sector as well as the positive impact of micro- and small enterprises (MSEs) owned by women entrepreneurs. A good number of literatures are available on the effect of female entrepreneurship on the poverty eradication in society (De Vita *et al.*, 2014; Aidis *et al.*, 2008; Manaf *et al.*, 2017).

3. Social and national integration through women entrepreneurship in India

After the independence, Government of India introduced several steps such as Central Social Welfare Board for implementing welfare measures for women in the first five-year plan (1951-1956). Steps to bring more women into agricultural development programmes was also taken up during the second five-year plan (1956-1961). Third (1961-1966) and fourth (1969-1974) five year plans mainly focused on the aspects of education for women. Policy shift from welfare-oriented approach to development-oriented approach by the Government of India in 1970's brought priorities for women in all the sectors. Government bodies recognized self-employment and industrial endeavors as ways of improving the societal and economic status of women. In order to equip women to earn a steady income and protection from exploitation, the fifth five-year plan (1974-1979) started training programmes exclusively for the skill and capacity development of women. During this period, under the Ministry of Social Welfare, Women's welfare and Development Bureau was also set up. Sixth Five-Year Plan (1980-1985) and Seventh Five-Year Plan (1985-90) gave attention for improving the availability of resources, and training for women to gain skillsets to ensure gender equality and empowerment. The Eighth Five-Year Plan (1992-1997) concentrated on empowering women, through Panchayat Raj Institutions and Ninth Five-Year Plan (1997-2002) introduced Women's Component Plan for providing a good amount of funds in women related fields. National Policy of Empowerment of women implemented in 2001, and the following Tenth Five-Year Plan (2002-2007) concentrated on the schemes to ensure survival, protection and development of women and children (Nagaraja, 2013; Mokta, 2014; National Policy for the Empowerment of Women, 2014).

4. Women in agripreneurship

Though women make up nearly half of the agricultural workforce in developing countries, they own a mere one-third of small and medium-sized enterprises in emerging markets. The composition of women extension officers comprise only 15 percent of the world's agricultural extension agents. Evidences suggest that the economic empowerment of women often lead to reinvestment in their families and communities mostly by challenging stereotypic roles of women in agriculture. This encourage women to restructure their livelihoods which in turn can spur a ripple effect towards self-reliance, prosperity, food and nutritional security. It is reported from emerging economies that 65% of female entrepreneurs had a female role model in lighting their path towards success. This highlighted the need to recognize the challenges encountered so that a gender-equitable enabling environment for agripreneurship development is created.

4.1. Challenges for women entrepreneurs in agribusiness

In developing countries, women entrepreneurs are faced with both internal and external constraints. Evidences from African countries indicate that women entrepreneurs face issues related to external market challenges. Similarly, issues related to religious believes, family norms, traditions etc. create problems for women entrepreneurs in all most all the developing countries. Women entrepreneurs in all developing countries are also prone to challenges related to inadequate support for business, harsh business conditions, political uncertainty, unreliable economy, inadequate access to resources, lack of business acquaintance and work-family imbalances (Panda, 2018). Experiential training sessions in entrepreneurship, leadership, and financial management encourage girls to deploy innovative solutions to practical challenges faced. Key constraints which affects women entrepreneurs and functional solutions are discussed in the following sections.

a) Gender discrimination and work family conflict

Women face severe discrimination in all aspects of entrepreneurship as it is generally considered as a man's province. It is really hard for women to establish credibility and earn respect in this male dominated patriarchal sector. The traditional gender roles are home bound that stop them to pursue goals outside family. Moreover, the patriarchal societal norms obstruct their path towards success in entrepreneurship preventing them to capture their spaces in the business sector. Studies in the Middle East countries show that women experience strong resistance from family members and the pressure put them in precarious position that hinder their entry into entrepreneurship. These results suggest that the gendered roles in the society hinder women's business career in most of the developing countries. Even though the number of women who enter into industries and service sector has increased in recent years, there still exists a huge gap in the high risk entrepreneurship field. Women often struggle to make their work - life balance mostly to satisfy the social expectations. Moreover, the roles assigned by the patriarchal society to women, restrict them to invest their time in networking and mentoring, which are essential for business successes and hamper their performance (Panda and Dash, 2013; Danish and Smith, 2012; Jennings and Brush, 2013).

b) Financial constraints and lack of infrastructural support

Access to finance and financial independence are two major constraints faced by women entrepreneurs. As the business sector is prone to high risk, it is difficult to get loans for starting the firm or to expand the scheme. Studies show that there are hindrances for entrepreneur's opportunity to raise public equity capital. Though efficient alternatives such as venture capital and private equity opportunities are available, women face the following difficulties to raise capital:

a) Women's weaker credit record due to inconsistent work histories, lower salary and insufficient savings.
b) Preference to male-owned firms with respect to allocation of funds as loans by Government schemes.
c) Reluctance of women to take external capital as the entrepreneurship field does not favor women and as such women are inclined to start up with their personal savings in service sectors that are inexpensive and easier to start.

The effectiveness of women entrepreneurs are severely affected by a lack of supporting infrastructure such as access to technology and business services. Access to resources to perform market research is very limited for women. These constraints negatively affect the performance of the business firm. At the same time the gender stereotypes stops them from proper negotiation with vendors, clients and customers. The improper appreciation and acknowledgement from fellow entrepreneurs also demotivate women entrepreneurs (Jamali, 2009; Panda, 2015; Panda and Dash, 2014).

c) Unfavorable business, economic and political environments

Both developed and developing countries are facing issues such as unfriendly business environments due to economic recessions which happened several times in the last two decades. The prevalence of unstable political environments also pose constraints for entrepreneurs. Unnecessary regulations and requirements, bribery, inefficient bureaucracy, unbalanced tax system and inefficient public policies are acting as detrimental to all entrepreneurs especially women entrepreneurs in developing countries. Gender disparity reinforces the inclement political environment and leads to situations which are not suitable for the growth of women entrepreneurship (Panda and Dash, 2014; Danish and Smith, 2012).

d) Lack of entrepreneurship training and education

In order to gain momentum in the field of entrepreneurship, women require training on business capital management, marketing skills and relationship with customers and human resource management skills. Present systems are not providing training and education to women entrepreneurs to build these qualities. This has resulted in lack of essential skills among entrepreneurs especially women. Therefore, exclusive training programmes are recommended for women to enable them to adapt to newer technologies and up-gradation of the existing skills and resources to survive in the market (Adema *et al.*, 2013; Brush *et al.*, 2009).

e) Personality-based constraints and inter relationship among constraints

Historically societal norms play a huge influence on women's cognition and mental dispositions which created a doubt among women entrepreneurs about their self-confidence. This has lead to isolation affecting their performance. Comparatively less number of female role models also reduces the inspiration levels. In developing countries, a significant number of women endorsed entrepreneurship as recourse to unemployment. This has been a huge detriment in the functioning of the firm due to the lack of training and education with respect to the entrepreneurship skills. Moreover, most women started their business with small capital that mostly relied on their assets and income (Shelton, 2006; Danish and Smith, 2012).

All these constraints have to be ranked and considered to help policy makers take well-versed decisions related to allotting resources in priority fields. Studies show that women entrepreneurs face intense challenges and that these challenges are intensified due to adverse conditions in emerging countries. Under these circumstances, it is hard for women entrepreneurs to start and sustain businesses of their own. Therefore, it is essential that policy makers are conscious about these challenges and help to craft a favorable atmosphere to endorse women entrepreneurship (Minniti and Naude, 2010; Reynols *et al.*, 2002).

4.2. Closing the gender gap in agro entrepreneurship

Despite this mounting number of programs aimed to encourage and improve women's entrepreneurship in developing countries, gender gap in terms of the volume of business continue unabated. In order to close the identified gender gap, it is necessary to implement and practice the schemes which are designed by considering all the perspectives. In order to design schemes, collection and analysis of reliable gender-segregated data, extra-economic and normative contexts of entrepreneurship, work family interface etc are essential. Vossenberg (2013) reviewed several literatures on closing the gender gap in entrepreneurship by analyzing the ongoing programs and schemes based on the effects of different factors on the success of these programs. Significant number of programs to close the gender gap in agro-entrepreneurship are being carried out by stakeholders, and policy makers in international (public) institutions, NGOs, private firms, public-private partnerships etc. These programs for promoting women agro- entrepreneurship include arranging instruments and methodologies for skill training, business services, technical support, and funding (Minniti and Naudé, 2010; Horrell and Krishnan, 2007). Corporates are also involved in this using their corporate social responsibility (CSR) fund and collaborating with public sector to implement PPP programs. These have contributed to the reduction of the gender gap by providing increase in income levels and economic growth for women entrepreneurs. Projects in partnership with the Cherie Blair Foundation and Millicom International Cellular by The U.S. Agency for International Development that provide trainings and education to improve women entrepreneurship in Tanzania, Rwanda and Ghana formed an example for such corporate initiatives (USAID, 2012). Another scheme to reduce the gender gap is the Women's Entrepreneurship Development programme of the International Labour Organization (ILO-WED). This scheme aim to empower women entrepreneurs, by creating suitable conditions for inclusive, resilient and sustainable economies. Third Billion Campaign is an example for Public Private Partnership (PPP) from La Pietra Coalition aimed to influence and strengthen the collective effect on women as drivers of global economic progress, (ILO, 2012; ILO, 2008; ILO, 2003).

Assessing impact of women entrepreneurship promotion is another important aspect which requires to study the efficacy of the schemes implemented for the promotion of women in entrepreneurship. A good number of studies have been carried out by researchers about the impact of these schemes and programs. Results from these studies varies from no noticeable impact to significant impacts in the economic aspects (Karlan and Valdivia, 2011; Beath *et al.*, 2013). As per the study of Duflo (2012), it is suggested that we should aim for women entrepreneurship development as it helps to achieve gender equality, social transformation and economic growth for the society. It became necessary to ask few questions such as whether these women entrepreneurship promotion and development schemes bring a noticeable change with respect to creation of more women employment opportunities and changed the gender biased context in favor of gender equality. Even though some observations can be drawn from case studies, it is highly recommended to implement inventory and critical analyses of the programs proposed and endorsed by different stakeholders. It has been observed that the mere execution of a lot of programs and policies that consider market engagement and generation of income may not necessarily lead to women's economic empowerment (Minniti, 2010). Impact should also be measured by considering the social changes for which more studies that examine the role of popular women entrepreneurs' in political and economic context are required.

Only few key points of entrepreneurship can effectively be educated and these alone cannot make a successful entrepreneur as the success depends on other variables as well. There are promising studies which consider the perceptual variables that can empower the skills possessed (Henry *et al.,* 2005). Policy outlines must be diverse and focus on normative attributes of entrepreneurship. The main objectives of the policies should be the assistance to provide resources, provide technical assistance and education and to make sure the policy framework endorse societal outlooks favorable to women's entrepreneurship and suitable business environment (Kelley *et al.,* 2012). As per ILO (2012), women's entrepreneurship is efficiently pushed through policy frameworks that safeguard, nurture and regulate business start-up and enhance their social inclusion. ILO also suggests that such a policy outline should reduce the risks women entrepreneurs encounter and establish a better enterprising culture for them along with the support services and policies by the governments.

An alternative approach to close the gender gap in entrepreneurship is based on institutional theory. It proposed the method of social marketing to increase the legitimacy of women's endeavors to become entrepreneurs in sub-Saharan Africa. It also recommended promotional activities for women's political leadership to make modifications in laws to increase the social status of women (Amine and Staub, 2009).

4.3. Agricultural extension reforms with special reference to women in agribusiness

Agriculture extension also helps in enhancing agricultural production and are provided by governments, NGOs and other organizations. The list of services offered in agriculture extensions are advice, training and information related to different fields of agriculture such as processing, manufacturing or livestock production and marketing etc. These services are designed to increase farmers' capability to increase productivity and revenue. The delivery of agriculture extension activities include meeting, industrial visits, providing information and communication technologies, providing demonstrations on model farms etc. Emphasis is given to offer education to enable knowledge exchange and impact assessment studies to bring better outcomes. Agricultural extension services help to reduce constraints such as unavailability of the right advisory methods, inadequate training and advisory agents, poor management system etc. (Birner *et al.,* 2009).

It is documented that gender inequality is connected to food insecurity. The number of undernourished women and girls is much higher than that of men. As per FAO, if the access to the resources remain same for women and men, the productivity is predicted to be increased by 20-30%. Recently, a significant number of efforts were taken by different governments, to design and implement social protection policies related to food security. Under these policies efforts were made to introduce tools such as public employment programmes and subsidies and transfers.

These social protection schemes provided for the following

a) Protection related to stable income even during economic crisis situations

b) Reduce the impact of economic and market shocks

c) Helps in productivity by providing financial and technical supports

d) Removal of the gender barriers and constraints for women empowerment

In maximizing the productivity and bringing transformation among women entrepreneurs, it is necessary to address different vulnerabilities faced by women. The recent developments in integrated programs support women not only to tackle economic and domestic vulnerabilities but also to bring a positive transformation about future protection (Jones *et al.,* 2017). Social protection interventions which are essential for transformation for food security and productivity of women in agricultural sector are depicted in Figure 1.

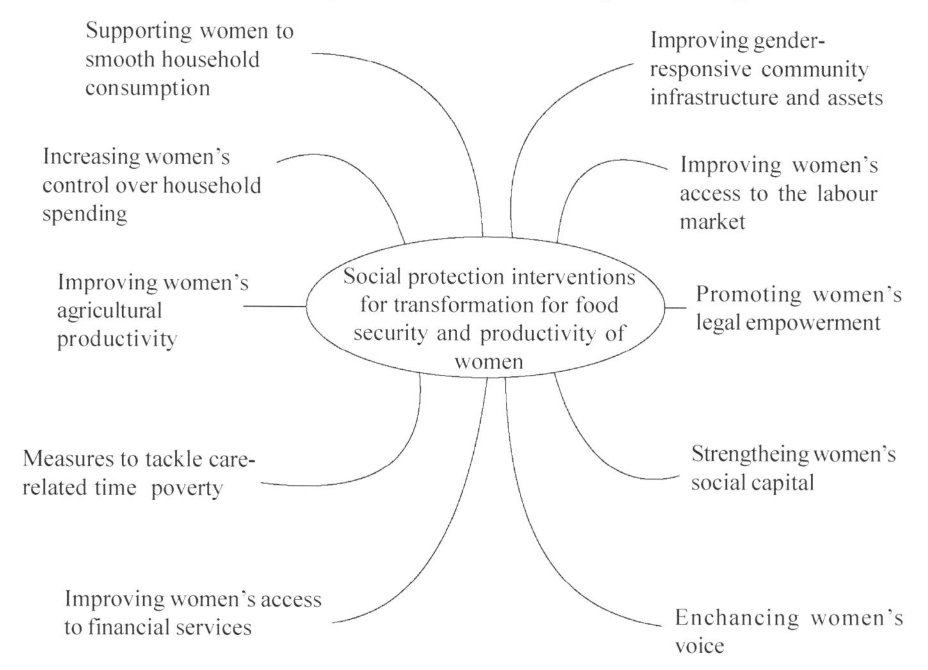

Fig. 1. Social protection interventions which are essential for transformation of women entrepreneurship

A study that focused on the approaches to increase the gender-responsiveness of agricultural extension systems based on extension reforms in India, Uganda, Venezuela, and Ethiopia considered the following aspects as reported by Ragasa (2014).

i) Whether the gender of the extension agent affects the efficacy of the services
ii) Whether everyone receive extension services irrespective of their gender and
iii) Whether the gender affects the delivery of extension services

ATMA (Agricultural Technology Management Agency Model in India: An example for gender sensitive extension system)

The ATMA model got attention due to its innovative nature with respect to decentralized extension service. The ATMA is a semiautonomous organization comprised of different stakeholders in agricultural sector. In this model several provisions were made to reduce the gender barrier by ensuring the allotment of 30% of the resources on beneficiary-oriented programs for women entrepreneurs. Policy guidelines for the providing more funds for women has been done for the "Support to States for Extension Reforms" based on Agricultural Technology Management Agency (ATMA) Model launched in 2006-07 (Policy Commission, 2007). It introduced gender-sensitization attributes in the training of trainers and also ensured representation of women in all its boards and groups at the district level. Even though the ATMA model showed an impact on women farmers in India, the outcome on improvement of agricultural production or the marketing practices were limited (Asante *et al.*, 2010).

4.4. Measures to strengthen the efficiency and participation of women in agro-entrepreneurship in India

The rate of progression of women entrepreneurs with time is substantial in spite of the not so supportive socio-economic factors in the Indian society. Governments initiated several schemes and implemented several policies to improve women agri-entrepreneurship in India. Various schemes for women entrepreneurship in India are collected from various sources and compiled in Table 1. Various agencies and centers and their activity towards providing infrastructure and insurance for women entrepreneurship are collected from various resources and compiled in Table 2. (Ministry of Micro, Small & Medium Enterprises, 2021; Ministry of Commerce and Industry, 2021; National Agriculture Infra Financing Facility, 2021, Department of Agriculture, Cooperation & Farmers Welfare (2021) and Indifi Technologies Business Loan Blogs, 2021).

Table 1. Various schemes for women entrepreneurship development in India

Name of scheme	Objective	Agency
Bharatiya Mahila Bank (BMB)	Scheme is to provide women entrepreneurs business loans for setting up of manufacturing firms	State Bank of India, Government of India
Mahila Udyam Nidhi Scheme	Scheme offers financial assistance of up to Rs. 10 lakh to set up a small scale enterprise	Small Industries Development Bank of India (SIDBI)
Mudra yojana scheme for women	Financial support to women who wish to start a small scale business	Nationalized banks
Trade Related Entrepreneurship Assistance and Development (TREAD) Scheme for Women	Scheme provides financial support, training, and counseling for economic empowerment of women	Ministry of Micro, Small and Medium Enterprises, Government of India
Dena Shakti Scheme	A scheme for financing women entrepreneurs in micro-credit, manufacturing, agriculture & retail sector.	Dena Bank
Udyogini Scheme	Scheme for the welfare and development of Indian Women Entrepreneurs	Women Development Corporation (GOI)
Cent Kalyani Scheme	Scheme envisages empowerment of women to start new project or expand or modernize the existing unit.	Central Bank of India
Mahila Kisan Sashaktikaran Pariyojana	Scheme aims to improve the status of women in Agriculture, and to increase the opportunities for the empowerment	Ministry of Rural Development

Source: (author compilation)

Table 2. Activities of agencies involved in providing infrastructure and insurance to women entrepreneurs.

Agency/Centre	Activity 1	Activity 2	Activity 3
Agriculture Technology Management Agency (ATMA)	Support for Women Food Security Groups (FSGs), Support for Gender Coordinator	Provision of Seed money / Revolving Fund, capacity building, skill development and support services	Representation of Women farmers in decision making bodies
Agri-Clinics & Agri-Business Centers (ACABC)	Back-ended Composite Subsidy	Compared to men, more back ended composite subsidy is given to women towards cost of the project.	—
Mass Media Support to Agricultural Extension	Outreach to women	—	
Mission for Integrated Development of Horticulture (MIDH)	Target the empowerment of SC/ST, and women beneficiaries through assistance for horticulture mechanization (for individual & self help groups)	Procurement of Agricultural machinery & equipment at higher subsidy rates to women groups	—
National Mission on Oilseeds & Oil Palm (NMOOP)	Objective is to promote women groups and self-help groups/ cooperative societies	Farmers associations, women groups, self help groups etc are eligible for getting assistance with respect to equipment related to oil processing and pre-processing and extraction.	—
Integrated Scheme for Agricultural Marketing (ISAM)	Storage infrastructure under Agriculture Marketing infrastructure (AMI) for Registered FPOs, Panchayats, Women, Scheduled Caste (SC)/ Scheduled Tribe (ST) beneficiaries or their cooperatives/ Self-help groups.	For Infrastructure projects other than storage Infrastructure for Registered FPOs, Women, Scheduled Caste(SC)/ scheduled Tribe (ST) beneficiaries or their cooperatives	—
National Food Security Mission (NFSM)	At least 30% allocation of the funds is for women farmers.	—	

National Mission for Sustainable Agriculture (NMSA)	Soil & water conservation; Water use efficiency; Soil health management and Rain-fed Area Development	At least 50% of the allocation is to be utilized for small, marginal farmers of which at least 30% should be women. —
Sub-Mission on Agricultural Mechanization (SMAM)	Training on Gender friendly Equipment for Women farmers are to be conducted by Farm Machinery Training & Testing Institutes.	More subsidy to women farmers —
Agricultural Insurance	Ensuring maximum coverage of SC / ST / Women farmers with budget allocation and utilization for these category of farmers to be in proportion to their population in the respective state	
Modified National Agricultural Insurance Scheme (MNAIS)	Insurance protection for notified food crops, oilseeds and annual horticultural / commercial crops.	
Weather Based Crop Insurance Scheme (WBCIS)	Weather based Crop Insurance Scheme (WBCIS) provides protection to the insured cultivators if they face loss in crops yields due to adverse weather incidents	
Coconut Palm Insurance Scheme (CPIS)	The Coconut Development Board, provide insurance schemes to assist small and medium coconut growers.	

Self-help groups (SHGs) the women-entrepreneurship in India

Nowadays Women's groups are acting as a source for social, political, and economic empowerment. In the last three decades in India, SHGs are working towards in the reduction of poverty and improvement of livelihood. At the initial period, governments helped these SHGs to solve the credit constraints through linking SHGs to banks and microcredits. For an example, Swarnajayanti Gram Swarojgar Yojana (SGSY), by the Ministry of Rural Development (MoRD) from 1999 to 2013 helped SHGs to overcome the financial constraints faced by the SHGs. Later SHGS at the state level, have expanded the volume of their business through social mobilization and responsibility, knowledge of rights and improvement of health and nutrition. Examples of these schemes are SERP (the Society for Elimination of Rural Poverty) in Andhra Pradesh, JEEViKA in Bihar and Kudumbashree in Kerala. Now National Rural Livelihoods Mission (NRLM) was launched by the State level programs such as SERP (Kumar *et al.*, 2019).

Kudumbashree: A successful initiative from Kerala

Kudumbashree is the poverty eradication and women empowerment program started in 1997 by the State Poverty Eradication Mission (SPEM) of the Government of Kerala. The implementation of this programme was associated devolution of powers to the Panchayat Raj Institutions. Kudumbashree consist of Neighbourhood Groups, Area Development Societies and Community Development Societies. Kudumbashree as a programme has been recognized as the State Rural Livelihoods Mission (SRLM) under the National Rural Livelihoods Mission (NRLM by Ministry of Rural Development (MoRD), Government of India. The mission of Kudumbashree include the expansion and promotion of the community network and through them, it emphasizes empowerment of women both in social and economic aspects by financial and technical assistance.

Nutrimix: A Success story of Kudumbashree in food processing

Nutrimix is a cereal-based powder mix containing wheat, soya, coconut sugar, jaggery, Bengal gram, groundnut etc. developed by Central Plantation Crops Research Institute (CPCRI), Kasargode. Through Nutrimix units, Kudumbashree is producing Nutrimix since 2007. Under the Integrated Child Development Scheme, Nutrimix is provided as free take-home ration to children in the age group of six months to three years through anganwadis. Nutrimix's popularity has increased over the years for the quality of the product and its capability to meet the demand with sufficient supply.

Aadrics Agro: The triumph of women in Coconut Processing

Aadrics agro is a venture by Ms. Chithra K Kumar. Chithra started her business journey with the production of virgin coconut oil. To increase the profit, Chithra introduced byproducts from the production of virgin coconut oil such as Coconut Based Masalas (Chicken, Meat, Fish, Sambar with tamarind), Coconut water Lemonade, VCO based Capsules, Mouth wash, Baby Oil, Virgin Coconut Oil (Hot Processed), Copra Pressed Coconut Oil etc. Chithra was awarded with Rupees 12 lakhs as Grand-in-aid from RAFTAAR Agri Business Incubation scheme of RKVY, Government of India for the development of Lactose free coconut milk drink and coconut milk cubes.

5. Changes in the perception of women entrepreneurship

With the advent of industrialization and globalization the status of women also changed significantly. In the developed countries, women have achieved a significant position in the corporate and entrepreneurship field simultaneously. But this change is not so prominent in developing counties and underdeveloped countries. Ideas of social entrepreneurship and inclusive innovation are getting attention due to their significance in building and executing new ideas and prospects which are essential to uplift the excluded members in these societies. Many researchers in the gender and diversity pointed out that women are better suited in employing strategies and show effective leadership in social enterprises. This has been attributed to their gender specific characteristics such as compassion, empathy and emotional nature (Muntean and Ozkazanc-Pan, 2016). These results mostly from developed countries has led to a better socio-economic development perspective for women in these countries. However, more studies are required to be conducted in developing countries to derive a more realistic empirical relationship (Rosca *et al.,* 2020). Recently, a significant number of studies were attempted respect to women entrepreneurship in the context of class, sector, caste, culture and regions and how these factors affect the performance of women entrepreneurs (Yadav and Unni, 2016). These were significant in bringing to fore new concepts and technology handholding and networking services through Agribusiness Incubation (ABI) centers for women entrepreneurship development.

6. Agribusiness incubation centres (ABI) and technology handholding for women enterprises

Agribusiness incubation centres act as a platform for the speedy commercialization of the technologies and reinforcing of public private partnerships. It facilitates interfacing and networking mechanism between R&D institutes, industries and financial institutions, thereby contributing to a knowledge-based economy. The ABI serves as a gender friendly platform to commercialize the ventures of women entrepreneurs through different initiatives of institutional support. Some of the innovative women friendly technologies for the production of safe and healthy value-added products developed under Agri Business Incubator of Kerala Agricultural University are briefed below.

6.1. Small scale parboiling cum drying unit

Rice is rich in carbohydrates, vitamins and minerals. However, polishing of grain affects the nutritional quality by removing the superficial layers of caryopsis from the rice, concentrating the carbohydrate content and reducing the vitamins and minerals. Parboiling of rice resolves these problems to a certain extent since it helps to concentrate nutraceuticals, improves the palatability and extends shelf life. Parboiling is a traditional hydrothermal rice processing treatment intended to gelatinise the starch to enhance the head rice yield, to facilitate the penetration of nutrients present in the bran layer into the endosperm and to reduce their loss during milling.

Technological advancement in the industrial parboiling has resulted in enhanced process controls in the large-scale industrial parboiling units. But the enormous initial investments required for infrastructure, along with the time, labour and energy involved in paddy parboiling made it uneconomical for small scale milling and processing units to undertake parboiling. The small-scale parboiling cum drying unit consists of soaking cum steaming cylinder, LSU dryer, bucket elevator, steam boiler and heat furnace. The unit was fabricated using food grade material SS 304, to ensure safety. Soaking and steaming were carried out in the same chamber of capacity 100 kg/cylinder. The Steam boiler generates the steam at a pressure of 2 kg/cm^2 in the soaking cum steaming chamber. The bucket elevator with a capacity of 500 kg/h conveys the steamed paddy into LSU dryer and facilitates the drying process by continuous recirculation of paddy. The LSU dryer consist of inverted V channels which acts as air inlet and outlet ports. The hot air generated by burning of firewood in the furnace will heat up the interior ducts of the LSU drier.

6.2. Technology for the production of tasty Ready-To-Eat snacks

A "Ready-To-Eat" food product is the one which needs minimum processing procedures and is ready-to-eat as soon as the pack is opened. Nowadays, many consumers want to relax in the comfort of their own home rather than to spend time at a restaurant. Extrusion technology plays an important role in the snack and ready to eat breakfast food industry.

Kurkure type extruded products are becoming popular day by day due to its convenience to use and its palatability. Moscicki *et al.* (2012) states that the extrusion technology is very useful to retain the nutrients compared to other thermal processing methods (Moscicki *et al.*, 2012). Its main advantages are faster processing time and reduction in energy consumption over other cooking processes. Thus the cost of production is reduced. Considering the nutritional and health benefits of finger millet, corn, rice and yam, nutritious, convenient and safe to consume RTE snack food was developed at the agribusiness incubator attached to Kerala Agricultural University. The detailed process flow chart for the production of RTE snack is given in Figure 2.

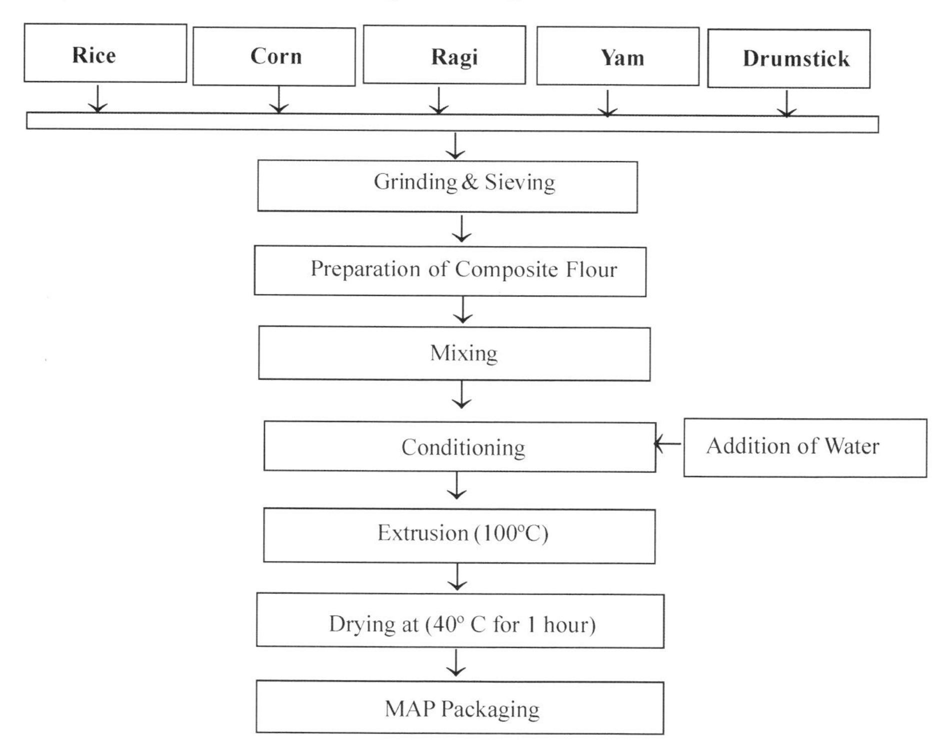

Fig. 2. Process flow chart of RTE product

Standardisation of the extruded snack food was conducted with different combinations of raw materials and at varying temperatures. The product was evaluated in agri incubation laboratory and the best combination was found to be the one prepared with 60% corn, 15% elephant yam, 20% purple yam, and 5% drumstick prepared at 100°C.

6.3. Technology for the production of healthy vacuum fried chips

Vacuum frying is a technology of producing healthy snacks with low fat content. Frying under atmospheric conditions absorb more oil, degrade the quality of oil to a considerable level and may lead to several health problems like obesity, cancer, heart diseases etc. Hence, vacuum frying is an alternate innovative technology to preserve the quality of oil, to lessen oil absorption and to keep the nutritional quality of the fried product.

During vacuum frying, the sample is heated under a negative pressure which drops the boiling point of oil and water in the sample (Troncoso *et al.*, 2009). So when the oil temperature reaches the boiling point of water, the unbound water in the fried food is rapidly removed. Moreover, this frying technology inhibits lipid oxidation and enzymatic browning in the absence of air and thus preserve the colour and nutrients in the sample. The oil used for vacuum frying can be reused several times without change in the quality when compared to atmospheric frying. The flow chart for vacuum frying of banana chips is presented in Figure 3. Three D view of vacuum frying system is depicted in Figure 4. The quality of edible oil can be determined by the percentage of total polar molecules (TPM) using TESTO 270 device and maximum allowable limit for the total polar molecule is 25%. The low value of TPM in vacuum frying is due to low temperature and low pressure conditions applied for frying when compared to atmospheric frying (Ranasalva and Sudheer, 2018).

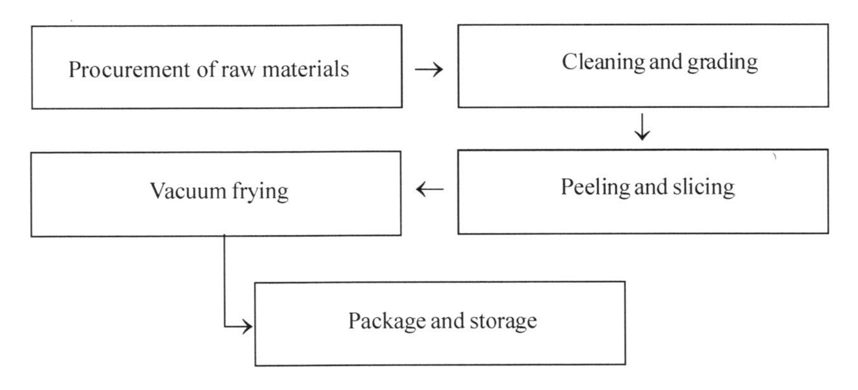

Fig. 3. Process flow chart of vacuum frying of banana chips

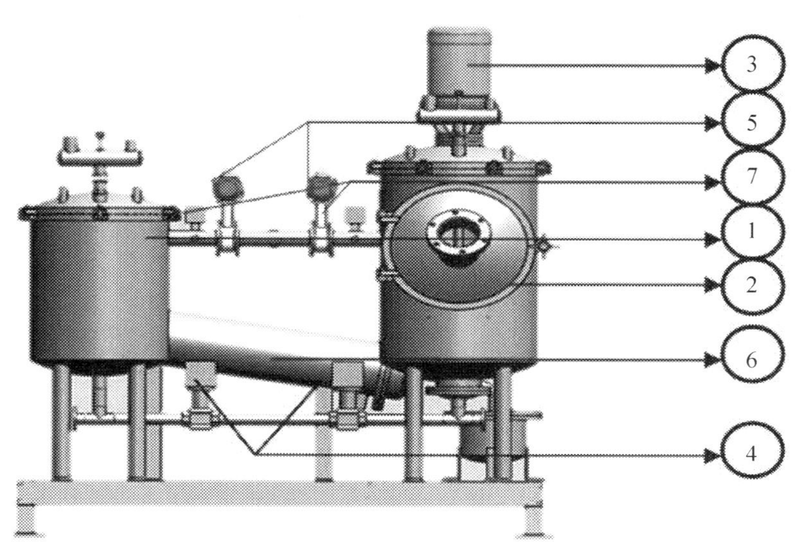

1. Oil storage chamber
2. Frying chamber
3. De-oiling motor
4. Oil flow control
5. Vacuum valve
6. Condenser
7. Nitrogen flow valve

Fig. 4. 3D view of vacuum frying unit.

The shelf life of vacuum fried banana chips was found to be more than 6 months when packed under active packaging (Nitrogen fill package MAP). This technology can also be used for the production of ripened jack fruit, bitter guard, carrot, beet root, ladies finger etc (Sudheer and Indira, 2018a).

6.4. Technology for the production of Intermediate Moisture Foods (IMF)

Osmotic dehydration is one of the food preservation technique that involves the partial removal of water from fruits and vegetables by dipping in aqueous solutions of high osmotic pressure *viz.*, sugar and salts (Pandharipande *et al.*, 2012). Osmotic dehydration is receiving much devotion in the food industry as it is a cost saving drying technology. Various perishable fruits such as banana, seasonal fruits like jack fruit and vitamin rich goose berries etc can be preserved and value added by this method. The drawbacks of traditional drying involves the spoilage due to insects attack, non-uniform drying chances of shrinkage and colour deterioration. Hence, a process protocol has been developed and standardized under agribuisness incubator at Tavanur to solve these hurdles and to ensure prolonged shelf life and quality retention of final product.

To produce Intermediate Moisture Foods (IMF), the selected fruits/vegetables are cleaned to remove undesirable foreign materials and graded based on their maturity and ripening. Cleaned fruits/vegetables are sliced in circular or longitudinal shapes using a slicer so as to get uniform thickness. Sliced fruits/ vegetables are then blanched before soaking in sugar syrup. Blanching is carried out mainly to retain the colour of fruits/vegetables after drying and to prevent enzymatic activity.

A blancher cum drier or a vacuum dryer is utilized for drying operation. Since, vacuum drying is a low temperature process; the quality of the product is maintained even after drying. From an industrial standpoint, blancher cum drier is more economical and less time consuming since both blanching and drying are done within the same machine (Sudheer and Indira, 2018b).

6.5. Microencapsulation technology for healthy ready to drink mix formulations

Conversion of food products into powdered form make it simpler, with reduced storage volume and transportation cost. Microencapsulation aids this by using spray drying to develop nutraceutical products from commodities with high medicinal and nutritive value. It also offers protection of sensitive food components against nutritional loss and also preserves flavour by coating tiny droplets with a suitable wall material. Spray drying will assist in converting the liquid slurry to

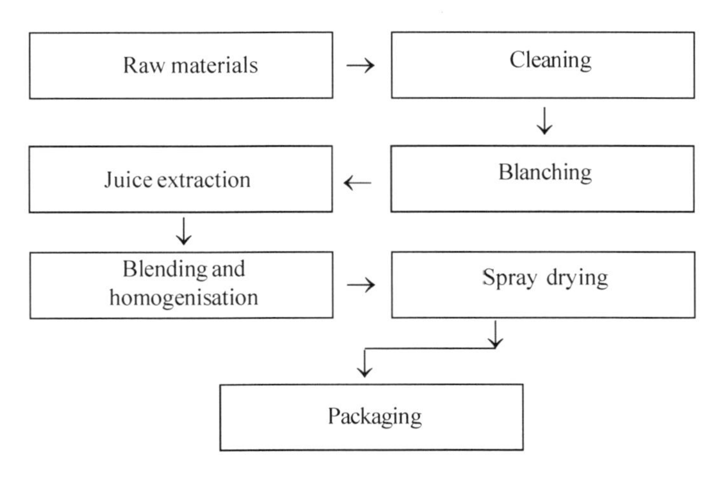

Fig. 5. Process flow chart for the production of microencapsulated healthy ready to drink formulations

powdered form which reduce the storage space, transportation cost and prolong shelf life.

Microencapsulated powders were developed at agri incubator from banana pseudo stem, horse gram, milk, whey, kokum. These are healthy ready to drink formulations that can compete with the fruit powders available in the market. In order to enhance the organoleptic properties, natural flavours like ginger juice, cardamom powder, mint extract etc were incorporated in the powder (Saranya

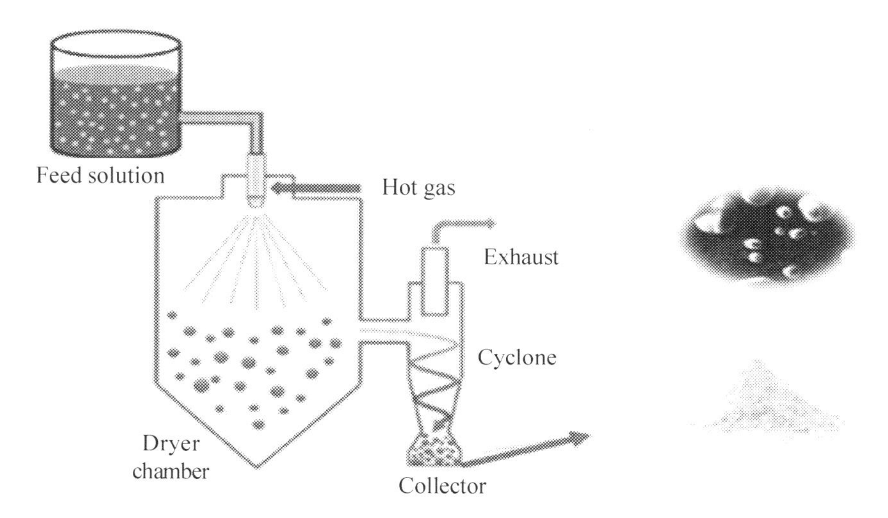

Fig. 6. A schematic diagram of spray drying process

and Sudheer, 2018). The detailed process flow chart for the production of microencapsulated powder is depicted in Figure 5. A schematic diagram of spray drying process is also given as Figure 6.

6.6. Technology for the production of nutraceutical pasta

Ready to cook food items are widespread in developed countries because of its flexibility in taste, convenience, ease of preparation and enticing nature. It is also cheaper, and fitting to the current life style. Gluten free pastas were developed using finger millet, corn, purple yam, elephant foot yam and drumstick as pasta flour to enhance the medicinal as well as nutritional value of final product.

An industrial model Pasta Machine is used for preparing cold extruded products (Figure 7). Process flow chart for the production of nutraceutical pasta is given in Figure 8. Different types of "die" could be attached to produce pasta of various shapes as per the requirement. The dough to be extruded is prepared just prior to extrusion with the aid of an industrial mixer. Once the dough of necessary consistency is ready, extrusion is carried out with greater ease and

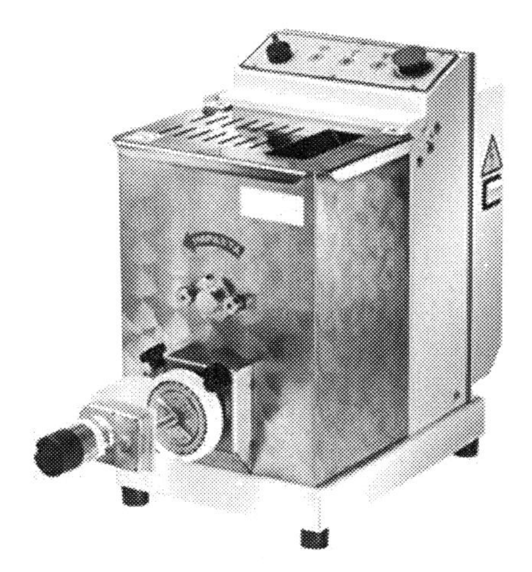

Fig. 7. Industrial pasta maker

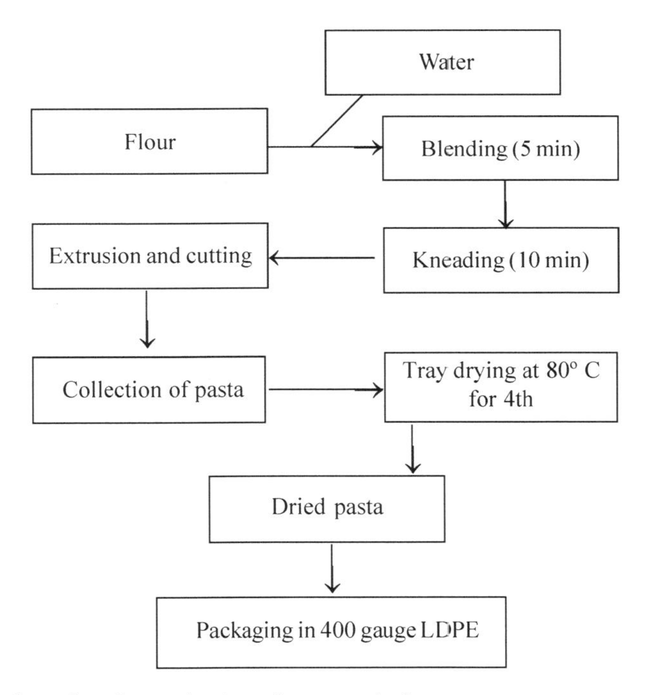

Fig. 8. Process flow chart for production of nutraceutical pasta

anticipated shape. A pasta cutter blade, connected at the outlet of the "die", is used to cut the extruded pasta to the required size. The pasta products were produced by the following the procedure prescribed by pasta machine manufacturer.

6.7 Retort processing technology for shelf – stable convenience food

During the last decades, life style and food habits have undergone drastic changes in India. Modern life style always prefer fast food culture either due to the time constraints or due to the aversion towards time consuming cooking practices. Hence, it is vital to design a safe alternative to fast food which could be both nutritious and shelf stable. The retorting or sterilization process confirms the stability of the Ready-to-Eat foods in retort pouches in the shelf and at room temperature. Retort thermal processing is the procedure used to retortable containers in a chamber with steam valves which assure the accurate temperature control to destroy bacteria. The injection of steam under pressure makes the temperature to reach above the boiling point of water inside each pouch within the chamber (Praveena and Sudheer, 2015). Retort processing will assist in the destruction of all viable micro-organisms especially the heat resistant pathogenic bacteria *Clostridium botulinum* to produce commercially sterile product. Flexible laminated pouch also known as retort pouches are used for retort processing, which can withstand thermal processing temperatures along with the advantages of metal cans and plastic packages. The multi-layer (4 layered) structure of retort pouch act as a good barrier for gas, moisture etc and thus prolong storage life.

Retort processing include filling of specific quantity food product in retort pouches, which can be achieved by using a filling machine. The pouches can be heat sealed using seal jaws and it should arrange in stackable pallets in retort chamber before retorting. After reaching the pre-set time, temperature and pressure sterilisation of product can be done by injecting steam or water into the retort. The detailed process flow chart for retort processing of *Ramasseri idli* is given in Figure 9.

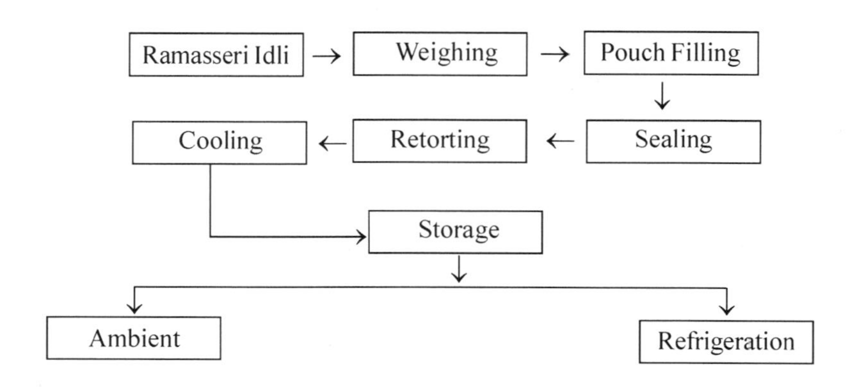

Fig. 9. Process flow chart of retort processed *Ramasseri idli*

7. Conclusion

Women's entrepreneurship is essential for sustainable economic development of a country. Women entrepreneurs face challenges such as gender discrimination, work-family conflict, financial constraints, lack of infrastructure and technical support, unfavorable business and political environments etc. Despite of all these obstacles, the number and scale of women entrepreneurship is increased due to the successful approaches from government and private entities in the forms of schemes and policies. For further growth of women entrepreneurship, implementation of women friendly legislation policies, public-private partnership, provision of extensive support for financial and infrastructural aspects, and eradication of gender gap in the entrepreneurship are required. Entrepreneurship has a conspicuous role in the economic development of the country. In the field of entrepreneurship, developing nations like India need a motivation and support to raise the per capita income of the farming community. Adoption of appropriate entrepreneurial sector plays an immense role in creating success rate and better utilisation of available resources. Enterprises/entrepreneurs should understand the fluctuating consumer styles and be able to provide high quality products with desired nutritional and sensorial qualities at reasonable cost. Compared to other sectors, food processing sector is a blooming sector having ample scope for entrepreneurship opportunities. Currently, our food market is ruled by multinational food companies and our diets are filled with unhealthy junk foods. This increases the chances of various diseases among different age groups especially youngsters who are addicted to fast food culture. Hence, the production of healthy products which are convenient and nutritional should be promoted to solve these hurdles. To fetch the market value and consumer acceptance, the products should be superior in quality as well as versatile in form, colour, attractiveness etc. Technological as well as mechanical support is mandatory for the production of better and safe foods. In this concept, Government has established agri-preneurship supporting centres *viz.*, agri incubators all over the country. Systematic research, monitoring and support of young entrepreneurs, technology development and transfer of developed technology are major activities of these incubators. Various technologies developed under the agri incubator in KAU, suitable for marketing are discussed in this chapter. We can hope that these technologies and agribuisness incubators will fill more colours to the food processing sectors, and farming community to boost the economy of our country.

References

Adema, W., Ali, N., Frey, V., Kim, H., Lunati, M., Piacentini, M. and Queisser, M. 2014. Enhancing women's economic empowerment through entrepreneurship and business leadership in OECD countries. [on-line]. Available: www.oecd.org/ gender/Enhancing %20Women %20Economic%20Empowerment _Fin_1 _Oct_2014.pdf (10 Feb 2021).

Aidis, R., Welter, F., Smallbone, D. and Isakova, N. 2007. Female entrepreneurship in transition economies: the case of Lithuania and Ukraine. Feminist Econ. doi:10.1080/ 13545700601184831

Alkire, S. Meinzen-Dick, R. Quisumbing, A., Seymour, G. and Vaz, A. 2013. The Women's Empowerment in Agriculture Index. World Dev. doi:10.1016/j.worlddev.2013.06.007

Amine, L.S. and Staub, K.M. 2009. Women entrepreneurs in sub-Saharan Africa: An institutional theory analysis from a social marketing point of view. Entrepreneurship Regional Dev. doi: 10.1080/08985620802182144.

Asante, F.A., Banful, A.B., Cohen, M.J., Gaff, P., Gayathridevi, K.G., Horowitz, L., Lemma, M., Mogues, T., Palaniswamy, N., Paulos, Z., Raabe, K., Randriamamonjy, J., Sekher, M., Sun, Y. and Tadesse F. 2010. World Bank/IFPRI (International Food Policy Research Institute) Gender and governance in rural services: insights from India, Ghana, and Ethiopia. Gender and governance author team. International Bank for Reconstruction and Development/World Bank, Washington, DC. doi:org/10.1596/978-0-8213-7658-4

Beath, A., Christia, F. and Enikolopov, R. 2013. Empowering women through development aid: Evidence from a field experiment in Afghanistan. Am. Polit. Sci. Rev. doi: 10.1017/ S0003055413000270.

Birner, R. et al., 2009. From Best Practice to Best Fit: A Framework for Designing and Analyzing Pluralistic Agricultural Advisory Services Worldwide. The J. Agric. Educ. and Ext. doi: 10.1080/13892240903309595.

Brush, C.G., de Bruin, A. and Welter, F. 2009. A gender-aware framework for women's entrepreneurship. Int. J. Gender Entrepreneurship. doi: 10.1108/17566260910942318.

Danish Y.A. and Smith, L.H. 2012. Female entrepreneurship in Saudi Arabia: Opportunities and challenges. Int. J. Gender Entrepreneurship. doi: 10.1108/17566261211264136.

De Vita, L., Mari, M. and Poggesi, S. 2014. Women entrepreneurs in and from developing countries: Evidences from the literature. Eur. Manag. J. doi:10.1016/j.emj.2013.07.009

Department of Agriculture, Cooperation & Farmers Welfare (2021), National Agriculture Infra Financing Facility. [online]Available: https://agriinfra.dac.gov.in/ (21 Feb 2021).

Duflo, E. 2012. Women empowerment and economic development. J. Economic Literature. doi: 10.1257/jel.50.4.1051.

Gupta, K.B. and Gupta, L.B. 2017. A framework for identification of opportunities for agribusiness and agripreneurship in India. J. Manag. Res. Anal. doi: 10.18231/2394-2770.2017.0005

Gupta, S., Vemireddy, V., Singh, D. and Pingali, P. 2019. Adapting the Women's empowerment in agriculture index to specific country context: Insights and critiques from fieldwork in India. Glob. Food Security. 23: 245–255. doi: 10.1016/j.gfs.2019.09.002.

Henry, C., Hill, F. and Leitch, C. 2005. Entrepreneurship education and training: Can entrepreneurship be taught? Part II. Educ. + Training. doi: 10.1108/00400910510592211.

Horrell, S. and Krishnan, P. 2007. Poverty and productivity in female-headed households in Zimbabwe. J. Dev. Stud. doi: 10.1080/00220380701611477.

ILO, 2003. Facilitating women's entrepreneurship: lessons from the ILO's research and support programs. Paper presented by Gerry Finigan at the OECD workshop on Entrepreneurship in a Global Economy: Strategic Issues and Policies

ILO, 2008. ILO Strategy on Promoting Women's Entrepreneurship Development. Published online, https://www.ilo.org/wcmsp5/groups/public/—ed_norm/—relconf/documents/meetingdocument/wcms_090565.pdf

ILO, 2012. Women's Entrepreneurship Development. Encouraging women entrepreneurs for jobs and development. [online] Available:https://www.ilo.org/empent/areas/womens-entrepreneurship-development-wed/lang—en/index.htm (20 Feb 2021)

Indifi Technolgies Business Loan Blogs, 2021. [online]Available https://www.indifi.com/blog/9-schemes-for-women-entrepreneurs-in india/#:~:text=Several%20schemes%20for%20women%20entrepreneurs,Annapurna%20Scheme (20 Mar 2021)

Jamali, D. 2009. Constraints and opportunities facing women entrepreneurs in developing countries: A relational perspective. Gender in Manag.: An Int. J. doi: 10.1108/17542410910961532.

Jennings, J.E. and Brush, C.G. 2013. Research on Women Entrepreneurs: Challenges to (and from) the Broader Entrepreneurship Literature?. The Acad. Manag. Annals. doi: 10.1080/19416520.2013.782190.

Jones, N., Holmes, R., Marshall E.P., and Stavropoulou M. 2017. Transforming gender constraints in the agricultural sector: The potential of social protection programmes. Glob. Food Security. doi: 10.1016/j.gfs.2016.09.004.

Karlan, D. and Valdivia, M. 2011. Teaching entrepreneurship: Impact of business training on microfinance clients and institutions. Rev. Econ. Statist. doi: 10.1162/REST_a_00074.

Kelley, D.J., Singer, S. and Herrington, M. 2012. Global Entrepreneurship Monitor: Global Report 2011. Babson College, Universidad del Desarrollo, Universiti Tun Abdul Razak.

Kumar, N., Raghunathan, K. Arrieta, A., Jilani, A., Chakrabarti, S., Menon, P. and Quisumbing, A. R. 2019. Social networks, mobility, and political participation: The potential for women's self-help groups to improve access and use of public entitlement schemes in India. World Dev. doi: 10.1016/j.worlddev.2018.09.023.

Manaf, N.A. and Ibrahim, K. 2017. Poverty reduction for sustainable development: Malaysia's evidence-based solutions. Glob. J. Social Sci. Stud. doi: 10.20448/807.3.1.29.42. 3(1): 29-42.

Ministry of Commerce and Industry, 2021. The Women Entrepreneurship Platform (WEP) [online]Available:https://www.startupindia.gov.in/content/sih/en/government-schemes/Wep.html (21 Feb 2021).

Ministry of Micro, Small & Medium Enterprises, 2021. [onlione] Available: https://msme.gov.in/women-entrepreneurs (Accessed: 21 February 2021).

Minniti, M. 2010. Female entrepreneurship and economic activity. Eur. J. Dev. Res. doi: 10.1057/ejdr.2010.18.

Minniti, M. and Naude, W. 2010. What do we know about the patterns and determinants of female entrepreneurship across Countries?. Eur. J. Dev. Res. doi: 10.1057/ejdr.2010.17.

Mokta, M. 2014. Empowerment of Women in India: A Critical Analysis. Indian J. of Public Administration. doi: 10.1177/0019556120140308.

Moscicki, L., Mitrus, M., Wojtowicz, A. and Oniszczuk, T. 2012. Application of extrusion-cooking for processing of thermoplastic starch (TPS). Food Res. Int. doi: 10.1016/j.foodres.2011.07.017.

Muntean, S.C. Ozkazanc-Pan, B. 2016. Feminist perspectives on social entrepreneurship: critique and new directions. Int. J. Gender Entrepreneurship. doi: 10.1108/IJGE-10-2014-0034.

Nagaraja, B.D.B.N. 2013. Empowerment of Women in India: A Critical Analysis. IOSR J. Humanities Social Sci. doi: 10.9790/0837-0924552.

National Policy for the Empowerment of Women .2014. Indian J. Public Administration. doi: 10.1177/0019556120140328.

Panda, S. 2015. Factors affecting capital structure of Indian venture capital-backed growth firms. Entrepreneurial Ecosyst. Perspectives from Emerging Economies. doi: 10.1007/978-81-322-2086-2_5.

Panda, S. 2018. Constraints faced by women entrepreneurs in developing countries: review and ranking. Gender in Manag. Emerald Group Publishing Ltd., 33(4): 315–331. doi: 10.1108/GM-01-2017-0003.

Panda, S. and Dash, S. 2013. Trust and reputation in new ventures: Insights from an Indian venture capital firm. Dev. Learning Organ. doi: 10.1108/DLO-02-2013-0003.

Pandharipande L., S., Paul, S. and Singh, A. 2012. Modeling of Osmotic Dehydration Kinetics of Banana Slices using Artificial Neural Network. Int. J. Comput. Appli. doi: 10.5120/7329-0188.

Planning Commission (India), 2007. Report. Report of sub-group on gender and agriculture submitted to the working group on gender issues, Panchayat Raj Institutions, Public Private Partnership, Innovative Finance and Micro Finance in Agriculture for the Eleventh Five Year Plan (2007–2012). New Delhi

Praveena N and Sudheer K.P. 2015. Optimization of blanching treatment for 'koozha' variety tender jack fruit (Artocarpus heterophyllus L.). Int. J. Agric. Sci. Res. 5 (6).

Ragasa, C. 2014. Improving gender responsiveness of agricultural extension. Gender in Agric. Closing the Knowledge Gap. doi: 10.1007/978-94-017-8616-4_17.

Ranasalva, N., and Sudheer K.P. 2018. Effect of Pre- treatment on quality parameters of vacuum fried ripe banana (Nendran chips). J. Trop. Agric. 55: 161-166.

Reynols, P.D., Bygrave, W.D., Cox, L.W. and Autio, E.H.M. 2002. Global Entrepreneurship Monitor 2002 Executive Report. Sci. Technol.

Rosca, E., Agarwal, N. and Brem, A. 2020. Women entrepreneurs as agents of change: A comparative analysis of social entrepreneurship processes in emerging markets. Technol. Forecasting and Social Change. doi: 10.1016/j.techfore.2020.120067.

Saranya, S. and Sudheer, K.P. 2018. Development of fortified banana pseudostem juice powder utilizing spray drying technology. J. Trop. Agric. Vol. 55.

Shelton, L.M. 2006. Female entrepreneurs, work-family conflict, and venture performance: New insights into the work-family interface. J. Small Business Manag. doi: 10.1111/j.1540-627X.2006.00168.x.

Sudheer, K.P. and Indira, V. 2018a. Entrepereneurship Development in Food Processing, 1st Edition. New Delhi: New India Publishing Agency. 381p.

Sudheer, K.P. and Indira, V. 2018b. Entrepereneurship and Skill Development in Horticultural Processing', 1st Edition. New Delhi: New India Publishing Agency. 432p.

Troncoso, E., Pedreschi, F. and Zúñiga, R.N. 2009. Comparative study of physical and sensory properties of pre-treated potato slices during vacuum and atmospheric frying. LWT - Food Sci. Technol. doi: 10.1016/j.lwt.2008.05.013.

USAID, 2012. [online] Available:https://geneva.usmission.gov/2012/12/20/women-entrepreneurs-africa/. (21 Feb 2021)

Vossenberg, S. 2013. Women Entrepreneurship Promotion in Developing Countries: What explains the gender gap in entrepreneurship and how to close it. Maastricht School of Management Working Paper Series. 8(1): 1-27.

Yadav, V. and Unni, J. 2016. Women entrepreneurship: research review and future directions. J. Glob. Entrepreneurship Res. doi: 10.1186/s40497-016-0055-x.

12

Gender in Household Nutrition and Community Health

*T.V. Hymavathi**

1. Introduction

Adequate nutrition is the fundamental keystone of any individual's health. It is especially crucial for women because it causes havoc not only on the woman's health but also on their children's health and in turn the community. Malnutrition causes a variety of risks to women. It undermines women's ability to survive childbirth, makes them more susceptible to infections, and leaves them with fewer reserves to recover from illness. Children of malnourished women are found to face impaired cognition, short stature, decreased resistance to infections, and an increased risk of disease and death throughout their lives. The increasing number of malnourished population in any country is a threat to its technological and economic development. Moreover, a chronically undernourished woman in every probability will give birth to a baby who is likely to be undernourished as a child.Thus malnourishment in women perpetuate the cycle of under-nutrition to be repeated over generations.In other words, women's nutritional status impacts the countries nutritional status.

Despite economic progress, India has failed to combat malnutrition that adversely affects the country's socio-economic progress. This is reflected in the hunger score of 27.2 which categorized India in 'serious' hunger category. The Global Hunger Index (GHI) rank of the country is 94 among 107 countries in 2020 and was 102 out of 117 countries in 2019. In fact, GHI is calculated based on the four indicators viz. undernourishment, child wasting, child stunting and child mortality. Estimates also suggest that more than one-third of the world's malnourished children are in India. Among the varied causes of malnutrition in the country, major ones include mother's nutritional status, lactation behaviour, women's education and sanitation. These in turn affect children in several forms and get manifested as childhood illness, stunting, and retarded growth. Therefore, addressing the nutrition of women population at all levels is essential to safeguard the future generations, and in turn the human capital and economic development of any nation.

*Author contact: hymasarathi@gmail.com

2. Importance of nutrition for women

More nutritional deficiencies are seen in women than in men, due to their distinguished biology, poverty, low social status, sociocultural traditions and lack of education. Differences in household work patterns can also increase women's chances of being malnourished. About 50 percent of the world's pregnant women are malnourished, out of which at least 120 million women are living in less developed countries. Underweight affects the productivity of women thus leading to high rates illness and mortality. Most of the underweight women are also stunted and below the median height for their age. The poor nutrition of the women lead to wasting in the children. Wasting is a nutritional deficiency state that carries severe health consequences, the most immediate being an intensified risk of mortality. A study in 2015 found thataround 13 per cent of worldwide deaths among children under 5 years of age were attributed to wasting. It accounted to 875000 preventable child deaths. In 2016, approximately 52 million children worldwide suffered from wasting, more than half of them resided in South Asia. The prevalence of wasting in South Asia is above the 15 per cent threshold which confirms child wasting as a 'critical public health problem.

These estimates are indicative of the importance of nutrition of women before and during pregnancy. It needs to be inferred that nutrition plays a significant role in reproductive health and is identified as being critical for optimizing pregnancy outcomes. The accessibility and supply of nutrients to the developing fetus lies on the status of maternal nutrition which in turn depends on her dietary intake, nutrient stores, and obligatory requirements. Nutritional and health studies show that the first 1000 days or first 2 years of life is a critical period for human growth and development. Evidence from around the world indicates that growth failure that initiates *in utero*, is noticeable during the first year of life, & continues, until around 2 years, however with lesser force (Victoria *et al.*, 2010). There may be modest catch up in length/height after 2 years, perhaps because of delayed maturation and a longer period of growth. Thus, the short stature that one observes in adults from many developing countries perhaps signifies growth failure before 2 years of life. Additionally the intrauterine growth period might also contribute equally to short adult height as in the first 2 years (Martorell & Zongrone, 2012).

3. Nutritional needs of females

Recently the ideal adult women weight has been changed from 50 kg to 55 kg (ICMR, 2020). Pregnancy, lactation, and menstruation increase women's requirements for various nutrients compared to their premenarcheal years. Additional protein of 9.5 and 22.0 g/day is required during pregnancy and lactation respectively against the regular requirement of 36 g/day. Calcium requirement increases during pregnancy, lactation and also in elderly years. Iron and zinc requirement also increase during adolescence (Table 1).

Table 1. Nutritional requirements of the females of different ages

Age group	Category of work	Body Weight	Protein (g/day) *	Calcium (mg/day)	Iron (mg/day)	Zinc (mg/day)	Iodine (μg/day)	Folate (μg/day)
Women	Normal	55.0	36.0	800.0	15.0	11.0	95.0	180.0
	Pregnant woman	50+10	+9.5 (2nd trimester) +22.0 (3rd trimester)	1000	27.0	14.5	250	570.0
	Lactation 0-6m 7-12m		+17.0 + 13.0	1200	23 .0	14.0	280	330.0
Infants	0-6 m*	5.8	8.0	300			100	25.0
Girls	6-12m	8.5	10.5	300	3.0	2.5	130	85.0
	10-12y	36.4	33.0	850	28.0	8.5	150	225.0
	13-15y	49.6	43.0	1000	30 .0	12.8	150	245.0
	16-18y	55.7	46.0	1050	26 .0	17.6	150	340.0
Women ≥ 60 Yrs			45.7	1200	19 .0	13.2	150	200.0

*0.83/kg body weight/day (*Source:* ICMR 2020)

Women have peak iron needs during the reproductive years. Iron deficiency anemia results from inadequate intake of iron-rich foods, as well as from excessive blood loss during events such as childbirth, hemorrhage, menstruation, and various parasitic infections. The consequences of iron deficiency anemia are severe. Folic acid is required to prevent neural tube defects, and essential fatty acids for brain growth. Iodine supplementation in pregnant and lactating women is critical for healthy brain growth in the fetus and young child. Woman's iodine requirements increase markedly during pregnancy to ensure adequate supply to the fetus.

4. Nutritional deficiencies in women

Nutritional deficiencies are caused not only due the lack of nutritious diet, but also due to other factors associated with the poor absorption. Mild-to-moderate deficiencies of iron, iodine, vitamin A, and energy may result in women from frequent reproductive cycling in a context of poverty and chronic deprivation. About half of all the global pregnant women are affected by anemia which is an important reproductive health issue. In India about 36 per cent of the population of women are underweight, and 56 per cent of women and 56 per cent of adolescent girls between the age of 15 and 19 years suffer from iron deficiency anemia due to undernourishment..

The prevalence of anemia among females of age 5-9 and 10-19 years are higher than in the males of same age group (CNNS fact sheet 16-18). In pregnant women the prevalence of anemia varies from 53.8to 90.2 per cent in developing countries and 8.3 to 23 per cent in developed countries (Bruno 2008). This is evident from the Indian state of Rajasthan, where majority of women are anemic. Anemia was high among pregnant and lactating women (80·7%) in the state. So also was severe anemia which was three-fold higher among pregnant and lactating women (4·1%).

Anemia continues to be a persisting nutritional problem among women for many years in India. The NHFS-4 reported that there was only 2 per cent decline of iron deficiency anemia among women from 55 per cent in 2005–2006 to 53 per cent in 2015–2016. It also indicated that there were 51 and 54 per cent of women with anemia in urban and rural areas of India, respectively. There were eight states in India which recorded over 60 per cent of women suffering from anemia (Aurino, 2016).

Vitamin A deficiency was higher among pregnant women (8·8%) than normal women (Singh *et al.*, 2009). More than 70 percent of rural and tribal population are unable to meet even 50 per cent of Recommended Allowances (RDA) of iron and vitamin A. This suggest a strong relation between dietary intake and the nutritional status. Compared to rural women the tribal women dietary intake is less and the nutritional status of tribal women is lower than that of rural women. In developing countries it has been shown that early marriage and childbearing affects women's nutritional status both directly and indirectly (Figure 1) as reported by Goli *et al.* (2015).

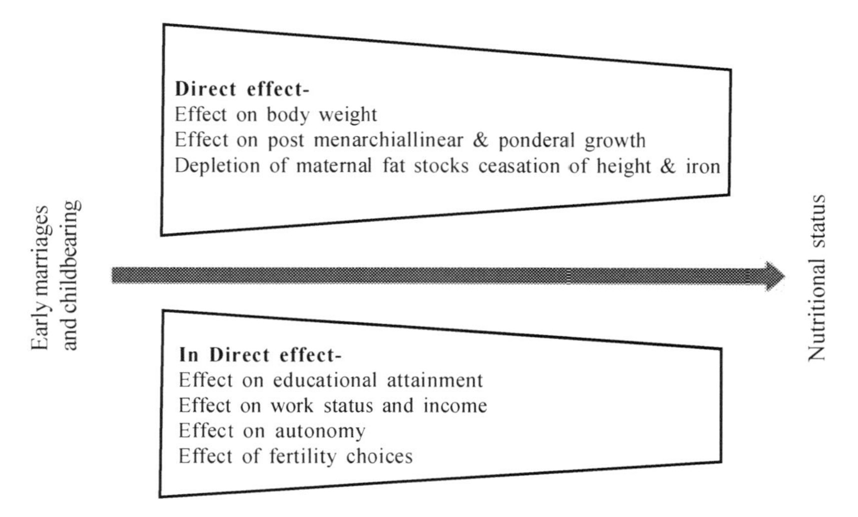

Fig. 1. Effects of direct and indirect pathways of early marriage and childbearing on women's nutritional status (*Source:* Goli *et al.*, 2015)

5. Causes of under-nutrition

The direct cause of undernutrition in women is lack of nutrients; however, there are many underlying factors that are related to socioeconomic, cultural and behavioral issues as discussed here.

5.1. Poverty and social vulnerabilities

The problem of under-nutrition is created through social relationships rather than any biological effects of gender. Economic power plays a large role in facilitating or preventing access to adequate food and health care. Thus an important underlying cause of malnutrition is poverty. Within the settings of poverty, females' social vulnerability-a product of their low social status-increases nutritional problems.

Women's main biological vulnerability for developing nutritional problems relative to men is their role in reproduction. However, reproduction alone will not compromise nutritional status. It is only in combination with the nutritional deprivation generated through socioeconomic relationships that conditions malnutrition. It will develop from frequent episodes of pregnancy and lactation. United Nations Population Activities (UNFPA) quoted the frequent pregnancies as too young, too old, too many, and too close.

5.2. Preference for male children

Preference for male offspring can express itself from birth onward through a reduced perceived need for, or delivery of food, health care, and education to the girl child, mainly when resources are scarce. As commonly seen in low-income countries, hard physical work may impose additional physiological stress and nutritional requirements in adolescence. In certain cultures, from infancy onwards, including adolescence, girls are at particularly high risk because of gender discrimination. For instance, in several countries of Asia, women's lower nutritional status becomes apparent during adolescence, with a delay in maturation (Waslien and Stewart, 1994).

This gender difference results from gender inequality that is seen even in middle-income families, though it is more prevalent in poorer families. Adolescent girls are less likely to consume foods rich in proteins, vitamins, and micronutrients compared to boys of the same family. Children are at risk for many nutritional deficiencies. Their high growth rates, small stomach capacities, and higher illness rates due to less fully developed immune systems are partly responsible. The energy and nutritional demands are the same for both the genders up to the age of nine years. However, the discriminatory behavior of the families and the society towards girls especially when the resources are highly constrained makes them more vulnerable to malnutrition.

5.3. Education of Women

In developing countries, women's education plays a prominent role in reducing malnutrition. Education of women is directly proportional to the reduction in the percentage of malnutrition and early marriage (NFHS-3). Education has significantly contributed to the reduction of child underweight, from 43% to 15.5% in developing countries.

5.4. Women's status

Women's status impacts child nutrition since women with higher status have better nutritional status themselves, have better care, and provide higher quality care to their children. However, the strength of influence of women's status and the pathways through which it influences child nutrition differ considerably across regions. In South Asia, increases in women's status have a strong influence on both the long- and short-term nutritional status of children, leading to reductions in both stunting and wasting. The human costs of women's lower status in the region are high. It was estimated that if women and men have equal status, the under-three child underweight rate would drop by approximately 13 percentage points, which means 13.4 million fewer malnourished children in this age group alone (Smith *et al.*, 2003).

6. Adolescent girls under nutrition

Adolescents' nutritional status is an important health issue because the growth during this period is rapid in the entire life except in infancy. Growth during this phase of life help overall development and provide adequate stores of energy for pregnancy and healthy adulthood. In South-East-Asia region, an enormous number of adolescent girls experience the ill effects of chronic malnutrition and anemia, which unfavorably impacts their wellbeing and improvement. The high rate of adolescent malnutrition contributes to increased risk of morbidity and mortality linked with pregnancy and delivery and the increased risk of delivering low birth-weight babies and contributes to the intergenerational cycle of malnutrition. Age between 10-19, 15-24 and 10-24 years are referred to as adolescence, youth, and young people, respectively (WHO, 2016). The United Nations classified young people as those aged 10-24, early adolescents as those aged 10-14 years, and late adolescents as those aged 15-19. Out of the 1.8 billion global adolescents, 90% are residing in low- and middle-income countries and forms the largest generation of young people in our history. As per the NFHS 4 2015-16 reports, one fourth of India women of reproductive age are undernourished, with a body mass index (BMI) of less than 18.5 kg/m. Girls are most vulnerable to the influences of cultural and gender norms, which often discriminate against them. Dietary patterns and physical activity, schooling, and socio-cultural norms for early marriage influence adolescents' health and nutritional wellbeing.

7. Overnutrition

The present day world is also faced with the irony has been the problems of obesity resulting from over-nutrition to coexist with the burden of persisting under-nutrition. Consumption of unbalanced diets or diets contaminated with potential toxins effect the nutritional security. These conditions are considered malnutrition in the true sense of the word's roots (bad nutrition) and each has been shown to potentially reduce human development. Over-nutrition involves over-consumption of nutrients and food to the point at which health is adversely affected. Over-nutrition can lead to obesity, which increases the risk of serious health conditions, such as type-2 diabetes, hypertension, cardiovascular disease and even cancer. For many years, over-nutrition had been viewed as a problem that affected only developed nations, but currently over-nutrition is a growing problem worldwide. In urban Latin America and Caribbea, more than 50 per cent of women are overweight (BMI $\geq$ 25) or obese (BMI $\geq$ 30). In Africa and Asia-Pacific the probability of being overweight or obesity rises with rising wealth in urban areas. In Eastern Europe, this pattern is reversed as reported by WHO.

Obesity negatively impacts the women's health in several aspects. Overweight or obese women are more prone to the risk of diabetes and coronary artery disease. Obese women have a higher risk of lower back pain and knee osteoarthritis. Obesity badly affects both contraception as well fertility. Maternal obesity is linked with increased rates of cesarean section as well as higher rates of high-risk obstetrical conditions such as hypertension and diabetes. Maternal obesity negatively effects the pregnancy outcomes (increased risk of neonatal mortality and malformations). Maternal obesity is also associated with a decreased interest to breastfeed, decreased initiation of breastfeeding, and decreased duration of breastfeeding.

India is currently suffering from the duel burden of over-weight and under-weight. In fact there is a brisk rise in the proportion of overweight and obese population particularly among adult women in the country. In the context of huge socio-economic heterogeneity across the states of India, the inter-state scenario of overweight and obesity differs considerably. For instance, in overweight states'(Kerala, Delhi and Punjab, where overweight is the prime concern) the overweight problem has started extending from urban and well-off women to the poor and rural people, while the rural-urban and rich-poor gap has disappeared. On the other hand in 'underweight states' (Bihar, Orissa and Madhya Pradesh, where underweight proportion is predominant) overweight and obesity have remained socially segregated and increasing manifested among urban and richer section of the population (Sengupta *et al.,* 2015).

8. Addressing women's nutrition

Addressing women's malnutrition has a series of potential positive effects because healthy women can fulfill their multiple roles such as generating income, ensuring their families' nutrition, and having healthy children more effectively and thereby help advance countries' socioeconomic development. Women are mostlyresponsible for producing and preparing food for the household, so their knowledge about nutrition can influence the health and nutritional status of the entire family. Promoting greater gender equality, including increasing women's control over resources and their ability to make decisions, is crucial in this. Improving women's nutrition can also help nations achieve three of the Millennium Development Goals, which are generally recognized as a framework for measuring development progress.

9. Recommendations for improving nutritional status of women

Many studies have reiterated the fact that malnourished women give birth to malnourished children. Therefore, it is feasible to take action to improve nutrition through generations (Branca *et al.,* 2015). In order to improve nutrition of women, children, and adolescent a range of policies, programmes, and interventions at different stages of life are required. Some of the important strategies are discussed.

1. **Promoting dietary diversity**: Diversification of diets increases the variety and intake of micronutrient rich food (Allen *et al.*, 2006). Dietary diversity can be achieved by promoting nutrition sensitive agriculture involving women. It is particularly important for understanding nutritional anemia that results, most commonly, from an inadequate intake and/or absorption of iron. Kitchen gardens serve as efficient tools in ensuring
2. **Improve adolescent girls' nutrition:** Preventing adolescent pregnancy and encourage pregnancy spacing, efforts are required to ensure that pregnant and lactating teenage mothers are adequately nourished.
3. **Improve child nutrition:** Implementing maternal nutritional support coupled with exclusive breast feeding (Practicing of giving an infant breast milk only for the first six months of life, with no other food or water)
4. **Improve women's nutrition**: Top priority is female empowerment and women's full and equal access to, and control over, social protection and resources such as income, land, water, and technology.

5. **Nutrition across the life course**: Universal access to functioning and resilient health systems and the scaled-up delivery of interventions can improve nutrition (Figure 2).

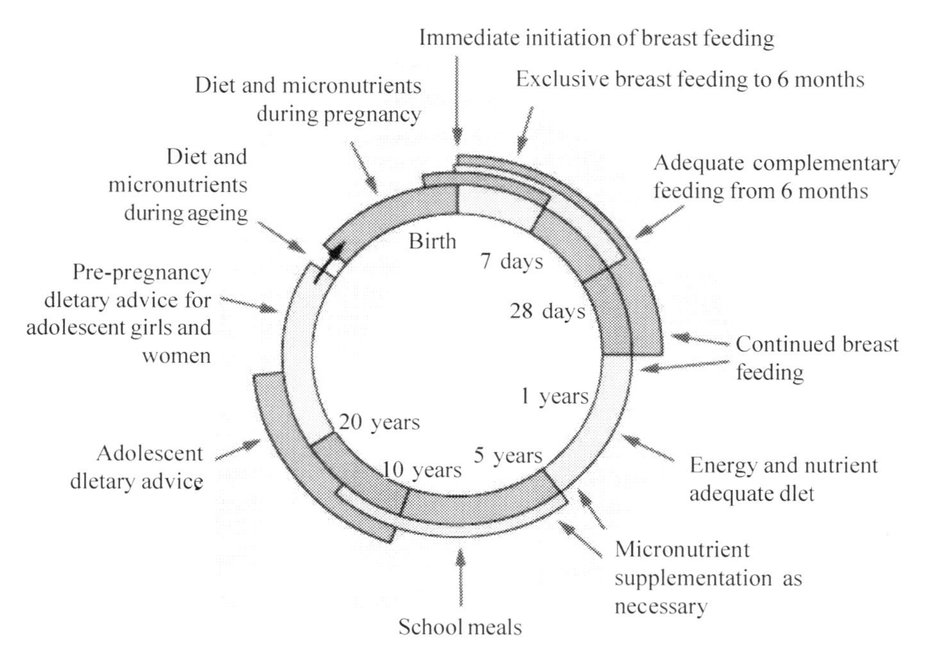

Fig. 2. Improving nutrition throughout the life course (*Source: BMJ* 2015)

10. Policy and programme initiatives

Among developing countries, India has been in the forefront for developing national food and nutrition databases and undertaking research studies and surveys documenting agriculture, food, and nutrition transitions. The country has also used the evolving knowledge and invested in nutrition intervention programmes to (i) improve food and nutrition security of its citizens, (ii) ensure that the ongoing food supplementation programmes provide sufficient food to meet the energy/nutrients gap in vulnerable segments of population, and (iii) improve ongoing nutrition interventions, converge existing programs and introduce newer ones, all aimed at preventing, early detection, and effective management of child under nutrition in the country.

The nutritional status of women is important for the wellbeing of the society and thus country. It is essential to address the women's nutrition though nutrition and non-nutritional approaches. Some of the non-nutritional approaches are education for females, improvement of societal status, economic empowerment etc., are found to have a high impact on the nutritional status of the women. Focusing on adolescent nutrition in developing countries as a pathway to improve the health of women and future generations of children on the basis of understanding that interventions targeted at adolescents enable for the interventions to have the maximum impact on optimizing health in the years ahead, including the health of women during future pregnancies and hence also the related health of the next generation.

11. Conclusion

Nutrition of women is crucial for community health and nutritional status improvement. Multisectoral approach is requiredto improve the nutritional status of women. Individual, household and policy level changes are required. Improving women's nutrition can also help nations achieve three of the Millennium Development Goals, which are generally recognized as a framework for measuring development progress.

References

Allen, L., Benoist de, B., Dary, O., Hurrell, R. (Eds.), 2006. Guidelines on Food Fortification with Micronutrients. WHO, FAO, Geneva; Rome.

Aurino, E. 2016. Do boys eat better than girls in India? Longitudinal evidence from young lives. Working Paper. Oxford: Young Lives, University of Oxford

Branca, Francesco and colleagues, 2015. British Medical J. 351:h4173.

Bruno, B. Mclean, E., Egli, I. and M. Cogswell, World Prevalence of Anaemia 1993–2005. WHO Global Database on Anaemia, World Health Organisation, Geneva, Switzerland, 2008.

Goli, Srinivas, Rammohan, Anu and Singh, Deepti, 2015. The effect of early marriages and early childbearing on women's nutritional status in India. Matern Child Health J. 19: 1864–1880.

Indian Council of Medical Research (ICMR)-National Institute of Nutrition, 2020. Nutrient requirements for the Indians, Recommended Dietary Allowances. Government of India.

Martorell, R. and Zongrone, A. 2012. Intergenerational influences on child growth and undernutrition. Paediatr Perinat Epidemiol. Jul 26 Suppl 1:302-14. doi: 10.1111/j.1365-3016

Sengupta, Angan; Angeli, Federica; Thelakkat, S. Syamala, Dagnelie, Pieter C. and Schayck, C.P. van. 2015. Overweight and obesity prevalence among Indian women by place of residence and socio-economic status: Contrasting patterns from 'underweight states' and 'overweight states' of India. Social Science & Medicine, 138: 161-169.

Singh, M., Fotedar, R.; and Lakshminarayana, J. 2009. Micronutrient deficiency status among women of desert areas of western Rajasthan, India. Public Health Nutrition 12(5).

Smith, Lisa C. Ramakrishnan, Usha; Ndiaye, Aida; Haddad, Lawrence and Martorell, Reynaldo, 2003. The importance of women's status for child nutrition in developing countries. Research Report Abstracts. 131. International Food policy Research Institute

Victora, C.G., de Onis M., Hallal, P.C., Blossner, M. and Shrimpton, R. 2010. Worldwide timing of growth faltering: revisiting implications for interventions. Pediatrics. 125: e473–e480.

Waslien, C.I., Stewart, L.K. 1994. Nutrition of the Asian adolescent girl. Asia Pac. J Public Health. 7: 31–33.

WHO, 2016. Strategic action plan to reduce the double burden of malnutrition in the South-East Asia Region 2016–2025. New Delhi: WHO-SEARO.

13

Gender and Livestock Development

*Reeja George P.**

1. The Context

Poverty continues to be a major issue the world over. Over the last decade there has been a significant reduction in the number of the extremely poor. Extreme poverty rates fell from 10.1 per cent in 2015 to 9.2 per cent in 2017 (https://blogs.worldbank.org/opendata/global-poverty-reduction-slowing-regional-trends-help-understanding-why). However, despite these achievements, we still have 689 million people living on less than two dollars a day and estimates by the World Bank in 2020 indicate that if one takes an account of higher poverty lines we have 24.1 percent of the world living on less than $3.20 a day and 43.6 percent on less than $5.50 a day in 2017 (http://www.worldbank.org/en/topic/poverty/overview). The disturbing aspect of this analysis are the facts that half of this poor are children and in most areas where poverty is encountered women represent a sizeable chunk of them. Poverty in real life is translated into lack of access to education and consumption of diets that are both quantitatively and qualitatively deficient in crucial micro nutrients and vitamins, placing them at higher risks on account of protein energy malnutrition, vitamin A deficiency and the all pervading problem of anaemia. The social consequences of such diets and implications for human capital of the world in the later part of this century would be devastating.

Poverty among women has huge repercussions for the citizens of tomorrow. It is in this context that the role of livestock products assumes crucial importance. Milk, eggs and meat have some of the highest bio-availabilities of crucial protein, iron and vitamin A. Berrazaga *et al.* (2019) observed that though plant proteins are beneficial in terms of the advantages that they offer from environmental and health points of view, the anabolic effect of plant proteins is generally lower than that of animal proteins. This is on account of their lower digestibility, relatively lower contents of essential amino acids such as leucine and the lack of sulphur amino acids such as lysine. The authors further observed that food protein quality was assessed by digestibility, net protein utilisation and biological value and these

* *Author contact: reeja@kvasu.ac.in*

indicators have so far been better for animal based protein sources. Another indicator used to assess the quality of proteins was the Protein Digestibility Corrected Amino Acid Score which takes into account the essential amino acid composition and scores for plant based proteins sources - except for some soy protein isolates- were less than 100 per cent. Other issues that limit plant protein sources include their lesser digestibility when compared to animal proteins on account of the secondary structure of plant proteins that is characterised by a high content in the beta sheet and a correspondingly lower alpha helix amount when compared to animal proteins. The beta sheet structure of the plant proteins tend to be hydrophobic thus resulting in lesser digestibility. All these factors are significant advantages for addressing nutritional imbalances in resource poor populations which also have significant implications for improving their overall health.

The other important contributions of livestock to food security include the capacity of livestock to transform very poor quality plant products, by products and residues into valuable food products that are useful to human life. This is not to mention the valuable manure contributed by livestock which is used by all traditional resource poor farmers in developing countries both to fertilise their own fields and as a source of income through sale of manure to others. Addressing poverty in third world countries is a priority for most governments, and this is possible only if the economy is robust and growing. Equally important is the issue of equity in development. Equitable spread of incomes is important in accelerating economic growth while inclusive growth can be achieved by making all citizens party to the economic activities that contribute to this phenomenon.

Livestock remains the mainstay of the household economy for rural families the world over. Over 750 million of the world's poor are engaged in this vocation that ensures food for nations as well as a source of income and asset building (FAO, 2012). Rural societies and rural life in most parts of the world are organised based on the broad canvas of local culture and traditions. Traditional customs and norms continue to shape the assignment of socially constructed roles and responsibilities to men and women. Many at times socially constructed gender roles tend to favour some over others; and generally women are at the receiving end. Despite the significant role that women play in the management of traditional livestock resources, this dominance is not reflected in their access to extension services, resources and most importantly in their involvement in decision making. The consequence of policies that ignore the basic principle of gender equity has detrimental implications on ecosystems and sustainable development. As the world moves through periods of immense social, economic and environmental changes as a result of aberrations in climates, there is a growing acknowledgement of the crucial links between gender, equality and sustainable development. More often than never, disastrous changes in the environment affect women and girls to a greater extent than men and boys on account of socially constructed roles. Capitalising on the substantial knowledge about sustainable use of natural resources that women are party to is crucial in acknowledging the role of women as central players in the world's journey towards sustainability. Policy measures that encourage women to pursue their capabilities and facilitate the enjoyment of their rights as well as those that promote their equal and engaging participation in all spheres of society are crucial in this regard. Such measures would have a catalytic role in pushing forward nations towards the attainment of economic, social and environmental sustainability.

2. The concept of gender into the centre stage

The World Development Report (WDR, 2012) elaborated on what gender is by defining this concept in terms of norms and ideologies that are socially constructed and that have a defining role in determining the behaviour and actions of men and women. These norms establish socially accepted distributions of relationships and determine the operation of power centres. It also determines the access of individuals to resources and the extent to which individuals participate in the decision making process. The actual concept of gender however emerged in the 1970s as a sequel to Ester Boserup's significant work where demands for an acknowledgment of the often invisible work of women and a redefining of the role of women in the development process from one of being mere beneficiaries to a more active role as women in the development process was voiced. Progress along these lines were forthcoming paving the way for other schools of thought to emerge and prominent among the newer proposals was the Gender and Development approach (GAD). It emphasised the need for a more holistic understanding of the realities of the actual lives of women navigating their socially constructed roles that were based on gender differences aswell as a deeper investigation into the implications of such a state of affairs on society in general. The GAD approach emphasised the need for a greater need for undoing the fetters of gendered systems and institutions thus transforming social systems into ones that operated based on gender equality. Subsequent approaches and initiatives by nations as well as international bodies such as the United Nations followed bring gender into the centre stage of development. But the most significant among these were the efforts by these institutions in affirming their endorsement and firm resolve to nullify and eliminate all forms of gender inequality through the implementation of the Millennium Development Goals 3 (MDGs). It gave a clarion call for embracing the essence of gender equality and empowerment of women in the fullest possible sense. The post 2015 framework envisages a more holistic view by accommodating various facets of women's lives ranging from completion of education to ensuring environmental sustainability and the economic empowerment of women. It is in this context that the role of livestock assumes a greater significance.

Gender thus can be viewed as the collective multitude of socially constructed roles, responsibilities and ascribed statuses of men and women. These are highly variable from nation to nation and within nations, between classes, ethnicity, religions and the point that an individual occupies on the human life cycle. Gender relates thus to the relationship between men and women rather than being concerned exclusively with women. The nature of these relationships and how they treat both men and women can be construed to mean equity or gender equity. Gender equality on the other hand accepts the freedom of individuals to pursue their personal choices without limiting themselves to socially constructed stereotypical frameworks while at the same time accepting biological differences and the consequent differing physical capacities of the sexes.

3. Gender equality: Implications for the economy

Gender equality exerts a significant influence on the growth of economies as well as poverty reduction strategies. Growing evidence in recent years suggests that in situations where equality of the genders is the norm, the growth of the economy and movement of the poor out of poverty are also enhanced (FAO, 2011; World Bank, 2012). The various facets of gender relations in society from disparities in access to power and resources to biases in education and extension all act symbiotically to stall economic growth, undermining the quality of life of millions across the globe. The brunt of this state of affairs is borne by women, who are further pushed down their already dismal situation. Poverty poses significant risks to individuals in terms of their empowerment, their security and the opportunities that they can harness. In order to ensure higher outputs and productivity of livestock, thus enhancing the income of farm families and boosting the rural economy, it is vital that gender be integrated into livestock projects and programmes.

3.1. Gender and livestock: Introducing crucial links

The role of livestock is complete when viewed against the broader canvas of the important issue of food security. Gender issues and livestock rearing cannot stand alone, divorced from mainstream issues affecting the broader issue of food security but should be viewed from within this canvas. The world's problems of lack of food have to a great extent been countered. Production and improvements in technology and accessibility to extension services have ensured that the world has enough food to go around. But the issue here is that despite these advances, it is estimated that nearly 690 million people remain hungry which represents 8.9 per cent of the world population. The projections based on these figures indicate that nearly 60 million people would be in this category in five years to come (FAO *et al.*, 2020). In addition to these figures of those who are actually hungry there are also people who are affected by severe food insecurity which is another indicator that comes close to hunger. This also shows an increasing trend and in 2019 one in ten people in the world, which works out to nearly 750 million- were in situations of severe food insecurity. It is in this context that other factors operate especially issues relating to impoverishment, so that despite the availability of food, people still lack food due to economic, physical or social factors. Despite gains in production, it remains a reality that the attainment of food security is hindered due to unequal availability, access and utilisation by all individuals in society irrespective of any intervening criteria. Women face an increased vulnerability in the food production chain from farm to fork on account of added inequalities that operate in social systems that are scripted on gender roles. This defeats the very purpose of sustainable livestock development programmes that seek to achieve food security and the consequent economic and social development of society.

4. Gender and food security

The concept of food security essentially revolves around the three pillars of availability, access to and utilisation of food. The importance of each of these pillars is contingent upon the level of aggregation that it addresses. A critical analysis of the various facets of food security at the individual level is most pertinent on account of the dynamics of cultural and social norms that operate along with inherent gendered power equations that dictate who consumes what. It is at this point that food utilisation becomes important – how are the nutritional needs of various categories of people in households defined? Does this operate based on scientifically based standards for various categories of individuals or is it dictated by socially constructed and culturally prescribed standards that endorse the preferential allocation. This is of specific relevance with respect to the distribution of crucial foods of livestock origin such as milk, eggs and meat. Experience from various cultures indicate that these livestock products are generally not freely accessible to all in many cultures and especially so in resource stretched households of the developing world. Despite the fact that women spend a considerable part of their lives in building a resilient system that sustains their families both economically and nutritionally, her access to and consumption of valuable nutrient rich livestock products are many at times limited and dictated by cultural norms steeped in inequalities of the gender.

5. Nutritional security: Interlinking roles of livestock products and women

The role of women on the food and nutritional security scene is widely acknowledged as being crucial, on account of their important role in making the products of their homestead available to their families in each meal. Research evidence from all over the world has consistently highlighted the crucial role that women play in ensuring that their homes are secure, both from the food availability point of view as well as from the nutritional point of view. Livestock products have unique characteristics that make them very valuable sources of easily digestible essential amino acids and vitamins. It plays a crucial role in the nutritional security of resource poor families where the effects of inequalities will have implications for generations. It is at this juncture that the crucial decisions made by women that determine what will be put on the table for the family and especially for the children who represent the human capital of a nation's tomorrow. The singular advantage that the livestock enterprises offer for women has been its capacity to serve as a continuous and diversified source of income as well as employment within the domestic domain. It enable her to dictate the terms of involvementwhich remains the important characteristic of livestock enterprises that make it a double edged tool for empowering women financially and psychologically. Besides these the potential implications these effects would have on curtailing intergenerational transmission of poverty and the associated addons also have to be considered. Empirical research world over, has reiterated the important consequences of livelihood options that offer women an income that they control. Such options are crucial determinants of food security as they directly translate into meeting the dietary needs of the family, especially children (Quisumbing *et al.*, 1995).

6. Current trends in engendering livestock projects

Gender affects development projects because of its socially constructed nature. The way in which this construct would affect or impact projects is contingent upon the place of implementation though there could be recurrent themes over geographical regions and social situations. Exploring social contexts and engaging the community in participatory learning exercises that are best suited to the embedded social constructs that relate to gender inequality could be a starting point for livestock development programmes. Research evidence from around the globe point to the fact that successful projects including those that are livestock based are those that address gender differences. This could be accompanied by steps to create a data base on gender disaggregated data using indicators that reflect gender based participation in programmes (Rubin *et al.*, 2010). The data could be the starting point to understand contextual and social aspects of gender influences on project outcomes.

6.1. Livestock based programmes to address asset ownership inequalities

The role of asset ownership in poverty reduction has gained increased importance in recent years (Naschold, 2008). Property rights are however complex and generally refer to a bundle of rights over that asset (Schlager and Ostrom, 1992). Although social norms influence the way in which assets are distributed in society, these norms are in many cases not the last word on the matter. Johnson *et al.* (2016) observed that continental disparities with regard to gender differences were less favourable for women in South Asia. Asian women were mostly married and had less opportunity to occupy positions of leadership in the household than African women (except Burkina Faso). The possibility of shifting gendered bias in asset ownership has been attempted by including direct distribution of assets to women in livestock programmes. A sensitisation phase was held concurrently to avoid any attitudinal hurdles in the process. Differing levels of access to various forms of property were also observed by Johnson *et al.* (2016). He reported that in many situations though women exerted control over milk for home consumption, the control of income from the sale of milk was vested with male members of the family. The impact of livestock projects such as the Building Resources Across Communities-Targeting the Ultra Poor (BRAC-TUP) programme in Bangladesh clearly demonstrated that both sole ownership as well as joint ownership of assets such as cattle, goats and poultry by women increased as a result of the programme when compared to the households that served as control group (Das *et al.*, 2013). The programme also had a positive impact on improving specific rights of women to sell, rent out or control products. This general trend of livestock programmes acting as instruments of change through promotion of control and ownership of assets could not be expected as a general consequence of such projects. But it was the result of separate *modus operandi* integrated within such projects to move explicitly in the direction of working on the norms that govern the acceptability of women having control over assets.

6.2. Livestock based programmes to address inequalities in decision making

Strengthening of women's control over assets resulted in enhancement of the role of women in the decision making process (Johnson *et al.*, 2016). It centred more on non-financial aspects of the dairy enterprise such as those involving decisions on feeding of animals and purchase of inputs (Quisumbing *et al.*, 2013). Though the ultimate aim of development projects would be to enhance women's decision making with respect to buying, selling or leasing of their animals, even lesser achievements of bringing about changes in their role with respect to caring of animals could have positive impacts on the productivity of the animals.

Conclusion

Livestock projects or any development project for that matter should be analysed in terms of that way they impact individuals in society. This is extra important for people who live on the brink of survival such as the traditionally marginalised groups which in most communities over the world include women as the majority. Therefore, the analysis should be interms of the traditional roles of men and women, their access and control over assets, their role in decision making and power relations as well as specific needs and priorities. Besides the integration of these concerns pertaining to men and women, livestock products should also be strengthened with specific sub programmes that seek to empower women. The approach that policy makers take could depend upon the extent to which attitudinal changes are possible in a society. In extreme situations where prejudices and stereotypes are deep rooted it may be required to go in for an approach that accommodates existing gender relations and accepts the existing state of affairs. But at the same time they need to be designed in such a way that measures to reduce the difficulties of the disadvantaged gender- in most cases women- are accounted, often known as the gender accommodation approach. Another most radical course of action that could be adopted in more receptive or sensitised societies could be a gender transformative approach where in imbalances in gender relations are addressed through more radical interventions that directly target improving the access of women to assets as well as their control over them.

Addressing gender disparities with respect to crucial indicators such as the access and control over resources, inputs and services could have significant impact on optimising yields of productive enterprises by 20 to 30 per cent. Besides it also raised the total outputs by 2.5 to 4 per cent and reduced the number of hungry people by 100 to 150 million (FAO, 2011). Control over assets has been reported to have a direct positive impact on family welfare as women tend to spend up to 90 per cent of their income on their families when compared to men who spend between 30 and 40 per cent (Hausmann *et al.*, 2009). Besides this, income into the hands of women increases her status within the family, improves her bargaining power as well as her ability to make her opinions heard in the family decision making process (Quisumbing and Pandolfelli, 2008).

Lack of access to productive capital such as land and the dominant role that men tend to play with respect to the income are significant areas for intervention by livestock development agencies. These have been associated with a lack of commitment on the part of women in tending to animals as they are not sufficiently motivated due to lack of returns for the jobs that they perform (Udry *et al.*, 1995). Interventions that seek to ensure equitable control on various spheres from the production of milk to its sales and receipt of income could be instrumental in boosting the income available to families as well as household food security.

The acknowledged position of livestock as the most ubiquitous non land asset on the rural farming canvas is on account of various factors. Significant advantages such as the higher expected returns from the sale of progeny coupled with availability of nutrient rich products for consumption, use on agricultural fields as draft animals have contributed to this. The easy liquidity of this asset for farm families especially in times of financial distress forms an added advantage (Bundervoet, 2006). Livestock as assets are also relatively easier for women to acquire when compared to other assets such as land. Hence the issue to be addressed by livestock development projects are twofold. On one hand, the programmes should be directed in such a way that communities are encouraged to push women forward as the sole owners of livestock. Such mandates with appropriate incentives should be made part of programmes. On the other hand, it is also necessary to address situations where in policy makers and planners themselves decide what is best for women, in many cases unilaterally deciding that women would rear only small ruminants and poultry. Though in certain agro ecological zones this may be the case, it should not be construed to be the accepted norm everywhere as variations in the preferred species is a function of culture and ecology (Yisehak, 2008).

References

Berrazaga, I., Micard, V., Gueugneu, M. and Walrand, S. 2019. Role of the anabolic properties of plant-versus animal based protein sources in supporting muscle mass maintenance: a critical review. Nutrients. 11(8): 18-25.

Bundervoet, T. 2006. Livestock, activity choice and conflict: Evidence from Burundi. HiCN Working Paper no. 24. Household in Conflict Network. Brighton: IDS.

Das, N., Yasmin, R., Ara, J., Kamruzzaman, M., Davis, P., Behrman, J., Roy, S. and Quisumbing, A.R. (2013) How Do Intrahousehold Dynamics Change When Assets Are Transferred to Women? Evidence from BRAC's Challenging the Frontiers of Poverty Reduction-Targeting the Ultra Poor Program in Bangladesh (December 1, 2013). IFPRI Discussion Paper 01317, Available at SSRN: https://ssrn.com/abstract=2405712 or http://dx.doi.org/10.2139/ssrn.2405712

FAO (Food and Agriculture Organization of the United Nations), 2011. Training guide: Gender and climate change research in agriculture and food security for rural development. Frederiksberg: Rome: FAO.

FAO (Food and Agriculture Organization of the United Nations), 2012. Livestock sector development for poverty reduction: an economic and policy perspective- Livestock's many virtues, by J. Otte, A. Costales, J. Dijkman, U. Pica-Ciamarra, T. Robinson, V. Ahuja, C. Ly and D. Roland-Holst. Rome, pp. 161 (WDR) 2012.

FAO, IFAD, UNICEF, WFP and WHO, 2020. The State of Food Security and Nutrition in the World 2020. Transforming food systems for affordable healthy diets. Rome, FAO.

Hausmann, R., Tyson, L. and Zahidi, S. 2009. The global gender gap report. Geneva: The World Economic Forum.

Johnson, N.L., Kovarik, C., Meinzen Dick, R., Njuki, J. and Quisumbing, A. 2016. Gender, assets, and agricultural development: lessons from eight projects. World Development 83: 295-311.

Naschold, F. 2008. Four Papers on Structural Household Welfare Dynamics. PhD thesis, Cornell University.

Quisumbing, A.R. Haddad,L. and Peña,C. 1995. Gender and poverty: new evidence from 10 developing countries. FCND DISCUSSION PAPER NO. 9. Food Consumption and Nutrition Division. International Food Policy Research Institute, Washington, D.C. U.S.A.

Quisumbing, A.R. and Pandolfelli, L. 2008. Promising approaches to address the needs of poor female farmers. IFPRI Note 13: 1-8.

Quisumbing, A.R., S. Roy, J. Njuki, K. Tanvin, and E. Waithanji, 2013. Can dairy value-chain projects change gender norms in rural Bangladesh? Impacts on assets, gender norms, and time use. Discussion Paper No. 1311. Washington, DC: International Food Policy Research Institute.

Rubin, D., Tezera, S. and Caldwell, L. 2010. A calf, a house, a business of one's own: Microcredit, asset accumulation, and economic empowerment in GL CRSP projects in Ethiopia and Ghana. Davis: Global Livestock Collaborative Research Support Program.

Schlager, E. and Ostrom, E. 1992. Property-Rights Regimes and Natural Resources: A Conceptual Analysis Land Economics 68(3): 249-262.

Udry, C., Hoddinot, J., Alderman, H. and Haddad, L. 1995. Gender differentials in farm productivity: Implications for household efficiency and agricultural policy. Food Policy 20(5): 407-423.

World Bank, 2012. World development report: Gender equality and development. Washington, D.C.: The World Bank.

Yisehak, K. 2008. Gender responsibility in smallholder mixed crop–livestock production systems of Jimma zone, South West Ethiopia. Livestock Research for Rural Development 20: 11.

14

Gender Roles in Fisheries and Aquaculture

Tanuja S.*, H.K. De, G.S. Saha, I. Sivaraman, S.K. Swain and S.K. Srivastava

1. Introduction

Fisheries is an important sector of food production which besides providing nutritious food to millions also serve as a livelihood option for many. It helps to bring about people centered rural development that protects the environment. Conventionally, fishing has been considered a male domain, because of the assumption that social, cultural, or religious taboos prevent women from participating in fishing activities. In fact, the involvement of women in fishing and related activities has remained invisible and unrecognized till the late 1980s. However, the publications and policy reports during the period highlighted the valued contributions by women in fisheries sector to economies and food security around the world (Harper *et al.*, 2020). Several high-level fisheries reports and policy instruments that followed added to this momentum and emphasized the importance of principles of gender equality in ensuring sustainable coastal livelihood security. Present day estimates indicate that there are around 60 million people who are directly engaged in the primary sector of fisheries and aquaculture globallyof which 14 per cent are women (FAO, 2018) Asia tops the list with the highest number of fishers and aquaculture workers (85 per cent of the world total), followed by Africa (9 per cent), America (4 per cent) and Europe and Oceania (1 per cent each) (FAO, 2018). However, despite growing attention to women and gender in fisheries there exits glaring gap in the gender disaggregated data from fisheries sector that is skewed against women worldwide. These warrant the need for the implementation gender balanced fisheries policies and management strategies that support women empowerment in the sector.

Women empowerment, is an interactive process between the individual and the environment, in the course of whichthe sense of worthlessnessof a self gets transformed into a proactive and assertive selfwith socio-political abilities (Kieffer,

**Author contact: tanujasomarajan@gmail.com*

1984).It is one of the eight millennium goals to which world leaders agreed at the Millennium Summit held at New York in 2000. However, it does not mean just economic betterment but entails deeper societal, institutional and individual changes, and is a long winding process (Chooi and Williams, 2014). Some studies report that income under women's discretion has greater positive impact on child nutrition and food security than men's (Smith *et al.,* 2003). Its impact has high visibility in terms of improved nutrition of mothers, their children, and other household members. Hence empowerment among women is considered a crucial component in alleviating poverty and has been endorsed through gender mainstreaming to reinstate their position in society in all sectors including fisheries and aquaculture in the past few decades.

2. Fisheries and aquaculture sector in India

The bounty of fisheries resources in India is evident from its long coastline of 8129 km, 2.02 million sq.km of Exclusive Economic Zone, and 1.95 lakh km of rivers and canals. Water sources in the country also include 3.15 million ha of reservoirs, 0.9million ha of brackishwater and lagoons, 0.2 million ha of floodplain wetlands and 2.41million ha of ponds. These resources contribute to the countries' nutritional security, earning foreign exchange through exports and provide employment to almost 14 million people either directly or indirectly. India produced 13.76 million tons of fish in 2018-19 which is 7.58 per cent of global fish production. Indian fisheries contribute to 1.24% of the country's GVA and 7.28% of its agricultural GDP. With a total fish production of more than 10 million metric tons, India stands 2nd in the world aquaculture fish/shell fish production, 3rd in inland capture fish production & 7th in world marine fish production (FAO, 2016).

3. Women in fisheries sector

Historically women have been playing a crucial role throughout the fish value chain. In India a total of around 16 million people are involved in fisheries activities (Both marine and Inland) of which 35% are women. When the sector is taken as a whole, 27% of women are involved in full time fisheries activities and 19% are involved in part time activities. Bihar has the highest number of people engaged in fisheries (23%) followed by Assam (16%) and West Bengal (14%). Bihar also has the highest number of women involved in fishing activities (32%) followed by West Bengal and (10.6%) and Tamil Nadu (9%). Tamil Nadu has the highest number of women in full time fishing activities (33%) followed by Kerala (24%) and Andhra Pradesh (10%). West Bengal has the highest number of women in part time fishing activities (28%) followed by Assam (26%) and Karnataka 15%. (Anon., 2019). As per the National Marine Fisheries Census, 2010, the marine fisheries activities in India are spread in approximately 1511 fish landing centres and 3288 fishing villages among the nine maritime states and the union territories of Puducherry and Daman & Diu (Chakraborty *et al.*, 2013). In Indian coastal artisanal fishing, women mend nets and manage the smaller boats and canoes that go out fishing. Tamil Nadu has the highest number (21%) of coastal women fishers in the country followed by Kerala and Andhra Pradesh. Of the total coastal women fisher population 34.76% are involved in fishing and allied activities whereas in the case of men, the participation is 86.2%. The exact nature of the work in which women are involved differs with culture, region and also between rural and urban areas. Their area of involvement can be divided into fishing and preharvest activities, processing and marketing, labour in processing and aquaculture. Women have intimate knowledge about seasonality of different species of fishes, and effects of lunar cycles, winds and other natural phenomena on marine species. They also have very close observations of the behaviour of various species. Hence with simple tools they can perform a variety of fishing related activities.

3.1. Women in fishing and pre harvest activities

Venturing out into the sea or offshore fishing activities have always been considered as a male domain. This is not only because of the vigorous work involved, but also because of women's domestic responsibilities and social norms. Regionally, female participation rates in fishing activities were estimated to be highest in Oceania, while the lowest participation by women in fishing activities is estimated for Western Asia and Eastern Europe. The data collection in fisheries is largely limited to commercial fisheries leaving behind the small scale fishing and gleaning activities mostly carried out by women. This is quite evident in shell fish fisheries. Overlooking their contributions leads to marginalization of fishers especially women although they contribute significantly to food and livelihood security of millions in the world. It is also important in terms of policy advocacies in favour of coastal women for the well being of their families and also for the sustainability of the resources that supports them. Many organizations have been working towards highlighting the women's contribution in the field of fisheries. The notable ones are, Gender in Aquaculture and Fisheries Section (GAFS) of the Asian Fisheries Society (www.genderaquafish.org), the International Collective in Support of Fishworkers (www.icsf.net/), the Pacific Community (www.spc.int/), the Food and Agriculture Organization of the United Nations (www.fao.org/fishery/topic/16605/en), World Fish (www.world fishcenter.org/) etc. Even then, the absence of appropriate gender indicators in fisheries is resulting in not highlighting the representation of women in the unorganized sectors like small scale fisheries, rural aquaculture and small scale processing.

Globally, women catch approximately 3 million tons per year of fish and invertebrates with a value of USD 6 billion or 12% of total landed value of small-scale fisheries catches (Harper, 2020). The expenditure pattern of men and women differ with women' income spend disproportionately on household provisioning, children's health and education compared to men. Catches by women were found to be highest in Asia, estimated at over 1.7 million tons per year suggesting the importance of small scale fisheries activities on the livelihood and food security of these communities. Catches attributed to fishing by women represents approximately 5 per cent of the overall landed value of marine fisheries globally (including all small- and large-scale sectors). Along with the substantial contribution it gives to the food and livelihood security, its economic valuation must also be considered (Harper, 2020).

In small-scale coastal fisheries, women are generally responsible for skilled and time-consuming onshore tasks like mending nets, baskets and pots, baiting hooks and providing services to the fishing boats. But in several developing countries like Ghana and Uganda women also invest the proceeds from their trading in boat and gear (Britwum, 2009). The fisherwomen of Cambodia, Democratic Republic of Congo, Thailand and Latin America actively participate in boat fishing. In the Indianstate of Maharashtra, the entire fishing economy revolving around Mumbai are controlled by women. Also in states like Kerala, women net prawns from backwaters and are also actively involved in the collection of bivalves and their marketing to ornamental dealers and lime collectors. In west Bengal, it could be noted that the main occupation of women fisherfolk is net making or repairing where around 72% of the women are involved. In Tamil Nadu, seaweed collection is carried out by fisherwomen. However, their work and income are rendered highly vulnerable by increasing levels of pollution, destruction of coastal habitats, reclamation of backwaters etc. Moreover, these activities are ridden with occupational health hazards such as backache, headache, myalgia, and anemia due to negligence about diet.

3.2. Women in fish processing and marketing activities

Since decades, women of fishing communities in India have been playing important roles in marketing of fresh fish, and processing surplus catch for sale at a later date in the form of cured fish. In Western Africa and in Asia, up to 80 per cent of seafood is marketed by women. The gender-based profile from nine fishing states and two Union Territories of mainland India provides further evidence that marketing of fish is primarily a women's domain, with Orissa and West Bengal as exceptions (Sharma, 2010). The estimates suggest that 81.8% of the fisherfolks engaged in marketing and 88.1% of the fisherfolk engaged in curing and processing were women (CMFRI, 2006). Also around 90-95% of coastal fisherwomen of Odisha are active dry fish producers and vendors and they contribute around 20000 INR to their family's annual income through dry fish trade. Majority of the peripatetic vendors who walk from place to place to sell their fish are usually women who purchase fish directly at auctions held at wholesale markets/landing centres. They sell fish door-to-door, travelling on foot, and carrying fish as headloads. They form a major source of fish supply to consumers within, and close to, coastal areas (Sharma, 2010). Inherent problems of low literacy, restricted mobility, limited access to training and information sources, lack of organized women groups, social and cultural issues limit their partnership in decision making. As such they lack knowledge on modern processing techniques like hygienic curing, good practices in handling and preparation of diversified and value added fish products that impede their progress. In their interaction with the public and the law, they are often forced to deal with inbred prejudices, gender discrimination and other sector specific problems detailed as follows.

1. **Longer distances and lack of basic facilities at harbours and landing centres**: With greater mechanization and motorization of fishing crafts, the location of harbours and fish landing centres have become more centralized. Women vendors thus have to travel long distances to access fish and may have to spend overnight at harbours and landing centres, in order to participate in early morning auctions. Transportation to landing sites/habours by public transport is difficult for women vendors because of the social stigma associated with it. Lack of basic facilities (toilets, storage, lights, waiting areas, night shelters) is a common phenomenon in all the harbors and landing centres of the country. Because of these reasons they are often prone to sexual abuse and harassment.
2. **Poor access to credit coupled with exorbitant interest rates by money lenders:** Women have poorer access to credit and capital, and hence cannot compete with large-scale traders, and commission and export agents. Relying on money lenders for credit often makes them fall into the vicious trap of debt because of the exorbitant interest rates.
3. **Lack of ice and proper storage facilities:** Fish is a highly perishable commodity that need to be preserved immediately for longer shelf life. The limited supply of ice and the high cost associated with it usually cannot be borne by women vendors. Hence they are forced to sell off the fish at a much lower price at the end of the day resulting in huge economic loss.
4. **Problems at marketplaces:** Because of the absence of legitimate vending zones, women are forced to vend fish on pavements and other areas and are thus considered as encroachers on public spaces resulting in legal complications.
5. **Poor market infrastructure:** In most of the fish markets around the country, basic facilities for storing, processing, and selling fish; clean toilets; access

to potable running water; and adequate waste disposal measures are usually not available. Such facilities are essential for the hygienic handling of fish, for the health and wellbeing of vendors, for consumer health, and for enabling women to engage in their occupation in a dignified manner (Sharma, 2010).

3.3. Women as workers in processing plants

The displacement of women from fish vending have landed them as wage earners in fish processing plants in peeling and grading of prawns, processing of cephalopods, bivalves, filleting and packing of fish and related activities. It is reported that as high as 90 per cent of all workers in secondary seafood activities, such as processing, are women (FAO, 2012). In India, majority (89.6%) of the fisherfolk engaged in peeling of prawns are women. But there is a huge disparity in terms of wages paid to them women that amount to almost 30% less compared to the payments received by men. Mostly migrant women between the age group of 18-25 are preferred as laborers in these units. Hence these women are forced to stay away from their family for longer periods, making it difficult for them to fulfill their domestic roles. They also experience poorer working conditions when compared to men. They are usually housed under very unhealthy conditions with very long working hours of 12-15 hours. There is no guaranteefor their employment as it is highly seasonal. They are also prone to a long array of health hazards and drudgery issues. Workers of most of the pre processing plants usually do not wear personal protective devices like gloves, gumboots or respiratory masks.They are either not provided with these safety gadgets by the plant owners or because of the lack of awareness(Nag and Nag, 2007). Some of the health risks associated with working in fish processing plants, are safety risks (mechanical and electrical accidents), excessive noise levels, low temperatures, bacterial and parasitic infections and the presence of bio-aerosols (which contain seafood allergens, microorganisms and toxins), exposure to refrigerant leakage etc. which lead tofatal or nonfatal injuries and occupational diseases. Occupational diseases often reported from the sector include frostbite, hearing loss because of continued exposure to loud noise, skin infection and sepsis, allergic respiratory diseases, musculoskeletal cumulative trauma disorders, and stress related health problems. The poor ergonomic practices of long hours of standing or awkward floor sitting postures also result in musculoskeletal pain and discomfort, mostly localized in the lower back, followed by knees, upper back, calf, shoulder, and other areas (Nag and Nag, 2007). The workers often get cut and stab wounds because of the use of cutting tools with poorly designed handle grips and finger guards and also due to loss of dexterity of hands caused by exposure to low temperature in the plants. Most fish factory employees are grossly deprived of health care services although periodic checkup for allergies or disease of workers are mandatory for all the processing factories.

Exhaustive studies on the working conditions of women in the sector, wages and existing gender biasness, the changing nature of employment, the impact of changes in technology and markets on their livelihood are essential in order to advocate welfare measures for women employed in this sector.

3.4. Women in aquaculture

Women have assumed a leading role in the rapid growth of aquaculture around the world with their participation along the aquaculture value chains. Active participation in aquaculture can empower women, notably by facilitating women's decision-making on the consumption and provision of nutritious food to their families. Historically aquaculture started as a backyard activity for recycling waste and as a source of family income. Women were involved in fish feeding and harvesting (Williams *et al.*, 2005; Napati *et al.*, 2016) considering it as an extension of their domestic work. Eventually women began taking up aquaculture, seaweed culture, mussel culture etc in ponds, pens, cages and around shores (Yap, 1999; Sumagaysay, 2013). As time progressed, this complementary role of women has changed and women involvement in these activities has become a full-time occupation. The type of work a woman has to do makes it necessary for her to be close to a pond where she has to wash, bathe, collect water and perform other household tasks. There exists, therefore, a natural condition for women to explore the possibilities for fish cultivation. These ponds are also used for vegetable cultivation, ensuring the supply of much needed household nutrition. This help families to meet the requirement of fish in their daily diet and also to meet extra family expenses.

Globally, the role of women in aquaculture and the need to consider gender issues in aquaculture development was first recognized by the FAO-NORAD sponsored workshop on *Women in Aquaculture* in 1987 (Nash *et al.*, 1987). Imparting required skills to women to carry out the activity and making available adequate credit were identified as the most important components to trigger aquaculture development (Meetei *et al.*, 2017). The scope and magnitude of women's participation in aquaculture production are influenced to a large extent by the level of advancement in aquaculture technology in a particular country and more importantly, the role and status of women in that society. This could be explained in terms of instances fromChina, Thailand, and the Philippines,which are the leading aquaculture producers of the world. These countries have a large pool of trained and skilled women fish farmers, technicians, extension workers, and professionals who are directly or indirectly involved in various capacities in fish production through aquaculture (Baluyat, 1987). In Japan, 51 per cent of the workforce in mariculture are women of which 31% are in freshwater aquaculture (Brugere and Williams, 2017). This can be attributed to their higher literacy rates and liberal legal systems that exists among and for women in these countries. But in developing countries like, Malaysia, women are only 10% of the total aquaculture workforce, mainly in freshwater aquaculture and hatchery operations. In Sri Lanka, women workers constitute from 5 per cent in shrimp farms to 30 per cent in ornamental fish culture (FAO, 2012).

Similar work conditions exist in aquaculture in Indian context as well. Although women of Manipur, Assam and West Bengal do participate in sustainable aquaculture, by engaging in pond fertilization, nursery rearing, fish feeding and harvesting, they are excluded or are not participating in skill involving practices like feed formulation, water quality management, disease management etc. (Meetei *et al.*, 2017). It is reported that in northern India only 13 per cent of women are involved in aquaculture activities. Their role is negligible in states like Andhra Pradesh and Punjab where carp culture has made rapid progress and has achieved a higher level of production compared to the other states. Reasons for nonparticipation are mostly traditional beliefs, social taboos, priorities for household chores, lack of ownership of pond and lower access to creditas well as technical knowhow and lack of women friendly technologies. In north eastern states and West Bengal, the participation is much higher at around 55 per cent. However, in these states, aquaculture still remains largely a subsistence activity and the higher involvement could be attributed to a higher number of fish eating population than the North Indian states.Moreover, in general the involvement of women in income generating activities in these states are also higher compared to men. Although Andhra Pradesh occupies the first position in fish production, the average monthly percapita consumption is only around 0.13 kg. In Punjab also the fish consumption is very negligible to the tune of 0.001 kg/ month/per capita. But the average monthly per capita fish consumption in West Bengal, Assam and Tripura are 0.92 kg, 0.72 kg and 1.27 kg respectively indicating a much higher demand and potential for fish production in these states (NSSO, 2014).

Aquaculture activities in which high women participation was recorded in India are listed as follows

1. Nursery pond preparation (disilting/drying/liming),
2. Seed rearing, seed stocking,
3. Manure and fertilizer application
4. Feeding of fishes
5. Weed clearance
6. Watch and ward

7. Net weaving and mending
8. Sorting and grading of harvest
9. Weaving of bamboo baskets and traditional tops
10. Collection/catch of mussels
11. Air breathing fishes
12. Engagement in shrimp hatchery
13. Sorting, packaging and transport of shrimp seed
14. Cleaning and drying of feed bags
15. Preparation of feed

Other activities that are potential areas for women entrepreneurs are ornamental fish breeding and culture, catfish and anabas breeding and culture, prawn breeding and culture, freshwater pearl culture, Spirulina cultivation, production of bio fertilizers like Azolla, net weaving and mending and preparation of fish feed as a cottage industry. Meetei *et al.* (2016) reported significant difference in index values of empowerment of rural women before and after taking up fisheries activity. Several studies conducted in Bangladesh have shown that overall levels of fish production, productivity and utilization can be strengthened when women fishers and managers of aquaculture technologies access resources effectively together with men.

4. Constraints to women's participation in aquaculture

Rural women, in particular, who live in poverty, with no purchasing power, and who suffer from malnutrition due to low protein intake has taken the lead in small scale aquaculture for improvement of their social status and economic power (Nwabueze, 2010). However, women's participation is hindered by several socio-economic constraints. Besides, there are certain policy related and infrastructural issues that come in the way of enhancing women's roles in aquaculture. A brief account of the most frequently reported constraints and their suggested remedies is presented here.

Constraints to women's participation in aquaculture

Constraint	Literature reports
Low literacy	Nwabueze (2010); Felsing *et al.* (2001); Ashaletha *et al.* (2002); Sultana (2002); FAO (1996); Gätke (2008); Murshed *et al.* (2008); Van (1997); Little and Edwards (2003); Minh *et al.* (1996); Bueno (1997); Samet (1997)
Limited access to modern technology	Sultana (2002); Mukherjee *et al.* (2002); FAO (1996); Kusakabe (2003); Van (1997)
Lack of government policies for addressing gender issues	Bueno (1997); Sujatha and Dixitulu (1998); Survarna *et al.* (1998); Nwabueze (2010)
Lack of access to credit and finance	FAO (1996); Kusakabe (2003); Van Crowder (1997); Harrison *et al.* (1994); Hoa (1998)
Lack of women's organization, women extension worker	FAO (1998); Hourihan (1986); Murray *et al.* (1998)
Male dominant society	Felsing *et al.* (2001); Ibrahim and Yahaya (2011)
Lack of access to resources	Nwabueze (2010); Van Crowder (1997)
Limited access to market and rural infrastructure	FAO (1996)

Source: De and Pandey, 2014

5. Gender mainstreaming initiatives

A review of the past efforts of Central Institute of Freshwater Aquaculture(ICAR-CIFA) in involving women in aquaculture reveal that women have successful partnership in aquaculture. A host of aquaculture ventures such as carp breeding, seed production, grow out carp polyculture, ornamental fish farming, integrated fish farming etc. have been popularized as part of ICAR initiatives. These interventions have benefited the weaker section of the society even in remote and backward areas of the country (De and Pandey, 2014). Rural women have also been trained in the rearing of carp seeds by the Central Institute of Women in Agriculture (ICAR-CIWA). Fry production was found to be a good proposition for farm women as the demand for good quality seed is always high. The constraints like inability of women to procure fry from long distance, inconvenience in transporting large numbers of fry over long distance, high mortality of fry during transportation, high cost of transportation and packing, unavailability of fry of desired species at the right time could be overcome by adoption of scientific practices. The domestic role of the woman in improving her family's health through more nutritious food and to increase her family's meager income has encouraged more women to be involved in homestead small-scale aquaculture.

ICAR-CIWA has also taken steps to popularize the integrated fish farming techniques by giving priority to farmer's choice and resource availability. Women were eager to adopt poultry cum fish farming as they could utilize the kitchen waste as feed for poultry as well as fish. The location of the poultry shed and pond near to their household helped them to keep watch and ward of the resources thus reducing the cost of labour. The continued sale of eggs and meat decreased the economic burden of their family and ensured their nutritional security. ICAR-CIWA has also made its mark in the field of fish processing and value addition. The process of creating awareness through training and demonstration on hygienic production of dry fish and several other value added fish products among the women dry fish producers of the coastal states of India has been well accepted. As a result of the hand holding support under these programs, rural women have started their own value added fish products enterprises. They were also facilitated with market linkage and procurement of licenses.

ICAR-CIFA has successfully implemented an innovative cluster farming approach involving women called 'Ornamental Fish Village' in three villages of Barkote block in Deogarh district of Odisha. The approach leveraged the group farming and formed active trade links with the traders, hobbyist and visitors because of mass scale production and availability of ornamental ûshes in these villages. The success spread to Saruali and Nuagaon adjacent villages where 21 farmers replicated the technology by constructing cement tanks in their backyards. It is more preferred by women since less drudgery is involved and they are able to earn by utilizing their leisure time.

Despite these efforts, social and cultural stigma continue to restrict women participation in aquaculture to negligible levels and following suggestions hold the potential to improve the situation.

- Recognition of the important contributions of women to the development of the family and the fishing community.
- Sensitization of donor agencies and the scientific community to the problems of rural women:
- Implementation of dvelopment projects involving woman/ women / group of women in fish farming andlinking the participatory stakeholders with credit, technology, infrastructure, training and trade.
- Development project involving women farmers can be successful if it is strongly supported by enabling institutions that provide education, skill, confidence, materials, inputs and right kind of environment.
- Promotion of education among women especially farm women is very important. Training programmes for women should be need based that adopt flexible timings and approaches.

- Develop women-friendly aquaculture technologies so that they can involve themselves in seed rearing, composite carp culture, integrated fish farming and design special nets convenient for women farmer to harvest fish.
- Appointment of more women trainers / extension workers would give better access to technologies by women.
- Success stories of women need to be documented and more research efforts on gender in aquaculture are required.
- Continuous monitoring of the impact of policies and programs on rural women is also advocated.

6. Fisheries enterprises to mainstream rural women

In freshwater aquaculture, culture of ornamental fish in the backyards of households, carp seed production, carp culture, murrel culture, magur culture and integrated fish farming are some of the technologies which could be adopted by rural women. These involved low capital investment and optimum utilization of natural resources available in the proximity of their households. Rural women inhabiting brackish water areas could indulge in aquaculture activities like shrimp farming, crab fattening, milk fish culture, bhekti culure etc. Mariculture technologies that possess potential for women's participation include mussel farming edible oyster farming, pearl oyster farming and pearlproduction, clam culture, lobster farming and fattening, sea cucumber culture, marine finfish culture, ornamental fish culture, seaweed culture, open sea cage farming etc. Another major arena of involvement of women is in fish processing. With proper skill and entrepreneurial training they can easily venture into value addition of fish which includes selling of dressed fish, hygienic fish drying, fish pickling, battered products preparation like fish cutlets, fish balls etc. Making of fish manures, fish feed and pet feed from fish waste also provide prospective vocations for women. In a study conducted in Kerala on the Theeramythri units supported by Society for Assistance to Fisherwomen (SAF), 10% of the SHG units (n=750) were involved in clam processing and fish pickling while 9% of them were involved in fish vending/selling. Aquatourism and fish aggregating devices are another promising fields where women can take part to improve their income (Vipinkumar *et al.*, 2017).

7. Conclusion

In order to make empowerment of women in fisheries a reality, policies and programmes must recognize the role and meet the needs of women in the fisheries sector. These suggest that studies on the nature of work and role of women in landing centres, the problems and competition they face, their involvement in organizations and how things have changed over time, would help in understanding the adapting ability of women. Organization of women into co-operative societies, Self Help Groups and federations (eg. Kudumbasree in Kerala, Samudram in Odisha etc.) have brought about a great change in facilitating them to carry on with the fish trade. It is necessary to have a deep understanding of the role women play in fish marketing, value chains and the drudgery involved in the process, problems they face in transport, in accessing market facilities, credit, etc. which would provide the needed information for better policy recommendations. The programmes should have a gendered approach with focus on increasing awareness of gender issues in fisheries and ensuring that these approaches improve the quality of their life. Inorder to bring equity in the sector, easy access to resources, technology, market and credit, technical trainings, encouraging formations of women's group and organizations are all pre requites.

References

Anonymous, 2019. Handbook of Fisheries Statistics, Government of India Ministry of Fisheries, Animal Husbandry and Dairying Department of Fisheries Krishi Bhavan, New Delhi, 175p

Ashaletha, S., Ramachandran, C., Immanuel, S., Diwan, A.S. and Sathiadas, R. (2002). Changing Roles of Fisher women of India- Issues and Perspectives. In:Kohli,M. P. S. and Tewari, M.S.R. (Eds.)Women in Fisheries. Indian Society of Fisheries Professional, Mumbai,p. 21-43.

Baluyat, E. 1987. Women in aquaculture production in asian countries. (Eds Colin E. Nash,, Carole R. Engle and Donatella Crosetti). Proceedings of the ADCP/NORAD Workshop on Women in Aquaculture. Rome, FAO, 13-16 April 1987

Britwum, A.O. 2009. The gendered dynamics of production relations in ghanaian coastal fishing. Feminist Africa, 12: 69-85.

Brugere, C. and Williams, M.J. 2017. Profile: Women in Aquaculture. https://genderaquafish.org/portfolio/women-in-aquaculture/

Bueno, P. (1997). Gender issues on the participation of women in fisheries development. In: Nandeesha, M. C. and Hanglomong, H. (Eds.) Women in Fisheries in Indo-China Countries. PADEK, Bati Fisheries Station, Phnom Penh, Cambodia, 21-38.

Chakraborty, K., Vijayan, K.K., Pananghat, P. and Mohanty, B.P. 2013. Marine Fishes in India: Their importance in health and nutrition. CMFRI Special Publication No 110 Central Marine Fisheries Research Institute. 115pp.

Chooi, P.S. and Williams, M.J. 2014. Avoiding Pitfalls in Development Projects that Aspire to Empower Women: A Review of the Asian Fisheries Society Gender and Fisheries Symposium Papers. Gender in Aquaculture and Fisheries: Navigating Change Asian Fisheries Science Special Issue 27S: 15-31.

CMFRI, 2006. Marine Fisheries Census 2005, Part 1. DAHD &F, Ministry of Agriculture, Govt of India, New Delhi and CMFRI, Kochi. 105p.

De, H.K. and Pandey, D.K. 2014. Constraints to Women's Involvement in Small Scale Aquaculture: An Exploratory Study. International Journal of Agricultural Extension 02 (01): 81-88.

FAO (1996). Gender issues in the fisheries sector and effective participation, Report of the Workshop on Gender Roles and Issues in Artisanal Fisheries in West Africa, Lomé, Togo, 11-13 December 1996.

FAO (1998). Asia regional paper on Rural Women and food security in current situation and prospective, FAO, 36-47.
FAO, 2012. The state of world fisheries and aquaculture. Food and Agriculture Organisation of the United Nations, Rome, 207pp.
FAO, 2016. The State of World Fisheries and Aquaculture 2016. Contributing to food security and nutrition for all. Food and Agriculture Organisation of the United Nations, Rome. 200 pp.
FAO, 2018. The State of World Fisheries and Aquaculture 2018 - Meeting the sustainable development goals. Rome.
Felsing, M.,Brugere, C. Kusakabe, K. and Kelkar, G. (2001). Women in Aquaculture. Asia Pacific Economic Cooperation, Project FWG 03/99. 60pp.
Gätke, P. 2008. Women's participation in community fisheries committees in Cambodia. Department of Environmental, Social and Spatial Change, Roskilde University.
Harper, S., Adshade, M., Lam, V.W.Y., Pauly, D. and Sumaila, U.R. 2020. Valuing invisible catches: Estimating the global contribution by women to small-scale marine capture fisheries production. PLoS ONE 15(3): e0228912. https://doi.org/ 10.1371/journal.pone.0228912
Harrison, E., Stewart, J. A., Stirrat, R. L. and Muir, J. (1994). Fish Farming in Africa - what's the catch? London, Overseas Development Administration (ODA), 51.
Hoa, P.T. 1998. The role of women in mangrove planting and protection in Nghia Hung district, Nam Dinh province- measures for life and environmental improvement. In: Hong, P.N. and Dao, P.T.A. (Eds.) National Workshop on the Socio-Economic Situation of Women in Mangrove of Coastal Area- Trend to Improve Their Life and Environment, Centre for Natural Resources and Environmental Studies (CRES), Vietnam National University, Hanoi and Action for Mangrove Reforestation (ACTMANG), Japan, Hanio, Vietnam, 66-69.
Hourihan, J.J. 1986. Women in development; guidelines for the fisheries sector. Report to the Asian Development Bank, Manila, The Philippines.
Ibrahim, H.Y. and Yahana, H. 2011. Women participation in homestead fish farming in North Central Nigeria. Livestock Research for Rural Development 23(2) Retrieved May 18, 2012, from http://www.lrrd.org/lrrd23/2/ibra23032.htm.
Kieffer C. 1984. Citizen empowerment: a development perspective. Prevention in Human Services, 3: 936-942.
Kusakabe, K. 2003. Women's Involvement in Small-scale Aquaculture in Northeast Thailand. Development in Practice. 13(4): 333-345.
Little, D.C. and Edwards, P. 2003. Integrated livestock-fish farming systems. FAO, (2003).
Meetei, W., Saha, B. and Roy, A. 2017. Constraints, issues and suggestive framework for fishery based strategic empowerment of rural women: a study in Manipur. Journal of the Inland Fisheries Society of India, 49(2) : 55-62.
Meetei, W.T., Saha, B. and Pal, P. 2016. Participation of women in fisheries: A study on gender issues in Manipur, India. International Journal of Bio-resource and Stress Management, 7(4): 906-914.
Minh, L.T., Huong, D.T. and Tuan, N.A. (1996). Women in Cantho City are profitably involved in fish nursing activities. Aquaculture Asia, 1: 40-41.
Mukherjee, M., Banerjee, R., Datta, A. and Sen, S. 2002. Socio-Economic Status of Fisherwomen of Peri-Urdan Areas of Calcutta. In: Kohli, M.P.S. and Tewari, M.S.R. (Eds.),Women in Fisheries. Indian Society of Fisheries Professionals, Mumbai, 84-87.
Murray, U., Sayasane, K. and Funge-smith, S. 1998. Gender and Aquaculture in Lao PDR: A synthesis of a socio-economic and Gender Analysis of The UNDP/FAO Aquaculture Development project LAO/97/007. FAO, Rome.

Murshed, E., Jahan, K., Beveridge, M.C.M. and Brooks, A.C. 2008. Impact of long-term training and extension support on small-scale carp polyculture farms of Bangladesh. Journal of the WorldAquaculture Society, 39(4): 441-453.

Nag, P.K. and Nag, A. 2007. Hazards and health complaints associated with fish processing activities in India: Evaluation of a low-cost intervention. International Journal of Industrial Ergonomics 37: 125–132.

Napati, R.P., Sefil, A.S., Serofia, G.D., Peralta, E.M., Palmos, G.N and Yap E.E.S. 2016. The Role of women in blue swimming crab (*Portunus pelagicus*) fisheries in the Philippines. In: GAF6. Available from https://genderaquafish.file s.wordpress.com/2016/06/13-napata.pdf

Nash, C.E., Engle, C.R and Crosetti, D. 1987. Women in aquaculture. Proceedings of the ADCP/NORAD workshop on women in aquaculture, 13-16 April, Rome, FAO.

NSSO, 2014. Household consumption of various goods and services in India. NSS report no.558. National Sample Survey Office, Ministry of Statistics and Programme Implementation, Government of India, 69pp.

Nwabueze, A.A. 2010. The role of women in sustainable aquacultural Development in Delta State.Journal of Sustainable Development in Africa, 12(5): 284-293.

Samet, H. 1997. Women in Cambodia and their contribution to agriculture/fisheries. In: Women in Fisheries in Indo-China Countries, pp.42-47. PADEK, Bati Fisheries Station, Phnom Penh, Cambodia.

Sharma, S. 2010. Women Fish Vendors in India: An Information Booklet. International Collective in Support of Fishworkers, Chennai, India, 53p.

Smith, L.C., Ramakrishnan, U., Ndiaye, A., Haddad, H. and Martorell, R. 2003. The importance of women's status for child nutrition in developing countries.Research report 131, International Food Policy Research Institute, Washington, DC Department of International Health, Emory University, 155p.

Sujatha, K. and Dixitulu, J.V.H. (1998). Role of women in changing trends of fisheries industry- a case study of selected coastal village in North coast of Andhra Pradesh, India. Paper presented at the Fifth Asian Fisheries Forum: Fisheries and Food Security Beyond the Year 2000, Chiang Mai, Thailand.

Sultana, M. 2002. Status of women in coastal aquaculture sector in India and avenues for gainful entrepreneurship. In: Kohli, M.P.S. and Tewari, M.S.R. (Eds.),Women in Fisheries. Indian Society of Fisheries Professional, Mumbai. p. 62-69.

Suvarna, M., Bhatta, R. and Rao, K.A. 1998. Gender perspectives on fisheries and livelihood in coastal Karnataka, India. Paper presented at the Fifth Asian Fisheries Forum: Fisheries and Food Security Beyond the Year 2000, Chiang Mai, Thailand.

Sumagaysay, M.B. 2013. Work spaces for women and girls in the mussel industry value chain: promoting small-scale entrepreneurship. In: GAF4. Available from: https://genderaquafish.org/events/gaf4-2013-yeosu-korea/ gaf4-tentative-program/).

Van, C.L. 1997. Women in agricultural extension and education.FAO. Rome (Italy). Research Extension Division. Retrieved May, 1, 2006, from http://www.fao.org/sd/EXdirect/EXan0016.htm

Vipinkumar, V.P., Narayanakumar, R., Johnson, B., Swathilekshmi, P.S., Ramachandran, C., Shyam, S,.S, Reeta, J., Shinoj, S. and Aswathy, N. 2017. Gender Mainstreaming and Impact of Self Help Groups in Marine Fisheries Sector, Central Marine Fisheries Research Institute, Kochi, Project Report. p 235.

Williams, M.J., Agbayani, R., Bhuket, R., Bondad-Reantaso, M., Brugere, C., Choo, P.S., Dhont, J., Glamiche-Tejeda, A., Ghulam, K., Kusakabe, K., Little, D., Nandeesha, M.C., Sorgeloos, P., Weeratunge, N., Williams, S. and Xu, P. 2012. Expert Panel Review 6.3: Sustaining aquaculture by developing human capacity and enhancing opportunities for women. In:

Subasinghe, R.P, Arthur, J.R., Bartley, D.M., De Silva, S.S., Halwart, M., Hishamunda, N., Mohan, C.V. and Sorgeloos, P. (eds) Proceedings of the Global Conference on Aquaculture 2010 : Farming the Waters for People and Food. FAO, Rome and NACA, Bangkok, pp. 785-922.

Yap, W.G. 1999. Rural Aquaculture in the Philippines. FAO, Bangkok RAP Publication, Bangkok.

15

Gender Perspectives in Forest Management

Parvathy Venugopal* and Abha Manohar, K.

1. Introduction

Much of humankind has either at least interacted with forests or benefited from forests for their lives and livelihoods. Approximately 1.6 billion rural people worldwide, mainly in tropical regions, depend upon forests in one form or the other. People rely on forests or forest services as sources of food, shelter, energy, building materials, fiber, fodder, income, medicine, and inspiration (Gabay and Rekola, 2019). The earlier efforts of international forest development practices were focused only on the production and industrial sector. Nonetheless, in the late 1970s and 1980s, this focus shifted to better understanding the links between people and forests. The same period has witnessed the birth of community-led (social/participatory) forest management in many countries. Women's role in the forestry sector has also started to receive more attention and understanding during this time (FAO, 1986; Ginsburg and Keene, 2020; Kristjanson, 2020). Thus, gender roles have remained an area of interest in forest management, especially in developing countries, and the perspectives have evolved over the years. The agenda 2030 of the sustainable development goals (SDG) encompasses strategies designed to tackle gender inequities and protecting ecosystems, including forests, biodiversity, and oceans (UN, 2015). In fact, the interlinked trajectories of women and the environment have led to the emergence of the theory of ecofeminism and women environment and development (WED) in the late 1970s (Tyagi and Das, 2017). It integrates the ethics of ecology with feminism to explore the intangible interconnections between environmental degradation and sexist oppression (Warren, 1993). Though feminist discussions are not new to the forestry sector, attempts to generate gender-disaggregated data exclusively on forestry management perspectives have been scarce.

Sunderland *et al.* (2014) reported that historically men and women had played distinctive roles in the forest sector. Men are mostly involved in hunting and

**Author contact: pvparvathycof@gmail.com*

fishing while women collect edible forest plants, fruits, and medicines (Shackleton *et al.,* 2011). The cultural, social, economic, and institutional disadvantages have been attributed to women's restricted roles in forest activities. These restrictions made the role of women in community forestry and other related forestry practices often less pronounced. In earlier times, even the professional jobs in forestry were reserved for men, and it was generally uncommon for a woman to become a forester. Such patriarchal norms made the work and efforts of women in the sector mostly invisible. It also restricted the spread of innumerable traditional knowledge and practices concerning plants/trees/forest, of which women were the major custodians. However, such stereotypes have been broken in India with three women's induction into the national forest service in 1980. Ever since, many inspiring women personalities have adorned several positions in the multitier forest administration and academic institutions, both at the union and state levels in India.

However, studies about the problems faced by women in a conventionally male-dominated forestry sector are still few. They are comparatively more available from Europe and North America than developing countries like India (Follo 2002, Lidestav and Ekström 2000, Reed 2008, Teske and Beedle 2001, Thomas and Mohai, 1995). In a case reported from Norway, where forestry has traditionally been one of the most male-dominated occupations, women had to face negative responses upon joining it (Brandth and Haugen 1998, Brandth and Haugen, 2000). In Canada, women in forestry faced no welcome from both outside and inside forest officials (Reed, 2008). A study by the Society of American Foresters (Kuhns *et al.,* 2004) described that 65% of the women felt gender discrimination in their workplace, and 71% of all surveyed women did not have the same opportunities as men. In 2007, the first female graduate from the Central Forest Rangers College in Chandrapur, also had to face negative responses from senior forest officials (Sainath, 2007).

Despite these odds, most green movements against forest exploitation and related environmental degradation that originated globally and at the community level had women at their helm. A review of the significant roles played by women in forest management and environmental protection has been attempted here.

2. Forests as a livelihood base for women and their households

According to FAO (2018), 850 million rural poor still depend on forest resources for family consumption, of which 83% is women. Women who live in and around forests in developing countries collect the fuelwood, fodder, and wild vegetables for their households from the forests (Agarwal, 2009; Biran *et al.,* 2004; Nakro and Kikhi, 2006; Nilsson, 2006; Robinson and Kajembe, 2009; Roy, 2008; Tabuti *et al.,* 2003). They also serve as repositories of ecological knowledge related to their local environment base (Boer and Lamxay, 2009; Eyzaguirre, 2006; Johnson and Grivetti, 2002). Though men and women have highly specialised knowledge of trees and forests, women's knowledge is more directly linked to household food, nutrition, health, and culture. For example, tribal women in India use almost 300 forest species for medicinal purposes. Women's traditional knowledge of plants and other forest products is crucial for household survival, especially to cope with natural crises and food shortages (FAO, 2014). In Uttar Pradesh, 33 to 45% of women's income was generated from forests and common land, compared to 13% for men (Agarwal, 2001). A study from West Bengal indicates that the women's contribution in forest industries like mining, processing, marketing, and consumption was 75%, 100%, 67%, 50%, respectively, compared to men who contributed 25%, nil, 33%, 50% respectively (Agarwal, 2001). Forests are the only source of livelihood for the tribal women of Chhattisgarh, and they use most of the forest products for food, fodder, medicine, and fuelwood. Among them, forest management is undertaken by both men and women as partners.

Non-Timber Forest Products (NTFP) serve as the main source of income, well-being, and food security of many women-headed households in and around forest areas. Regarding NTFPs, women make significant contributions to value chains, specifically in collecting, processing, and marketing. Nevertheless, most of the activities they engage intended to have low returns as little attention is given to the potential of Non-Wood Forest Products (NWFP) for local markets. Moreover, though several reports reiterate the importance of roles played by women in forest product value chains, their contributions are yet to be well documented. Few studies from various Indian states highlighted the importance of organized efforts in making the forest-based livelihood more women inclusive. A study from Uttaranchal suggested that it was the migration of men to different works areas in search of employment that created the space for women in forest-based income-generating activities (Agarwal, 2001). In the Garhwal region, inspired by the Chipko movement and innovative programs like Mahila Samakhya, all women Van Panchayats and informal women's committees were organized to help forest-based livelihood activities of women. In Gujrat, women engaged in Kareya gum collection were getting low prices for the product. An intervention by Self Employed Women's Association (SEWA) helped them to win the right to sell Kareya in the open market (Carr *et al.*, 1994; SEWA, 2000).

However, women's roles in these value chains are not much supported by policymakers and development programs. Many of them have knowledge of trees and forests in terms of biological diversity, specifically medicinal plants and edible products, their sustainable management, use, and conservation practices (FAO, 2012). There need to be public support programs in empowering women in the forest sector to get significant benefits for the food security of rural, especially tribal households, and the sustainable management of forests. Enhancing women's participation in forest user groups, improving their access to fuel sources, and increasing market access have been found to make a significant difference in the livelihoods of forest-dependent people and their communities.

Worldwide about 3 billion people rely on forest wood for cooking and other activities (WHO, 2006). Mainly women and girls are involved in the collection of fuelwood (Sunderland *et al.*, 2014). In areas affected by deforestation, natural disasters, or various conflict, they had to walk for long hours for its collection (WFP, 2012). This is one factor that restricted their participation in education, paid work, and other productive activities. In addition to the drudgery involved in carrying fuelwood over long distances, health hazards induced by smoke and dust from the smoky fires add to problems of women who depended on forests. In a case study from Bangladesh, it has been reported that women around Chunati Wildlife Sanctuary used improved stoves, which caused less air pollution with better energy efficiency. This helped them to reduce the collection of forest fuelwood, compared to traditional stove users (Roy, 2008).

Another woman-dominated vocation related to forestry reported from India has been the cultivation of fodder and cattle rearing in forest lands. Women also use forest lands to collect fodder for cattle, and their earnings are spent for household purposes (Franzel and Wambugu, 2007). Common property resource management principles dictate the terms and conditions that ensure sustainable management of available resources.

In Tanzania, most leafy vegetables consumed in the area are from the wild, i.e., collected by women from fields, field margins, fallows, and agroforest. Proximity to the forest was the key determinant of vegetable consumption in the study area. The poor who lived away from the forest had to spend a significant time traveling to the forest to collect vegetables (Powell *et al.*, 2011). In Ethiopia, sorting and cleaning gum and resins is the primary income source for 96% of the women involved. However, in Burkina Faso, sorting gum arabica was reported as the most important source of income for 3–4 months in a year (Shackleton *et al.*, 2011).

3. Women in Joint Forest Management

National Forest Policy, 1988 and Joint Forest Management (JFM) program of the 1990s mandated that women comprise 33% of the Vana Samarakshana Samitis. To a large extent, the inclusion of women has resulted from the analysis of village community forestry groups in India. The success of these groups differed among states. It was observed that the exclusion of women from decision-making could negatively affect these initiative's long-term efficiency and sustainability. Women groups also have been found to handle forest regeneration, protection of biodiversity, and watershed management better than male-dominated groups (Pandolfelli, 2009; Coleman and Mwangi, 2012). The logic conclusively proved that deforestation could be better handled if the state Forest Departments worked out joint management agreements with local communities.

The Forest Rights Act (2006) provided a major opportunity to strengthen the economic and social security of tribes and other forest dwellers. It is suggested that the environmental policies should consider the economic impact of women in protecting and managing the forest resources (Agarwal *et al.,* 2006). For sustainable forest management, women should have access to the forest, the benefits from it, and the right to make decisions about it (Giri, 2012). Thus, women's participation in forest management involving decision-making are found to have a good impact on the outcomes of resource use. It has increased forest management effectiveness, ensured the protection of illicit harvesting of forest products, and regeneration of degraded forest (Agarwal *et al.,* 2006).

Nevertheless, it could be inferred from many studies that even though women use the forest to a greater degree than men, their perspectives seldom found a place in management decisions (Jewitt, 2000; Sunderland *et al.,* 2014). This resulted from the limited access and less control of rural women to shared economic resources than their male counterparts (Allotey *et al.,* 2008). Even in utilizing the knowledge acquired by women through their close association with forests, such as the efficiency of different types of firewood in cooking (Edmond 2008; Gbadegesin 1996; Godfrey *et al.,* 2010; Brouwer *et al.,* 1996) or availability of wild plant/foods in their forest and their uses, better management participation needs to be ensured (Daniggelis, 2003; Johnson and Grivetti, 2002). Thus, women being the major beneficiaries of the forests, gender differences must be addressed, and rural women's close relationship with forest resources must be utilized in effective and sustainable forest management (Fortmann and Rocheleau, 1985; Hoskins, 1980; Nhem and Lee, 2019; Tinker, 1994).

4. Role of women in forest and environment conservation

Across the globe, women, especially at the grassroots, relied on natural resources for fuel, food, and fodder. They served as custodians of these common property resources and have shown more resilience in protecting the forest, water, and land on which their livelihoods are intrinsically dependent. History has glowing annals of women who acted as a shield by struggling to protect forests from habitat destruction activities. The interlinkages between gender and environment are complex and mostly arose from the imbalances in power related to the control over land and natural resources, which remained skewed against women. Some of the major women-led movements in the forest and environment conservation from India and worldwide have been discussed.

4.1. Women-led environment movements in India

Bishnoi movement led by Amrita Devi Bishnoi to protect the Khejri trees (*Prosopis cineraria*) in 1731 is renowned as the first women-led environment movement from India. Along with her three young daughters and 360 other villagers, she had to lay down their lives in their efforts to protect the sacred grove of Khejri trees in the Khejarli village of the Marwar region of Rajasthan. However, their martyrdom brought public policy legislation by the Maharaja, who declared the whole region as a protected area for trees and wild animals. Amrita Devi Bishnoi Wildlife Protection Award instituted by the Ministry of Environment, Forests and Climate Change in 2013 honors the memory of India's first eco-feminist.

The Bishnoi movement later served as an inspiration for the rural women of the Chipko Movement (1973) led by Chandi Prasad Bhatt. It went down the history as the first environment movement of modern India. The movement saw resistance of village women from the Terai region of Uttarakhand in the foothills of the Himalayas who embraced trees to prevent its felling. It helped them to safeguard their livelihoods, which were dependent on the forests. Agarwal (2001) observed that deforestation could qualitatively alter all village residents' lives, but it was the women who agitated for saving the forests.Another remarkable forest regeneration effort by Indian women has been reported from Pargana district of Bihar where the forests had been degraded by human exploitation. The women who took an interest in working in the forest protection committee were instrumental in bringing this change (Agarwal, 2001). Environmental movements in India started gaining more significant momentum in the later decades of the twentiethcentury and continue to rise over the recent years. Jungal Bachao Andolan (1980s), Navdanya Movement (1982), Save Silent Valley movement (1970s), Narmada Bachao Andolan (1985) were all notable women-led environmental movements from India.

4.2. Global women-led environment movements

The publication of Silent Spring, the seminal work by Rachel Carson in 1962, has laid the foundation of the modern environmental conservation era. Rachel's lone voice brought to fore the danger of indiscriminate use of pesticides, particularly DDT on the environment and its severe effects on human health. Considered the mother of the environment movement, she paved the way for women to join environment conservation efforts.

Green Belt movement (1977) is one of the most showcased global movements from Africa led by Wangari Maathai. It focused on the restoration of Kenya's rapidly diminishing forests through a holistic approach that integrated environmental conservation, community development, and capacity building. It also facilitated the empowerment of rural women through environmental conservation, emphasizing planting indigenous trees by women (Agarwal, 2001). This grass-root indigenous NGO's efforts were globally recognized and honored with the Right to Livelihood Award (1986) and the Nobel Peace Prize (2004).

In Thailand, the Dhamma Raksha Reforestation Program was the response of Thai peasant women against the decline of forest area from 53% (1961) to 29% (1985). They restored forests and developed new sources of income linked to reforestation and sustainable agriculture.

Elinor Ostrom was the first to bring a gender perspective to natural resource management by providing a route map for just and sustainable resource management through collective action in Governing the Commons (1990). Amazon environment movement under Marina Silva continues the fight to save the Amazon forests in association with Brazil's women by promoting sustainable forest management and conservation.

5. Conclusion

Forests are a crucial livelihood support system, especially for the women living in forest fringe areas. They are highly dependent on forest resources as it plays a vital role in the viability and subsistence of their households. Women are more directly affected and influenced by forest degradation, making them more active participants in the protection of forests and the environment than men (Rao, 2012). Besides, women have emerged into leadership roles in movements that spearheaded conservation and enhanced the environment from communities worldwide. This conclusively proves that the women's movement and environmental movement share a close association, as implied by the theory of ecofeminism.

Moreover, women's need for forest products and forest management priorities also differs from that of men (Buchy and Subba, 2003; Paudel, 1999). Evidence of growing environmental crisis has brought more intensified focus on women as environment crusaders. Most of the environmental agencies and institutions around the world have acknowledged women's role in protecting and conservingthe environment. As custodians of traditional knowledge,women have critical functions to perform. Therefore, women's works should inevitably get recognition and support from authorities/ government institutions for the well-being of rural women and society concerning forest conservation and its resources.

The status of women and forests can't be improved by simply increasing the quota of women in decision-making. The government and policy makers should address the struggle women face in joining the existingmale-dominated institutions. More needs to be done to achieve gender equality and social inclusion through women empowerment. Women participation in forest management and conservation could be encouraged by training and educating rural women, ensuring theirland tenure and access rights, employing women in protection activities, or encouraging them to start small enterprises. Small efforts can make a big difference and ease forest conservation activities, thus leading to a sustainable future.

References

Agarwal, B. 2001. Participatory Exclusion, Community Forestry, and Gender: An Analysis for South Asia and a Conceptual Framework.World Dev. 29: 1623-48.

Agarwal, B. 2009a. Gender and Forest Conservation: The Impact of Women's Participation in Community Forest Governance. Ecol Econom. 68: 2785-2799.

Agarwal, B. 2009b. Rulemaking in Community Forestry Institutions: The Difference Women Make. Ecol Econom. 68: 2296-2308.

Agrawal, A.G., R.A Yadama & A. Bhattacharya. 2006. Decentralization and Environmental Conservation: Gender Effects from Participation in Joint Forest Management. CAPRi working papers 53, International Food Policy Research Institute (IFPRI).

Allotey, P., M. Gyapong & C.J.P. Colfer. 2008. The Gender Agenda and Tropical Forest Diseases. In: C.J.P. COLFER (ed.),Human Health and Forests: A Global Overview of Issues, Practice, and Policy. Earthscan: London. pp. 135-160.

Biran, A., J. Abbot & R. Mace. 2004. Families and Firewood: A Comparative Analysis of the Costs and Benefits of Children in Firewood Collection and Use in Two Rural Communities in Sub-Saharan Africa. Hum Ecol. 32: 1-24.

Boer, H. de & Lamxay, V. 2009. Plants Used During Pregnancy, Childbirth, and Postpartum Healthcare in Lao PDR: A Comparative Study of the Brou, Saek and Kry Ethnic Groups.J. Ethnobiol. Ethnomedicine. 5: 25.

Brandth, B.& Haugen, M.S. 1998. Breaking into a Masculine Discourse. Women and Farm Forestry. Sociol. Ruralis. 38: 427-442.

Brandth, B. & Haugen, M.S. 2000. From Lumberjack to Business Manager: Masculinity in the Norwegian Forestry Press. J. Rural Stud. 16: 343-355.

Brouwer, I.D., A.P.D. Hartog, M.O.K. Kamwendo & M.W.O. Heldens, 1996. Wood Quality and Wood Preferences in Relation to Food Preparation and Diet Composition in Central Malawi. Ecol Food Nutr. 35: 1-13.

Buchy, M. and Subba, S. 2003. Why is Community Forestry a Social- and Gender-blind Technology? The Case of Nepal. Gender Tech. Dev. 7: 313-332.

Carr, M., M. Chen & R. Jhabvala, 1994. Speaking out: Women's Economic Empowerment in South Asia (eds.). IT Publications:New Delhi. 252p.

Coleman, E. & Mwangi, E. 2012. Women's Participation in Forest Management: a Cross-Country Analysis. Glob. Environ. Change. 23: 193-205.

Daniggeli, S.E. 2003. Women and 'Wild' Foods: Nutrition and Household Security among Rai and Sherpa ForagerFarmers in eastern Nepal. In: Howard, P.L. (ed.), Women and Plants: Gender Relations in Biodiversity Management and Conservation. Zed Books: London. pp. 83-97.

Edmond, J. 2008. Incorporating Gender into PHE Strategies: Experiences from Conservation International. Conservation International and USAID. pp. 25.

Eyzaguirre, P.B. 2006. Agricultural Biodiversityand How Human Culture is Shaping It. In: Cernea, M.M. & Kassam, A.H. (ed.),Researching the Culture in Agriculture: Social research for international development. CABI: Oxford. pp. 264-284.
FAO (Food and Agriculture Organisation), 2014.Report Women in Forestry: Challenges and Opportunities. FAO: Italy. pp.11. http://www.fao.org/3/a-i3924e.pdf.
FAO, 1986. Monitoring and Evaluation of Social Forestry in India. FAO:Italy. p75. http://www.fao.org/3/an799e/an799e00.pdf.
FAO, 2012. Forests for Improved Nutrition and Food Security. FAO: Italy. P12. www.fao.org/docrep/014/i2011e/i2011e00.pdf.
FAO, 2018. The State of the World's Forests - Forest Pathways to Sustainable Development. FAO: Italy. p28.http://www.fao.org/3/ca0188en/ca0188en.pdf.
Follo, G. 2002. A Hero's Journey: Young Women Among Males in Forestry Education. J. Rural Stud. 18:293-306.
Fortmann, L.& D. Rocheleau, 1985. Women and Agroforestry: Four Myths and Three Case Studies. Agroforest. Sys. 2: 253-272.
Franzel, S. & C.Wambugu, 2007. The uptake of fodder shrubs among smallholders in East Africa: key elements that facilitate widespread adoption. In: M.D. Hare & K. Wongpinchet (ed.)Forages: APathway to Prosperity for Smallholder Farmers. Proceedings of an International Symposium, Faculty of Agriculture, Ubon Ratchathani University: Thailand, pp. 203-222.
Gabay, M. & M. Rekola, 2019. Forests, peaceful and inclusive societies, reduced inequality, education, and inclusive institutions at all levels: Background study prepared for the fourteenth session of the United Nations Forum on Forests. Department of Forest Sciences, University of Helsinki: Finland. p. 89.
Gbadegesin, A. 1996. Management of Forest Resources by Women: ACase Study from the Olokemeji Forest Reserve area, Southwestern Nigeria. Environ Conserv. 23: 115-119.
Ginsburg, C. & S. Keene, 2020. At a Crossroads:Consequential Trends in Recognition of Community-basedForest Tenure from 2002-2017. China Eco. J. 13: 223-248.
Godfrey, A.J. K. Denis,W. Daniel& O.C. Akais, 2010. Household Firewood Consumption and Its Dynamics in Kalisizo Sub-County, Central Uganda. Ethnobotanical Leaflets.14: 841-855.
Hoskins, M.W. 1980. Community Forestry Depends on Women. Unasylva. 32: 27.
Jewitt, S. 2000. Mothering Earth? Gender and Environmental Protection in the Jharkhand, India. J Peasant Stud. 27: 94-131.
Johnson, N. & L. Grivetti, 2002. Environmental Change in Northern Thailand: Impact on Wild Edible Plant Availability. Ecol Food Nutr. 41: 373-399.
Kristjanson, P. 2020. Closing gender gaps in forest landscape initiatives. Int. Forest Rev. 22: 44-54.
Kuhns, M.R., H.A. Bragg & D.J. Blahna, 2004. Attitudes and Experiences of Women and Minorities in the Urban Forestry/Arboriculture Profession. J. Arbor. 30: 11-18.
Lidestav, G. & M. Ekström, 2000. Introducing Gender Studies on Management Behaviour among Non-Industrial Private Forest Owners. Scand. J. Forest Res. 15: 378-386.
Nakro, V. & C. Kikhi, 2006. Strengthening Market Linkages for Women Vegetable Vendors – Experiences from Kohima, Nagaland, India. In: Vernooy, R. (ed.) Social and Gender Analysis in Natural Resource Management – Learning Studies and Lessons from Asia. Sage: India. pp. 65-98.
Nhem, S. & Y.J. Lee, 2019. Women's Participation and the GenderPerspective in Sustainable Forestry in Cambodia: Local Perceptions and the Context of ForestryResearch.Forest Sci Technol. 15: 93-110.
Nilsson, A.K. 2006. Gender work in Lao PDR: Women and the fuelwood collection. Kulturgeografiska institutionen, Umeå universitet.

Pandolfelli, L. 2009. Integrating Gender Analysis at CIFOR: Proposed Next Steps. Report submitted to CIFOR. CIFOR: Indonesia. p23.

Paudel, D. 1999. Distributional Impacts of Community Forestry Programmes on Different Social Groups of People in The Mid-Hills of Nepal. MPhil Dissertation, Department ofGeography, University of Cambridge, UK, Unpublished.

Powell, B.J. Hall& T. Johns, 2011. Forest Cover Use and Dietary Intake in the East Usambara Mountains, Tanzania. Int. For. Rev. 13: 305-324.

Rao, M. 2012. Ecofeminism at the Crossroads in India: A Review.Dep. 20: 124-142.

Reed, M.G. 2008. Reproducing the Gender Order in Canadian Forestry: The Role of Statistical Representation. Scand. J. Forest Res. 23: 78-91.

Robinson, E.J.Z. & G.C. Kajembe, 2009. Changing Access to Forest Resources in Tanzania. Environment for Development, Discussion Paper Series. pp.21.

Roy, B.C.S. 2008. Fuelwood, Alternative Energy, and Forest User Groups in Chunati Wildlife Sanctuary. In: J. Fox, B.R. Bushley, W.B. Milesand S.A. Quazi(ed.)., Connecting Communities and Conservation: Collaborative Management of Protected Areas in Bangladesh. East-West Center: Honolulu. pp. 209-226.

Roy, S.B., R. Mukerjee & M. Chatterje, 1992. Endogenous Development and Gender Roles in Participatory Forest Management. IBRAD: Kolkatha.

Sainath, P. 2007. A Forest Road Less Travelled. The Hindu. Retrieved 13 Oct. 2008, Jan. 23.

SEWA (Self Employed Women's Association), 2000. The Gum Collectors: Struggling to Survive in the Dry Areas of Banaskantha. SEWA Academy: Ahmedabad, India.

Shackleton, S., F. Paumgarten, H, Kassa, M. Husselman & M. Zida, 2011. Opportunities for enhancing women's economic empowerment in the value chains of three African non-timber forest products (NWFPs). Int ForestRev. 13: 136-151.

Sunderland, T., R. Achdiawan, A. Angelsen, R. Babigumira, A. Ickowitz, F. Paumgarten, V. Reyes-García& G. Shively.2014. Challenging perceptions about men, women, and forest product use: a global comparative study. World Dev. 64: S56-S66.

Tabuti, J.R.S., S.S. Dhillion & K.A. Lye, 2003. Firewood Use in Bulamogi County, Uganda: Species Selection, Harvesting and Consumption Patterns. Biomass Bionener. 25: 581-596.

Teske, E. & B. Beedle, 2001. Journey to the Top-breaking Through the Canopy: Canadian experiences. For. Chron.77: 846-853.

Thomas, J. & P. Mohai. 1995. Racial, Gender and Professional Diversification in the Forest Service from 1983-1992. Policy Stud. J. 23: 296-309.

Tinker, I. 1994. Women and Community Forestry in Nepal: Expectations and Realities. Soc. Nat. Resour. 7: 367-381.

Tyagi, N. & S. Das, 2017. Gender Mainstreaming in Forest Governance: Anaysing 25Years of Research and Policy in South Asia. Inter. Forest Rev.,19: 234-244.

UN (United Nations), 2015. Transforming Our World: The 2030 Agenda for Sustainable Development. New York (NY): United Nations (UN). pp. 41.

Warren, K.J. 1987. Feminism and Ecology: Making Connections.Environ. Ethics. 9: 3-20.

WFP (World Food Programme), 2012. Handbook on safe access to firewood and alternative energy (SAFE). World Food Programme: Italy. pp. 221. https://www.wfp.org/publications/wfp-handbook-safe-access-firewood-and-alternative-energy-safe.

WHO (World Health Organization), 2006. Fuel for life: household energy and health, by E. Rehfuess. WHO: Geneva, Switzerland. pp. 23.https://www.who.int/airpollution/publications/fuelforlife.pdf?ua=1.

16

Gender Friendly Tools and Equipments in Farm Mechanization

Suma Nair, Suresh A. and Ayisha Mangat*

1. The Invisible Gender

India, as is always said, is primarily an agrarian economy. Despite the forward leaps taken by our society in the different novel and pioneering areas of development, agriculture remains the force that supports rural India. According to the economic survey of 2020-21, the contribution of agriculture to the country's GDP has increased from 17.8% in 2019-20 to 19.9% in 2020-21. It is a sector that provides almost half of all the employment opportunities available in the world. The rural agricultural workforce all over the world comprises 37 per cent women and in low-income countries, it is as high as 48 per cent. Women also represent close to 50 per cent of the world's 600 million small-scale livestock managers and about half of the labor force in small-scale fisheries. They also have a highly significant presence in the post-harvest, processing, and small scale value addition sectors also. However, these numbers are pessimistic estimations as the women's contributions to agriculture are often unpaid labor and therefore their contributions are not fully captured (FAO, 2020).

They also have a highly significant presence in the post-harvest, processing, and small-scale value addition sectors also.According to the latest reports from Oxfarm Research, the agriculture sector employs 80% of all economically active women in India, which comprise 48% of the self-employed farmers, and 33% of the agriculture labor force.Barring few states like Punjab, Kerala, and West Bengal, most of the women in rural India are engaged in agriculture. It is estimated that 85% of the rural women in India work in agriculture yet own only 13% of the land. The modeled estimates of the International Labour Organization (ILO) shows a drop in females employed in agriculture from 74.38% in 2000 to 54.59% in 2019.However, under the cultural backdrop of Indian society, a significant number of women still exist as "invisible contributors." Their contributions to

**Author contact: suma.nair@kau.in*

domestic and agriculture activities go unacknowledged and they remain to be economically dependent on the male agents active in the agrarian sector, leading to their economic and social exploitation. Donning the role of family labor, they are often destined to form the invisible backbone of Indian agriculture.

In the agricultural sector, men are still the dominant decision-makers and the policy seekers of profit. The man was the prime user of tools developed during the evolution of the human race. When these became the tools used for agriculture the same scenario continued, as the anthropometric attributes of the males were more suitable to develop the power required for agricultural operations while using these tools. The male dominance prevalent in the ancient societal norms was also a contributing factor. Hence, women have been relegated to the background of the agrarian canvas from way back in time. Existing social norms in many restrictive societies do not favor the participation of women in income-generating agrarian activities including farm mechanization.

The agricultural sector as a whole has grown and evolved tremendously technology-wise and men have been significantly. The agricultural sector as a whole has grown and evolved tremendously technology-wise and as a result men have been significantly empowered. This evolution, however, was unable to uplift the role of women as an integral part of agriculture. Though two-thirds of the female workforce is employed in agriculture, their contribution to this sector is not quantified and the women are not given the status of employed labor. The status of women in a country is reflected in terms of indicators such as literacy, life expectancy, infant mortality, the participation of women in the development agenda, etc. These indicators are at a low in most developing countries. For these women, along with their major roles in household chores and rearing children, agriculture is a duty that has to be done routinely without any expectations of financial gain. If engaged as labor, their employment in agriculture is seasonal, scattered, and temporary, causing wide ups and downs in their income pattern also. The availability of cash in hand, which equals economic and some extent of social freedom, is hence unreachable. Despite this, however, a large number of women newly enter into this sector, especially in the rural communities, because agriculture still is an assured source of income (however small) and self-reliance for the rural woman.

2. Farm mechanisation for improved productivity

Agricultural landholdings in India are highly fragmented and about 78% of farm holders have land areas less than 2 ha. The lesser land area deters the farmers from mechanization and hence in these small farms, there is increased participation of women as agricultural labor. Traditionally women are engaged in the more monotonous and hard work, taking more time for completion, and also in works that require round the clock engagement such as processing, homestead farming, value addition, etc. However, various socio-cultural factors and economic reasons have hindered the adoption of farm technologies by women in India. Women often opt out of the outside-homestead agricultural economic opportunities.

The new technology involves the need for costlier inputs, intricate knowledge, and complex skills. As agriculture is being increasingly feminized, it is now becoming imperative that the women who are involved in agriculture embrace such new technologies to sustain and improve their income, improve their time use efficiency, reduce or tide over their stress and strain due to agriculture, to diversify and bring out quality products. To sustain an economic level of productivity in agriculture, it becomes imperative that farms and farming operations be mechanized. Mechanization in the Indian context should have a focus on improved and novel equipments and increased farm power availability. It should be a mix of mechanical power and human power for sustained mechanized farming.

Mechanization of agricultural operations on a farm reduces the drudgery, increases timeliness and efficiency of work, and increases profit. Agricultural machinery available today is more suited to male anthropometry. Operator comfort, safety, operation mechanisms, operator controls are all designed taking the male operator/labourer into account. The female agricultural workers are not even considered as participants in farm. The female agricultural workers are not even considered as participants in farm mechanization during the design process. The unsuitability of machines for comfortable use often deters women from operating these, and they often merge into the background, due to the socio-cultural perception that machines cannot be operated by females, and the lack of skill in operating machinery. When machines are introduced to reduce the drudgery of farming operations, women are thus displaced from their traditional livelihood opportunities, causing serious economic insecurities not only of the women per se but also of their families, because it has been established through various studies that the income of the farm woman is utilized fully for the benefit of her family.

A study in Thrissur district, on the changing gender roles in rice farming when new technologies are introduced revealed the following results.

Table 1. Major gender roles, technologies, and users in rice farming in Thrissur district, Kerala

Framing operation	**Traditional Actor**	**Technology Used**	**Changed Technology**	**The user of the Technology**
Ploughing	Men	Country plough	Tractor, Tiller	Men
Seeding	Women	Manual operation	Seed drill	Men
Transplanting	Women	Manual operation	Mechanical Transplanter	Men
Weeding	Women	Manual operation	Weedicide & sprayers, mechanical weeders	Men
Harvesting	Women and Men	Sickle	Harvester	Men
Threshing	Women and Men	Manual operation	Thresher	Men
Winnowing	Women	Manual operation	Winnower	Men
Irrigation	Men	Manual	Pumpset	Men
Processing/ milling	Women	Manual	Power-operated Mills	
Marketing	Men	Farmgate	Market	Men

(*Source:* Suresh Kumar *et al.,* 2005)

This pronouncedly shows the displacement of women from women-specific farming operations, as a social norm, when newer technologies are introduced. Apart from displacement from their traditional livelihood, lack of parity in wages, the ergonomic unsuitability of available machinery, and lack of social acceptance are severe hindrances to women in mechanized agriculture.

However, as women are now the major players in farming, it is high time that gender-specific interventions be made to increase the ease of participation of women in agriculture.

3. Ergonomic studies for drudgery reduction

Drudgery can be described as physical and mental strain, agony, monotony, and hardship experienced by human beings (Momin, 2009). Since they continue to be most affected by illiteracy, hunger, and unemployment, females tend to bear the brunt of drudgery. Women normally perform their economic activities with conventional tools, designed specifically for men taking into account their physical characteristics like height, hand strength, grip, a span of limb movements, etc. Along with unsafe, unhealthy, and long working hours, these mismatching tools often cause many health problems especially for women. Agriculture is considered as being one of the riskiest vocations due to the wide range of potentially detrimental health hazards encountered during the various cultural operations.

Farm workers are often unaware of the health hazards and risks they face in mechanized farming, due to a lack of information and education. As for the women involved, they take the pain resulting from agricultural operations as a normal part of their work. They often seek medical aid and care only when the condition becomes extremely worse. While considering human physical conditions from an occupational point of view, the parts like the cervical spine, head, neck, shoulders, elbow, and wrist joints play an interrelated aspect that affects human safety, design, comfort, and thus efficiency. Improper matches between operators' physique and machine cause serious health hazards such as musculoskeletal injuries, and preventative methods are to be ensured to overcome these problems in mechanized agriculture. One of the major drudgery-producing factors is the usage of un-ergonomic tools and types of equipment by farm women (Rehman, 2017).

The International Labour Organisation (ILO) defines ergonomics as, 'the application of human biological sciences, its conjunction with engineering sciences to the workers and his/her working environment to obtain maximum satisfaction and enhance productivity'. It is the study of a man at work and his environment. It is a man, machine, and environment system focusing on fitting the job to the worker.

Ergonomics is a new discipline in developing countries that is being applied to the assessment of workload in various activities performed by women at home and on the farm. From a physiological point of view, the workload refers to the demands placed on the cardiorespiratory system and is determined from the energy cost and cardiac cost of work (Chauhan, 1999). Studies conducted by CIAE, Bhopal have revealed that very few efforts have been made to design, develop and popularize suitable tools and implements for various agricultural activities like weeding, top dressing of fertilizer, sowing, etc. traditionally performed by rural women. Most of the designers of farm tools and machines are men, as a result of which the designers are biased towards a design to suit men. The occupational health and safety of farm women thus become the most neglected area in agriculture and allied sectors such as animal husbandry, fisheries, etc. The technological empowerment of farm women with occupational safety and better work output is needed for assuring the livelihood security of this vulnerable group. This paves the way for the introduction of gender-friendly equipment for reducing drudgeries in farm operations.

4. Gender friendly tools and machines in agriculture

The introduction of gender-friendly equipment and farm tools helps to reduce drudgery by giving more emphasis on safety and comfort to the worker. Generally, women-friendly and gender-appropriate tools are less used due to lack of awareness, unavailability, and being inadequate in number. However, a radical development was seen in the late 90's when many studies that emphasized more on ergonomic and gendered aspects of agricultural mechanization were conducted. The equipment used in earlier days was usually heavy, complicated, large, and due to these design parameters, hard to use by women. Resultant of several studies and researches, many easy-to-use implements have now been developed which have made farm operations easy for workers, especially for women. The introduction of gender-friendly tools potentially reduces the bio-mechanical drudgeries associated with agricultural tasks.

A few of the areas in which research is underway in the context of gender-friendly farm mechanization are illustrated below.

4.1. Weeding

Weeding is an important time consuming and drudgerous activity that has to be performed on time to assure crop productivity. This activity is mainly performed manually by women for almost all crops grown. Improved technological interventions are needed in this area to reduce the drudgery of the participating women as the operation demands high energy and a long duration of work in bending or squatting postures. Improved hand-operated weeders, *kutla,* and hoes have been developed to ease the drudgery of weeding operations. A study conducted in Himachal Pradesh and Uttarakhand assessed the physiological workload and musculoskeletal problems of the women farm labour who participated in the study while using both traditional and improved tools. Physical fitness level and physiological parameters were studied and the results showed that heart rate values were more than the acceptable limits when the operation was performed with traditional tools as compared to improved tools. Adoption of the improved tools relieved postural stress and severity of pain in the various body parts. It was clear that the improved tools with better ergonomic design could reduce the drudgery faced by farm women in manual weeding operation and their use is recommended (Kishtwaria & Rana, 2012).

4.2. Harvesting

Manual harvesting of crops is another labour intensive activity usually performed by women. This operation also involves long durations of work in uncomfortable postures, usually bending or squatting. Traditionally, the sickle is used for manual harvest. Improved models of sickles such as the Naveen sickle, Vaibhav sickle,etc. have been developed to ease the operator strain during harvesting. In a study conducted in Ratlam district of Madhya Pradesh, the local sickle was compared with the Naveen and Vaibhav sickles for harvesting wheat crops. Thirty farm women were the respondents of the study. The results of the study indicate that the improved tools are women-friendly as compared to the local sickle because the exertion experienced by the women workers was markedly less, more area could be covered in lesser time and drudgery and cost of operation was low. The availability of the same in rural areas is scarce, however, and this hinders effective transfer and use of technology (Sharma *et al.*, 2017).

Another study carried out in Keonjhar, Odisha also corroborated on the effectiveness of the improved Naveen sickle in drudgery reduction. Forty farm women were assessed for the impact of the improved sickle over the conventional one. Parameters such as drudgery index, RPE, degree of difficulty, musculo-skeletal problems, efficiency, and output, etc. were observed. It was found that farm activities that are time and labour intensive, monotonous, repetitive, and more drudgery prone are generally performed by women. However, the improved method reduced drudgery by 28 per cent, saved energy and time as well as minimized musculo-skeletal problems. A higher work output of more than 138 m^2 area indicated that the new technology was beneficial both to the labourers and farmers because an additional area of 286.7 m^2could be covered per day fetching a savings of Rs. 506/ha. Work could be completed faster and 3 h/ day could be saved, which could be utilized for other works. By this, it was pointed out that the farm woman could earn an additional income of Rs. 1450/ month (Pereira *et al.*, 2014).

4.3. Threshing

Threshing is another operation traditionally performed manually by women. Threshing of cereals, pulses, and millets are extremely drudgerous operations. These are traditionally the livelihood sources of rural farm women. The drudgery involved in the threshing is one of the reasons for the reduction in cropped area of millets in India. Processing of millets especially is very arduous and time-consuming, due primarily to the small size of the grain. A study on an efficient finger millet and barnyard millet thresher designed and developed at VPKAS, Almora found that the mechanized thresher was ergonomically better than traditional means of threshing. The new thresher helped reduce physiological ergonomics parameters i.e. Heart Rate (HR), Energy Expenditure Rate (EER), Blood Pressure (BP), Pulse Pressure, Total Cardiac Cost of Work (TCCW), Physiological Cost of Work (PCW), Blood Lactate Concentration and postural discomfort. These machines significantly reduce the workload and time spent for post-harvest processing of small millets and have been accepted by the cultivators as well as the development agencies (Joshi *et al.*, 2015).

4.4. Transplanting

The transplanting operation, especially in cereals, is traditionally performed by women. A very arduous operation, it requires the women to stand almost knee-deep in water in a bending posture for the whole period of the work. The drudgery involved has led to a large number of traditional workers abandoning the work and look for other avenues of work. Ojha and Kwatra (2014) assessed the drudgery perceived by women workers during manual hand transplanting and transplanting using power-operated eight-row transplanters. Ergonomic evaluation of twenty female subjects showed that the mean value of working heart rate was a maximum of 138.32±7.67 beats/ min in manual hand transplanting while it was 110.12±5.79 beats/min in eight- row paddy transplanter. The energy expenditure rate was lesser with paddy transplanter as compared to manual transplanting. The total cardiac cost of work (TCCW) and physiological cost of work was also found to be lower in the case of power-operated transplanters when compared to the traditional method. The economic cost of mechanized transplanting was 47% less than the conventional method.

Innumerable studies carried out in the ergonomic evaluation of machinery have indicated that the work performance output of women farm workers improves when ergonomically suitable machinery is introduced among them. Power-operated machinery and ergonomically designed tools and implements reduce the drudgery of farm operations and increase the physical well-being of the women's farm labour. The injuries due to improper usage of unsuitable machinery and repetitive farm operations are avoided.

5. Emerging Trends

Agriculture is going through a revolutionary phase. Traditional agrarian ways and means are not going to be sufficient to meet the food demands of the globe in the coming decades. The projected growth of population cannot be fed with the current land area available on the planet. Hence the focus is now on cutting-edge advanced technologies like digital farming, smart farming, using artificial intelligence (AI) etc., that have the potential to meet the growing challenge of providing food for all. However, as these technologies come to the fore, would this again pose a threat to the women in agriculture?

As the inclusion of women in agriculture is essential to ensure their livelihood, such new technologies should be developed keeping the gender perspective also in mind. The introduction of new technologies must not be the reason for pushing the women out of their livelihood. Technology for reducing the gender gap in work burden especially for women in agriculture can be achieved by Climate-smart agriculture (CSA). This gives a chance for women to assess the climatic conditions in high climate risk areas. Women responsive climate risk management is achieved by considering local crops, social conditions, agro-climatic conditions, etc. Practices like DSR (direct-seeded rice using zero tillage and minimum tillage), green manuring, laser leveling, system of rice intensification (SRI), smart irrigation methods, low weight –small machines,etc. also reduce drudgery in farm operations.

Apart from scientists and researchers, many farmers are also, interestingly, involved in developing gender-friendly, easy-to-use equipment.

A few of the gender-friendly equipment developed in the different scientific establishments across the country are presented below.

Conoweeder

Conoweeder, also called wetland weeder, is used to remove weeds from rice fields. The weeds are uprooted and buried into the soil between the crop rows. The equipment consists of a 'T-shaped handle' fitted to one side of a frame of 1.4m length On the other side, two conical rollers with serrated strips on the periphery are connected in tandem with opposite orientations. A float is provided at the front portion controls the working depth and prevents sinking of weeder in the puddle. It is operated by pushing and pulling action in a standing posture. Unlike in the traditional method of weeding with hands, this equipment prevents bending posture and thus reduces the drudgery of workers. In addition to weeding, trampling of green manure can also be done using conoweeders. It disturbs the topsoil for about 3cm depth and helps to increase aeration in the soil.

(*Source:* https://www.agri-companies.live/cono-weeder-1.html?category_id=71)

Drum Seeder

It is used for direct seeding of sprouted or pre-germinated paddy. The equipment consists of drive wheels with lugs, driveshaft, seed drums, and handle frames. The hyperboloid-shaped seed drums allow the free flow of seeds to the seed metering holes. A drum cap is provided to fill the drum with seeds. The number of seed drums may vary, but usually it is four (8 rows). Baffles are provided between the seed holes inside the drum to maintain a uniform seed rate. Ground wheels are made of plastics to ensure floating characteristics. Wheels are present at both ends for smooth operation. A swinging handle is provided for pulling the equipment. The seeder operates well in a shallow ploughed/ puddled field. The seeder can be operated at a walking speed of about 1-1.5 km/ hr. The wheel impression during the operation will help to keep the path for subsequent passes. Line sowing helps to maintain a specific row to row space which helps the subsequent operation of weeders like conoweeders in the field. The capacity of the seeder is considered to be around 920 sq.m/hr.

Keramithra- KAU Coconut Husking Tool

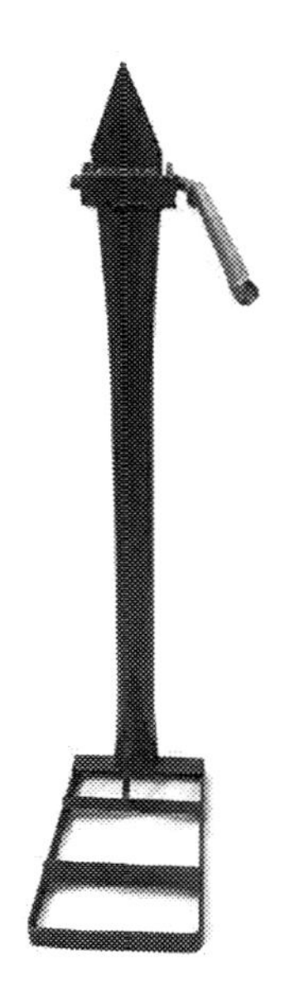

Keramithra is a manually operated coconut husking tool with a simple hand – lever principle developed by Dr. Jippu Jacob and Dr. Joby Bastin of Kelapaji College of Agricultural Engineering and Technology Tavanur under the Kerala Agricultural University. It consists of two wedges, a handle lever, and a base pedestal. One wedge is stationary and the other is movable. These wedges are upwardly positioned at the top end of the pedestal mounted on a stable base. The stationary wedge is fixed on a firmly fitted hinge seat and a movable wedge is joined to a long hinge pin, which enables the movement to about 1800. A hand lever is provided at the end of the hinge pin which is usually horizontal. The coconut to be husked is inserted on the closed wedges and the lever is lifted upwards. The action is repeated three to four times to completely remove the coconut husks (Varghese & Jacob, 2014).

Keradhaara- KAU Coconut Milk Extractor

Keradhara is used to extract coconut milk from grated coconut. The manually operated extractor is developed using a rotary pressing mechanism. It consists of the main frame with a tightening screw, extraction and perforated cylinders, screw shaft, pressure plate, and a handle. It is made of stainless steel. The mainframe is mounted on a table and the extraction cylinder is mounted on it. Through the closed end of the extraction cylinder, the screw shaft is inserted. The pressure plate and handle are press-fit at both ends of the screw shafts. Grated coconut is filled into the perforated cylinders (preferably three-fourth) and inserted into the extraction cylinder. By rotating the handle, coconut gets pressed and coconut milk is squeezed out of it through the holes of the perforated cylinders. The ejected milk is then collected through the slit provided at the bottom of the extraction cylinder. This simple, women-friendly equipment gives hygienic coconut milk and can be used for household chores and small scale processing and value addition.

Seed Treatment Drum

It is used to mix chemicals uniformly on seeds before sowing and consists of a cylindrical drum, frame, and a handle. The frame is a tripod angle frame on which the cylindrical drum is mounted. Three mild steel flat bars are welded inside the cylindrical drum for serving uniform mixing. Operators are recommended to wear hand gloves and a face mask before mixing chemicals until the completion of work. This is intended for health care and protection. After adding seeds and chemicals into the drum, a little water is sprayed and the lid is closed tightly. The lid is then opened after rotating the drum for 1-2 minutes (20-25 revolutions). The treated seed is taken out and collected in a separate container. It takes about 5-6 minutes for filling, treating, and emptying a batch of 20 kg of seeds. It is highly recommended to wash the hands, face, eyes, and legs after the completion of work. This equipment ensures safety to workers as direct contact with chemicals is avoided. It also provides uniform mixing of chemicals. Furthermore, the squatting or bending posture of the workers while treating the seed in the conventional method is avoided.

CRRI Two-Row Rice Transplanter

A machine used to transplant rice seedlings at 3-4 leaf stage old in two rows, it consists of a frame, floats, seedling tray, operating handle, fingers, tray drive unit, and depth control mechanism. A mat-type nursery 45 cm in length, 22 cm in width, and 1.5 cm thickness of the soil raised to operate the equipment. After sprinkling little water over the tray surface, seedling mats are loaded on it. Using the operating handle, seedlings are gently pushed down for transplanting. The operator has to walk in a backward direction for the operation of the transplanter. The bending posture of the workers in traditional methods is avoided by using this transplanter. Moreover, row sowing helps the usage of mechanical weeders and hence reduces human drudgery and the cost of weeding operation. The capacity of two-row rice transplanter is estimated to be around 61 m^2/h (*Source:* https://images.app.goo.gl/we1miraoEVq6tvNV8).

Pusa Wheel Hoe

It is used for weeding in between line crops. It is a foldable type weeder, which makes it easy to handle while transporting. According to the desired need, the angle of operation can be adjusted. It can be operated in a standing posture which helps to reduce fatigue as in the conventional method. It is operated by giving forward and backward motion. It is a cost-effective method of weeding that requires only one person for the operation. (*Source:* https://images.app.goo.gl/Ce4Xyi F8qnEWEAHA6)

IIHR Mango Harvester

It is an improved type of conventional harvester used to harvest mangoes with pedicel. It consists of a pole, a frame with a net, and a blade to separate the fruit from the tree. The fruit is lowered using the frame and is held between the pole and frame. The blade positioned between the pole and frame cuts the mango with a pedicel of 10-20 mm in length. This method of harvesting ensures no sap bleeding and the shelf life is increased by 3-4 days. The harvester is lightweight and needs less force for pulling. When the net is filled with 4-5 mangoes, it is emptied for easy handling (*Source:* https://images.app.goo.gl/fMrL3dY62WNDHD7c8)

Cassava Peeling Knife

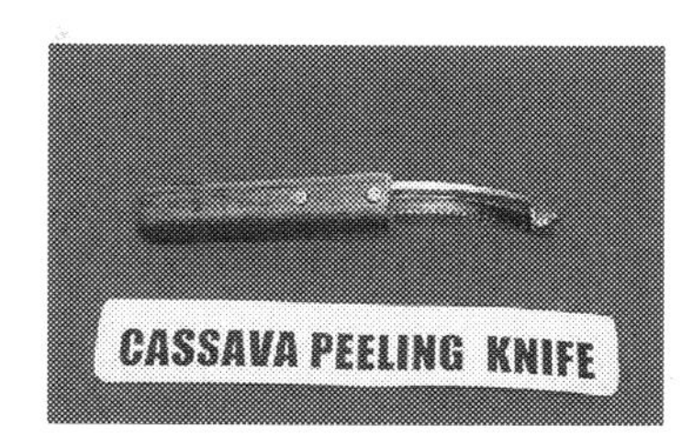

Peeling is the process of removal of corky skin alone or along with fibers. This is usually done manually by knives and is considered one of the most laborious and time-consuming tasks in process of starch extraction. The cassava peeling knife has been developed to address this problem of peeling off the skin from cassava. It is a novel design consisting of a knife with a curved blade. It reduces flesh loss during peeling. Flesh loss using the improved peeling knife is about 1.38% while that of using a traditional knife is 5.70%. The average output is considered to be 132 kg of cassava peeling in an hour. (*Source:* https://images.app.goo.gl/AZUTAqogrz5Tswh49).

Tender Coconut Cutter

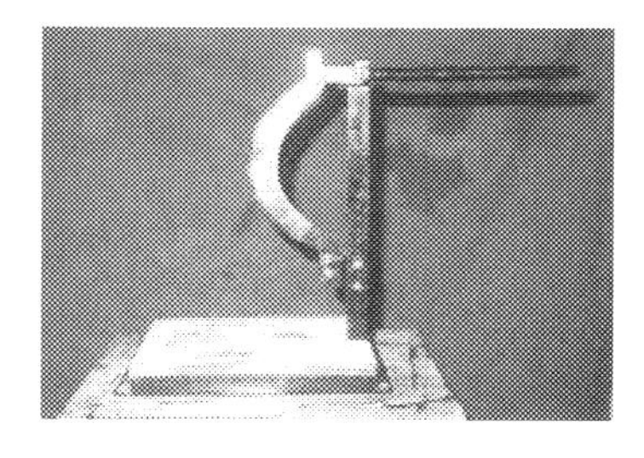

This equipment is used to cut the tender coconut into two halves to get the meat inside the kernel. It consists of a stand to which a serrated curved cutting blade is attached. The whole assembly is mounted on a firm base. A cutting blade is retained at a height of 150-200 mm. The blade is made of high carbon steel of 6mm thickness and is 450mm in length. A hand lever is pivoted to the other end of the blade through a horizontal hinge. On the lower side, a stopper is provided to apply concentrated load for easy cutting. (*Source:* https://images.app.goo.gl/gXky2LhY1vMR1khG8)

PAU Seed Drill

Used for line sowing of seeds like wheat, maize, soybean, etc., the PAU seed drill consists of a handle, hopper to keep seeds, a ground wheel, a fluted roller for seed metering, and a hook for pulling the drill. It is moved with the help of a chain and sprocket mechanism attached to the ground wheel shaft. It is operated in a well ploughed and prepared field by two workers; one for pulling the drill and the other for pushing and guiding. Pulling is done by using a rope tied to the hook in the front portion of the seed drill and pushing is done using the handle. The capacity of the seed drill is estimated to be 430 m^2/ h. PAU seed drill has been ergonomically designed for women workers considering anthropometric data. The main advantage of using seed drill is that the bending posture followed in the traditional method can be avoided which helps to increase the output by 18 times. Moreover, row sowing helps the usage of mechanical weeders and hence reduces human drudgery and the cost of weeding operation. (*Source:* https://images.app.goo.gl/cGv7rzawsEuQfrJz5).

Groundnut Decorticator

It is used for separating groundnut kernels from pods. It is a sitting type decorticator which is an oscillatory type device made of cast iron. It consists of a frame, handles for operating, shoes with projections attached to an oscillating arm, and sieves. A stool is provided for the women worker to sit and operate comfortably. The harvested groundnut pods are fed in batches of about 1.5 kg into the hopper. It should mark a level of half the capacity of the hopper so that, the oscillating arm can be adjusted and easily operated. Conventionally, groundnut was decorticated using hands or beating on stones, which was a time-consuming and drudgery-prone operation. The introduction of sitting type decorticator reduces the drudgery by 79% per unit of output. Women workers prefer sitting type decorticator due to its low requirement of force, less cardiac cost, and less chance of injury to fingers. Furthermore, the grains obtained could be used for sowing. (*Source:* https://images.app.goo.gl/t72mkJmXZgD39Vhm8).

Pedal Operated Paddy Thresher

The pedal operated paddy thresher consists of a cylinder attached with wood or aluminum strips. Wire loops are embedded or welded on these strips. Using the foot pedal, the cylinder is given a rotary motion through a power transmission system. Paddy bundles are held in the hands and kept over the rolling cylinder for threshing. This equipment helps to reduce drudgery by avoiding the bending posture. Also, arm rising above the shoulder level as in the traditional method is avoided by using a pedal-operated threshing machine. (*Source:* https://images.app.goo.gl/WUzcWHUoDFhaEwC86).

CRRI Paddy Winnower

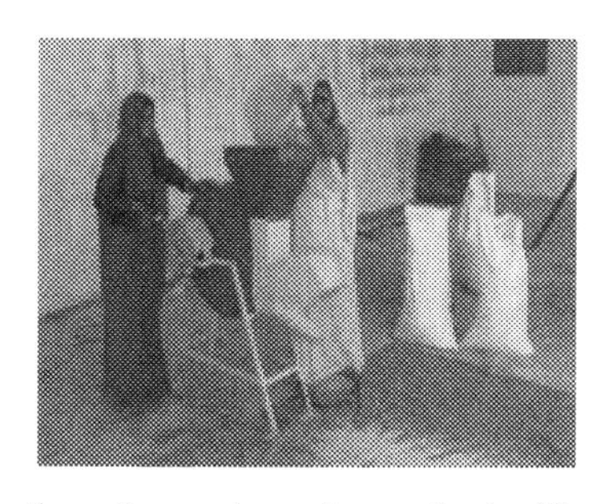

It is used for cleaning grains after harvesting. It consists of a frame on which handle, gear mechanism, volute case, a fan, hopper, outlets for clean grain, and separated chaff are attached. This is a useful and easily operated machine which requires two workers for the proper operation. One worker operates the machine and the other feeds the hopper and separates the cleaned grain and chaff. Operators can either sit or stand while operating. The capacity of the machine is considered to be 242 kg/h. The main advantage of using paddy winnower is that there is no need of waiting for natural airflow as in the case of the traditional cleaning method. Also, it can be operated indoors or under shade. (*Source:* https://images.app.goo.gl/ZKuEt57BjiU1NVpUA)

Wheel Barrow

Wheel barrow has now become a common equipment in farmsteads to carry all kinds of agricultural materials including seeds, fertilizers, manure, harvest etc. It consists of a frame on which ground wheels, a cart to carry the materials, and a handle are attached. They could have one or more wheels depending on the terrain it is used and the size of material to be transported. It is made of good quality mild steel sheet of 1.6 mm thickness. It is useful equipment to convey agricultural materials from one place to another, especially within the agricultural field. The attraction of this equipment is that the person does not have to exert much force in pulling or pushing the cart and the capacity is around 100 kg (*Source:* https:/ /images.app. goo.gl/GbF3qbc9tcXFRcmn6).

Scissor Type Tea Plucker

The main problem with the plucking of tea leaves in the traditional method is injuries occurring to the skin of fingers and hands due to the chemicals. This problem can be avoided by using a scissor-type tea plucker. It consists of a perforated steel pan attached to a scissor. The output is more when comparing with the traditional method of handpicking, which results in 32% cost saving and 40% time-saving.

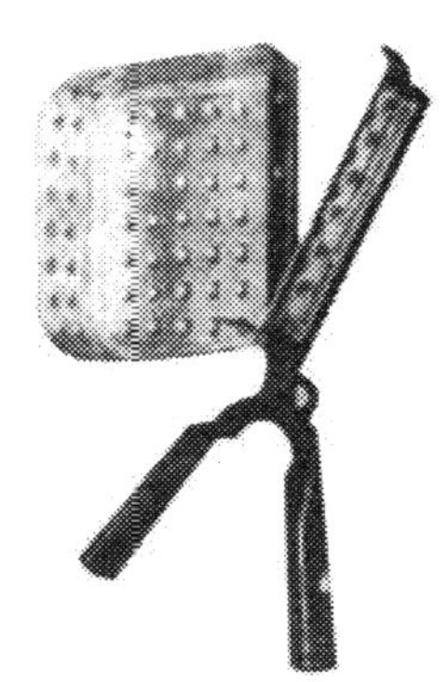

(*Source:* https://images.app.goo.gl/9AjixJHXjhM62rVq7, https://images. app.goo.gl/Q6g7BixuXNU87sWm7).

Rotary Arecanut Dehusker

It is used for dehusking arecanuts. It consists of a hopper, lead screw, cutting blade, and a handle for operating. Through hopper, nuts are fed to the lead screw. Due to the compression occurring between the lead screw and teeth on the cutting plate, the husk is peeled off and removed. The kernels are ejected through the leads of the lead screw. The chance of getting injuries to fingers and palms while dehusking in the traditional method is eliminated by using this arecanut de-husker. (*Source:* https://images.app.goo.gl/yunWH9ngTkXefQtr7)

IIHR Lime Harvester

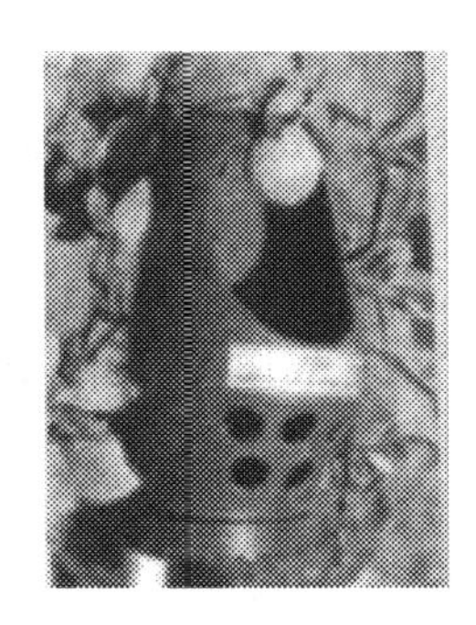

It is used to harvest lemon. The harvester consists of a small cylindrical box with 100 mm diameter and 200mm length, a hook, and a pole. The fruit is held in the hook and harvested by pulling. The harvested lemons are collected in the cylindrical box. The box is emptied when 8-10 fruits are collected in the box. (*Source:* https://images.app.goo.gl/aT5mE5LE2jZfSNXk9)

Bhindi Plucker

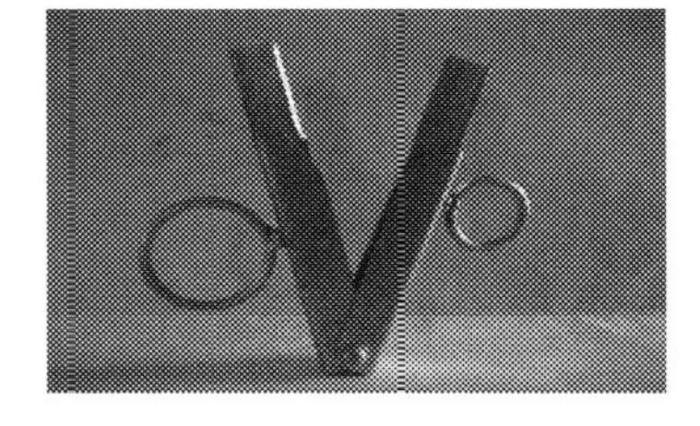

It is used to pluck bhindi without causing any injuries or discomfort to the workers. It consists of two arms with rings and cutting blades. The arms are hinged together and cutting blades are attached at the open ends of the arms. Rings are fit into the fingers: one in the thumb and the other in the index finger. By pressing the fingers, force is exerted and the pedicel is cut and harvesting is done. Usually, the thorny or chemical materials cause itching to the skin. This is eliminated by adopting bhindi plucker (*Source:* https://images.app.goo.gl/GG8UbUwp VdP5mdye9).

Sugarcane Stripper

This is a handy tool to strip leaves during harvesting of sugarcane. Also, it is used to de-top cane after harvesting, to push the leaf sheaths away from stalks, and to clean the roots. It consists of a knife welded on the stem of the stripper. It helps to reduce the chances of injury to workers during stripping operations. The capacity is around 46 kg/hr. (*Source:* https://images.app.goo.gl/BNduwLjJrqT8f6wb6).

6. Conclusion

Farm mechanization is necessary for the economic conduct of agriculture in the future. As women are a major stakeholder in agriculture, the development and adoption of women-friendly machinery should be emphasized, based on ergonomic analyses. A collaborative effort between the researchers, research institutes, industry, and extension agencies is required to be able to develop and popularise equipment suited to women in agriculture. Several types of equipment have already been developed by various research institutions and these can be used by women with very few minor modifications. Many equipment and machinery are still in the process of development and access to these are restricted. Whenever the new tools and implements are available, often they do not reach the end-user due to non-availability. Concerted efforts must be taken up by the extension personnel to ensure that viable technologies reach the identified target and constant skill training should be organized to disseminate the knowledge about the various available machinery. Farm women can be empowered in mechanized agriculture through capacity-building training and continuous hand-holding. Farming has a lot of job opportunities and efforts to build a congenial social climate will go a long way in supporting women entering the sector. It will also provide a means of financial self-reliance. Linkages among developmental agencies will help in popularising gender-friendly equipment. Sensitization of extension functionaries about drudgery-related health hazards is also needed to bring about a change in their perspective. These multi-pronged interventions will bring about a change regarding the role of women in the mechanized farming sector and women also will be able to earn a decent livelihood from their agricultural vocation.

References

Alagasundaram, J.T. Sheriff, Kanchan. K. Singh, G. Senthil Kumaran. Catalogue on tools and machinery (2016). ICAR.

Chauhan, M. 1999. Workload and health problems in some occupational activities Paper Presented in Advanced Training an Ergonomics At SNDT Women' Univerisy Mumbai, 22-27 Feb. 1999.

FAO, 2020. FAO Policy on Gender Equality 2020–2030.

Kishtwaria, J. and Rana, A. 2007. Drudgery of hill farm women due to weeding activity- some solutions, Women at work, Vol. II, Allied Publishers Ltd., pp. 43-49.

Kishtwaria, J. and Rana, A. 2012. Ergonomic interventions in weeding operations for drudgery reduction of hill farm women of India. Work, 41(SUPPL.1), 4349–4355. https://doi.org/10.3233/WOR-2012-0730-4349

Pereira, J.O., Schlosser, J.F., Souza, D. De, Silva, D.L., Angel, M., Opazo, U., Sales, E.J. De, & Severino, S. (2014). Anthropometric model and obesity index of agricultural tractor operators. 9(45): 3350–3359. https://doi.org/10.5897/AJAR2014.8948

Rehman, N. 2017. Significance of Ergonomically Designed Tools and Equipment in Drudgery Reduction among Farm Women/ : A Review. 17(1): 102–105.

Sharma, B., Verma, S. and Mustafa, M. 2017. Ergonomic Evaluation of Drudgery Load Faced by Farm Women in Wheat Harvesting. International Journal of Current Microbiology and Applied Sciences, 6(10): 3014–3022. https://doi.org/10.20546/ijcmas.2017.610.355

Sureshkumar, P.K., Geethakutty, P.S. and Nair, S. 2005. Women friendly farm mechanisation - status and scope in Kerala. Proc. Natl. Workshop on Role of Women in Mechanised Farming, 8th January 2005. NRCWA (ICAR)

Varghese, A. and Jacob, J. 2014. A study on the KAU coconut husking tool. 2014 Annual International Conference on Emerging Research Areas: Magnetics, Machines and Drives, AICERA/ICMMD 2014 - Proceedings, May 2016.

Websites

https://doi.org/10.1109/AICERA.2014.6908254

https://farmech.dac.gov.in/Women%20Friendaly%20Equipment-With%20photographs.pdfhttp://www.businessworld.in/article/Women-In-Agriculture-The-Potential-And-Gaps/05-01-2021-361877/

https://vikaspedia.in/agriculture/women-and-agriculture/tools-and-equipments-to-reduce-women-drudgery

https://farmech.gov.in/FarmerGuide/UP/6u.htm

https://vikaspedia.in/agriculture/women-and-agriculture/tools-and-equipments-to-reduce-women-drudgery

https://kvk.icar.gov.in/API/Content/PPupload/k0196_89.pdf

Part-III

Tools and Approaches in Gender Analysis

17

Gender Analysis: Strategies and Tools

Sabita Mishra*

1. Introduction

The gender gap in agriculture is perceived worldwide. It signifies that farm women have less access to resources, finance, capital and advisory services compared to men (FAO, 2011a). Further, women own only 1 per cent of the world's property in spite they dominate in world food production (50 to 80 per cent) and they own less than 10 per cent of land globally (Huyer *et al.*, 2015). But, in a healthy society, both men and women should enjoy the equal right, opportunity, accessibility and also should shoulder the equal responsibility in family building, society building or nation building. But till date, the equal status and opportunities of men and women has remained a desired goal. It happens so, as the livelihood needs of men and women are different due to their different roles, responsibilities, accessibility to resources and cultural norms. However, of late the researchers and policy makers have realized the importance of gender equality for national development and it has been one of the avowed targets in Sustainable Development Goals (SDG) for 2030. This has led to the recognition and conceptualization of the concept of 'gender analysis' in development research. Gender analysis frameworks provide a step-by-step methodology for conducting gender analysis (Ochola *et al.*, 2010). Gender analysis is relevant to education, although the frameworks used for development projects must be adapted to meet the needs of educational projects (Leach, 2003).

The chapter discusses the concept of Gender Analysis along with the tools and frameworks available in undertaking gender analysis. Each method is discussed in detail covering the context of use, merits and limitations. The data collection formats are also discussed.

---

* *Author contact: sabitamshra@rediffmail.com*

2. Gender Analysis-The concept

Gender Analysis is a tool to better understand the realities of the women and men, girls and boys whose lives are impacted by planned development. It refers to the gathering and examination of information on gender gaps. These include gender issues with respect to social relations, activities, access to and control over resources and needs. Therefore, to bring equal status, gender analysis is a must to understand the gender issues, their roles, responsibilities, needs, etc.

3. Gender Analysis Tools/Frameworks

A number of tools/frameworks are available for gender analysis, some of the popular analytical frameworks widely used by gender researchers and development policy makers are presented and described as follows.

1. Harvard Analytical Framework
2. Moser (triple roles) Framework
3. Levy (web of institutionalization) Framework
4. Gender Analysis Matrix (GAM)
5. Equality and Empowerment Framework (Longwe)
6. Capacities and Vulnerabilities Analysis Framework (CVA)
7. People Oriented Framework (POP)
8. Social Relations Framework / Approach (SRF)
9. SEAGA Approach of FAO

3.1. Harvard analytical framework

The Harvard Analytical Framework was one of the first frameworks designed for gender analysis. It was developed in USA and is widely used in agricultureand other rural projects. It helps to improve overall productivity and reduce poverty. It maps the work and resources of men and women in a community highlighting the main differences. The framework is used for collecting data at the micro-level. The Harvard Analytical Framework has four interrelated components as discussed below.

Harvard Tool 1: The Activity Profile: This tool identifies all the productive and reproductive activities of women and men, girls and boys by questioning who does what. Other data disaggregated by gender, age or other factors can also be included. It can record details of time spent on tasks specifying what per centage of time is allocated to each activity, and whether it is carried out seasonally or daily at home or elsewhere in the community. A sample of the activity profile used in data collection is given as **Annexure (I)**

Harvard Tool 2: Access and Control Profile-Resources and Benefits: This tool indicates whether women or men have access to resources, who controls their use, and who controls the benefits of a households/community's use of resources. In our society, it happens that women may have some access to resources but have little control over those to take decisions. Therefore, the person having control over the resources, ultimately make decisions about its use and sale. A sample data collection format of Harvard tool 2 is given as **Annexure (II)**

Harvard Tool 3: Influencing Factors: This tool identifies the influencing factors that shape gender relations and give an indication of future trends. It determines various present opportunities and constraints for men and women. These factors may be family/ community norms, cultural and religious practices, economic conditions, environment, infrastructure; training and education; etc. which affect women's or men's activities or resources. The field data collection needs to be precisely addressing these issues as presented in **Annexure (III)**.

Harvard Tool 4: Checklist for the Project Cycle Analysis: This consists of a series of questions or checklists. At each stage of project cycle, questions are to be asked covering identification, design, implementation and evaluation. This tool has been designed to examine a project proposal from a gender perspective like assessing women needs, impact on women's access to or control of resources and benefits, impact on women's activities, organisational structures and flexibility to meet the changing situations of women, operations and logistics, flexibility, women involvement in the collection and interpretation of data, fair allocation of funds for women, etc.

3.2. Moser (Triple Roles) Framework

The Moser Framework was developed by Caroline Moser at University of London, UK in the early 1980s as a method of gender analysis. Moser Framework aims for achievement of equality, equity, and empowerment. There are six tools in the framework that can be used for planning at all levels from project to regional planning. It can also be used for gender training. The Moser Framework includes gender roles identification, gender needs assessment, disaggregating control of resources and decision making within the household, planning for balancing the triple role, distinguishing between different aims in interventions and involving women and gender-aware organizations in planning.

Moser Tool 1: Gender Roles Identification / Triple Role: This tool includes making visible of the gender division of labour. It can be carried out by mapping all the activities of men and women in the household over a twenty four hours period. A triple role for low income women is identified by Moser as productive, reproductive and community roles. As defined by Moser, men mostly carry out productive activities which involves the production of goods and services for consumption. On the other hand, women mostly perform reproductive works such as caring household and family health-care, bearing and caring children, care for food, water and fuel, shopping and housekeeping. The productive activities can be performed by both men and women. But, women's productive work is often less valued than men's.

Moser Tool 2:Gender Needs Assessment: Women have particular needs which differ from those of men. According to Moser, meeting practical gender needs responds to an immediate perceived in a specific context. Women's practical gender needs may include water provision, health-care provision, opportunities for earning, provision of housing, distribution of food, etc. which are shared by all household members. On the other hand, women's strategic gender needs exist because of their subordinate social status. It is related to gender divisions of labour, power, control, legal rights, domestic violence, equal wages, and women's control over their own bodies. Meeting strategic gender needs helps women to achieve greater equality.

Moser Tool 3: Disaggregating Control of Resources and Decision: This tool asks the question; who controls what? Who decides what? How? It also identifies who has control over what resources within the household, and who has what power of decision-making?

Moser Tool 4: Balancing of Roles: This framework helps to examine whether a planned programme or a project will increase the workload of women in one of her roles. The users of this framework should see that women must balance their reproductive, productive and community responsibilities.

Moser Tool 5: WID/GAD Policy Matrix: The WID/GAD policy matrix provides a framework for identifying/evaluating the approaches that have been used to address the triple role, and the practical and strategic gender needs of women in programmes and projects. This is mainly a tool for evaluation. It helps to identify five different types of policy approaches such as welfare, equity, anti-poverty, efficiency and empowerment. It aims to meet women's practical gender needs with a purpose to promote equality for women with decision-making power. However, according to Khandker (2020) most of the work done by women are unpaid, and their contribution continues to be largely overlooked in agricultural policies and programs.

Moser Tool 6: Involving women, gender aware organizations and planners: Moser's framework users should involve women, gender-aware organisations, and planners in planning. The aim of this tool is to ensure that practical and strategic gender needs are identified by women so that the "real needs" are safeguarded.

Potential limitations of Moser's framework are: (i) radical agenda gets depoliticised by the language of planning, (ii) concept of triple role does not get fully captured due to the power imbalance between women and men, (iii) women's and men's separate activities are emphasised, rather than relationships between the two, (iv) the framework does not deal with other underlying inequalities, such as class and race, (v) division between strategic and practical is artificial, (vi) ignores men as 'gendered' beings and (vii) the change over time is not examined as a variable.

3.3. Levy Conceptual (Web of Institutionalization) Framework

The Levy conceptual framework is known as the web of institutionalisation and widely-recognised tool for analysing gender mainstreaming in institutions for development planning through a critical gender lens. The Levy Framework comprises 13 elements for effective gender mainstreaming which are organised into four spheres like: (i) Beneficiaries Sphere, (ii) Policy Sphere, (iii) Organisational Sphere and (iv) Delivery Sphere. Participants work in groups to know the elements of each sphere in terms of gender mainstreaming. Based on the time availability, the group may focus on only one sphere, or the whole framework.

3.4. Gender Analysis Matrix (GAM)

The Gender Analysis Matrix (GAM) was developed by Rani Parker for planning, designing, monitoring, and evaluating projects at a community level. It is a tool used for analysis about gender roles by community members themselves with equal representation of women and men facilitated by a development worker. It is simple, systematic, accessible, flexible and designed to accommodate changes over time. The tool fosters 'bottom-up' analysis, considers gender relations between women and men, includes intangible resources, encourages for support for those at risk, and includes men as gendered beings and quick data gathering. The GAM is to be used in addition to other tools of analysis such as monitoring tools, needs assessments, and so on. It is appropriate for use in transformatory gender training.

GAM Tool 1:Analysis at Four 'Levels' of Society: This tool analyses the impact of interventions on women of all ages, men of all ages, households, and community and so on.

GAM Tool 2:Analysis of Four Kinds of Impact: This tool looks at impact on labour, time, resources and socio-cultural factors.

GAM tool has some limitations such as: (i) needs a good facilitator, (ii) some factors can get lost because categories have many aspects, (iii) requires careful repetition in order to consider change over time, (iv) does not seek out the most vulnerable community members, (v) risk of misleading outcomes and (vi) difficulties defining a community.

3.5. Women's Empowerment (Longwe) Framework

The Women's Empowerment (Longwe) Framework was developed by Sara Hlupekile Longwe, a consultant on gender and development based in Lusaka, Zambia. Longwe defines women's empowerment as enabling women to take an equal place with men, and to participate equally in the development process. Gender equality denotes equal participation of women and men in decision-making, equal ability to exercise their human rights, equal access to control resources, equal share in benefits of development and equal opportunities in employment and in all other aspects of their livelihoods (FAO, 2009). This is a useful framework for planning, monitoring and evaluation.

Women's Empowerment Tool 1: The Longwe Framework indicates the extent of empowerment and equality of women with men. These levels of equality are hierarchical. It is focused on the concept of five 'levels of equality' like: (i) women's material welfare, (ii) access to the factors of production, (iii) conscious understanding of the difference between sex and gender, (iv) women's equal participation in the decision-making process, in policy-making, planning, and administration and (v) balance of control between men and women.

Women's Empowerment Tool 2: According to the Longwe Framework, women's empowerment must be concerned with both women and men. It identifies the extent of address to women's issues in project design in three different levels like: (i) negative level by making no mention of women's issues, (ii) neutral level by recognising women's issues and (iii) positive level by positively concerning with women's issues.

The Longwe Framework has some limitations such as: (i) it is static irrespective of the changesthat may occur in a situation over time, (ii) does not look at the complicated system which exists between men and women, (iii) can mislead the view of women as a homogeneous group, (iv) deals in very broad generalities and (v) does not examine the macro-environment.

3.6. Capacities and Vulnerabilities Analysis Framework (CVA)

It is like the People-Oriented Framework and was designed for disaster preparedness in emergencies to meet immediate needs, and to achieve social and economic development. The framework focuses on the existing strengths of individuals and the vulnerabilities exist before disasters to cope with the disaster.

CVA Tool 1: Categories of Capacities and Vulnerabilities: By using an analysis matrix, the CVA distinguishes three categories of capacities and vulnerabilities like: physical, social, and motivational including features of the climate, land, and environment. CVA identifies the ways in which both men and women are physically or materially vulnerable, and, who have access and control over the existing resources, what were people's beliefs and motivations before the disaster, how has the disaster affected them, and, do men and women feel they have the same ability?

CVA Tool 2: Additional Dimensions of 'Complex Reality': CVA matrix enables to map out on social differentiation, the level of wealth, political affiliation, language groups, age, and so on of a community. It can be used before an intervention and after to examine social change and impact in gender relations as the result of an emergency.

Limitations with respect to the use of CVA include (i) creates gender-blind analyses and responses, (ii) not designed to promote women's empowerment, (iii) offers a temptation to guess with a relatively superficial knowledge of the situation and (iv) the framework does not lend itself to participatory uses.

3.7. People Oriented Planning (POP) Framework

It is an adaptation of the Harvard Analytical Framework. The aim of POP is to reduce the disparities between the sexes. The POP Framework has three components as follows:

POP Tool 1: Refugee Population Profile and ContextAnalysis: The tool helps to assess the refugee population from a demographic perspective to find out the composition of the refugee group before they became refugees, and subsequent changes in relation to social, political, economic, institutional, religious and cultural context. These determinants are required to consider which ones affect activities or resources. It also helps to identify constraints and opportunities to consider while planning programmes.

POP Tool 2: The Activities Analysis: This tool is similar to the Harvard Tool 1 (Activities Profile) and helps to identify who does what, as well as when and where they do it. Also, it facilitates to find out what women and men were doing before, and what they are doing now, in the refugee situation.

POP Tool 3: Use and Control of Resources Analysis: Similar to the Harvard Tool 2 (Access and Control Profile) and helps to know how resources are distributed among men and women, who has control over their use and what resources must be provided for which refugees .

The limitations of POP Framework are: (i) too materialistic, (ii) over simplifies the concepts of access and control, (iii) changes over time are not taken into account, (iv)works best with homogeneous groups, (v) designed for use in planning for refugee situations and (vi) the question of control cannot be fully answered.

3.8. Social Relations Framework / Approach

The Social Relations Approach has been developed by Naila Kabeer at the Sussex University, UK. This approach can be used for training, project planning and policy development even at the international level. It aims to give a total picture of poverty and cross-cutting inequalities of class, gender, race, and so on. Evidence establishes that equal gender relations within agricultural households and communities lead to better development outcomes *viz.* farm productivity and family nutrition (World Bank, FAO and IFAD, 2015).

Social Relations Approach: Concept 1: Here, the development is not simply about economic growth or improved productivity but, primarily increasing human well-being with the goals of survival, security, and human dignity. Human wellbeing includes all those tasks which poor people carry out to survive; and which people perform in caring their environment.

Social Relations Approach: Concept 2: Kabeer uses the term 'social relations' to describe the structural relationships and systemic differences of different groups of people. It determines the availability of tangible and intangible resources to groups and individuals, the roles, responsibilities, claims, rights and control of own lives and those of others.

Social Relations Approach: Concept 3: This approach used for institutional analysis to know the causes of gender inequality which are not confined to the household and family, but are reproduced across a range of institutions, including the international community, the state & the market place. Here, an institution is defined as a framework of rules for achieving certain social or economic goals.

Social Relations Approach: Concept 4: Naila Kabeer classifies institutional gender policies into three types: (i) gender-blind policies where there is no distinction between the sexes, (ii) gender-aware policies which recognises both men and women as development actors and (iii) gender-neutral policy approaches used to overcome biases in development interventions and to meet their practical gender needs.

Social Relations Approach: Concept 5: While analysing a situation for intervention, this framework sightsees the instant, primary, and structural issues which cause the problems, and their effects on the various actors involved.

Social Relations Approach too has its inherent limitations in use such as: (i) gives an impression of monolithic institutions, (ii) women may get subsumed in many individual categories, (iii) unsuitable to use in a participatory way at community level and (iv) difficult to use fully in situations having lack of information.

3.9. SEAGA Approach of FAO

SEAGA technique has been developed by FAO and it stands for Socio-Economic and Gender Analysis. It helps to identify the priorities of women and men, the gender issues with respect to activities, access to and control over resources, decision making, needs and problems and also to formulate projects for gender mainstreaming in research and extension, and, to better understand the ground realities of the women and men. It is for analysis of the current situation and planning for the future. Broadly, all the tools are classified into three categories of gender analysis as follows.

A. Development context tools,

B. Livelihood analysis and

C. Stakeholders' priorities

A. **Development context tools:** Here, the centre of attention remains on present situation (What is) for learning economic, environment, social and institutional patterns that act as supports or constraints for development.There are six tools as listed belowwhich are used under the development context.

   i) Village Resources Maps

   ii) Transects

   iii) Village Social Map

   iv) Trend Lines

   v) Venn Diagrams

   vi) Institutional Profiles

i) **Village Resources Map:** This map focuses on available resources like roads, buildings, houses, water bodies, agriculture land, grazing land, forest area, shops, health clinics, educational institutions, religious institutions, bus stop, etc. Helps for learning about the environmental, economic and social resources in the community.

ii) **Transects:** It gives more details about environmental, social and economic resources in a community and provides a cross sectional picture of an area through direct observation. Helps for learning about the community's natural resource base, land forms, and land use, location and size of farms or homesteads, and location and availability of infrastructure and services and economic activities.

iii) **Village Social Map:** It gives a perceptional picture of resources existing in the community. It helps for learning about the community's population, local poverty indicators and number and location of households by type (ethnicity caste, female-headed, wealthy, poor, etc.).

iv) **Trend Lines:** It is a simple graph depicting change over time. It gives a picture of what is getting better and what is getting worse over time. It helps for learning about environmental trends (deforestation, water supply);

economic trends (jobs, wages, costs of living), population trends (birth rates, out-migration, in-migration), and other trends of importance to the community.

v) **Venn Diagrams:** Through this tool we can identify the potential conflicts between different socio-economic groups. It helps for learning about local groups and institutions and their linkages with outside organizations and agencies.

vi) **Institutional Profiles:** It helps for learning about the goals, achievements and needs of local groups and institutions

Field Techniques: A group of older men and women should be involved in discussion as they know more about the past events. Ask them about important changes in the community (may be better or worse) related to natural resources, populations and economic opportunities. A sample format for data collection for trend analysis over years has been included as **Annexure (IV)**.

B. **Livelihood Analysis:** Here, the focal point is on existing situation (What is) for learning the flow of activities and resources for living.Tools commonly used under Livelihood Analysis are listed below:

   i) Farming System Diagram
   ii) Benefits Analysis Flow Chart
   iii) Daily Activity Clocks
   iv) Seasonal Calendars
   v) Resource Picture Cards
   vi) Income and Expenditure Matrices

i) **Farming Systems Diagram:** It is a diagram to highlight the farming systems in family. It helps for learning about household members' on-farm (crop production), off-farm (fuel collection) and non-farm (marketing) activities and flow of resources to and from the home. It shows how livelihood depends upon various types of agro-eco-systems like forest, river, grazing land, etc which are in common use.

ii) **Benefits Analysis Flow Chart:** Through this analysis, we may be able to understand what the 'fruits' are from people's livelihood activities and who enjoys that. It also helps for learning about benefits use and distribution by gender. The bi-products are the result of any resource. Example, 'tree' as resource has bi-products like leaves, bark, fruits, seeds, fiber, fuel wood, fodder, etc, Here, who is the gender to enjoy these can be understood.

iii) **Daily Activity Clocks:** It gives a total picture of activities performed by gender in a day and who does more and also who does less. Helps for learning about the division of labour and labour intensity by gender and socio-economic groups. It helps to identify the workloads and leisure time for the community people including men, women, rich, poor, young and old. The clear picture comes that who works for longest hours and who does little activities. The field application format of the tool is given as **Annexure (V).**

iv) **Seasonal Calendars:** Helps for learning about the seasonality of women's and men's labour and seasonality of food and water availability and income and expenditure patterns and other seasonal issues important for the community. The calendars can be used to know the changes in income over the time and the work opportunity for the people at different periods of time.

v) **Resources Picture Cards:** Helps to know the gender based resource use and control within the household. This exercise facilitates us to know who is likely to be looser and who is likely to be gainer because of a particular development activity. It gives idea about who has access over the household resources (land, livestock, trees) and who takes decisions for its use. Furthermore, women are excluded from decision-making process on land use, accessibility and resources critical for their livelihood (FAO, 2011b).The World Bank (2001) documented that ignoring gender inequalities comes at great cost to people's well-being and countries' abilities to grow sustainably and thereby reduce poverty. A resource picture card format for field use is presented in **Annexure (VI)**.

vi) **Income and Expenditures Matrices:** Helps to find out about sources of income, sources of expenditures and changes in expenditure at crisis. Analysing their items of expenditure the priorities and limitations can be understood. It helps to understand the security or vulnerability of livelihood, meeting basic needs and saving if possible for rainy days.

C. **Stakeholders' priorities**: Here, the focus is on future prospect (What should be) for planning development activities based on women's and men's priorities.Tools under Stakeholders' Priorities are as follows.

i) Pair wise Ranking Matrix

ii) Flow Diagram

iii) Problem Analysis Chart

iv) Preliminary Community Action Plan

v) Venn Diagram of Stakeholders

vi) Stakeholders Conflict & Partnership Matrix

vii) Best Bets Action Plans

i) **Pair wise Ranking Matrix:** Helps to know the most important problems in the community, the priority problems of women and men and, of different socio-economic groups. Techniques for field application involve organization of two separate focus groups: one of women one of men with a mix of socio-economic groups. Ask the participants to list 6 problems important to them. Write the list of 6 problems on both the vertical & horizontal axis of the paper. Also write the problem in separate six cards, show the participants a pair of problem cards asking them the more important one with reasons of choice. Record their choice on the prepared matrix as depicted in the following case.

Problems	Cost of Inputs	Insect pest	Technical knowledge	Climate	Irrigation	Land
Cost of Inputs		Cost of Inputs	Cost of Inputs	Cost of Inputs	Irrigation	Cost of Inputs
Insect pest			Insect pest	Climate	Irrigation	Insect pest
Technical knowledge				Climate	Irrigation	Technical knowledge
Climate					Irrigation	Climate
Irrigation						Irrigation
Land						

The results from the above illustrated case can be documented in a tabular format and ranked as follows.

Problems	Number of times preferred	Rank
Cost of inputs	4	II
Insect Pests	2	IV
Technical knowledge	1	V
Climate	3	III
Irrigation	5	I
Land	0	VI

i) **Flow Diagram:** This analysis helps to identify about the causes and effects of their problems and can be used for possible solutions. This identifies the major problem in the community and decides which problem to be solved by the community, which can be solved by the external source and which has no solution like natural disasters.

ii) **Problem Analysis Chart:** It is used for bringing together the priority problems of all the different groups in the community, to explore local coping strategies and to identify opportunities to address the problems.

iii) **Preliminary community action plan:** It is helpful for planning possible development activities, including resources needed insider and outsider groups to be involved and timing.

iv) **Venn diagram of stakeholders:** Stakeholder is anyone who has interest in and is going to be affected in any developmental work. It helps us to know who is going to be affected by the proposed development plan. Gives a picture about the insider and outsider stakeholders for each action proposed in the Preliminary Community Action Plan. The extent of interest of a stakeholder is determined by the size of their stake in it.

v) **Stakeholders Conflict and Partnership Matrix:** This analysis helps for learning about conflicts of interests and common interests between stakeholder.

vi) **Best Bets Action Plans:** Facilitates for finalization of action plans for development activities meeting priority needs as identified by women and men of each socio-economic group.

Out of many available tools, it is found that, the **SEAGA** tool is very much appropriate for gender analysis in agriculture. Therefore, most of the users implement SEAGA tools for gender analysis in the field. But, based on the communities, priorities and needs, these tools for gender analysis can be used by the researchers with required modification.

4. Conclusion

All the stakeholders should go for gender analysis to get valuable gender disaggregated data as developmental aspects. Based on the situations and locations, the analytical, desired corresponding tools should be selected and used for its wider application. The researchers having the knowledge of PRA methods will find the gender analysis tools very useful in formulation of project/programmes for gender mainstreaming.

References

FAO, 2009. Bridging the gap: FAO's Programme for Gender Equality in Agriculture and Rural Development. Rome, Italy: FAO.

FAO, 2011a. The State of Food and Agriculture. Rome, Italy: FAO. (Available from http://www.fao.org/docrep/013/i2050e/i2050e00.htm).

FAO, 2011b.Women and Food Security. FAO FOCUS www.fao.org/ FOCUS/E/ Women/ Sustin-e.

Huyer, S., Twyman J., Koningstein M., Ashby J. and Ver-meulen S. 2015. Supporting women farmers in a changing climate: five policy lessons. CCAFS Policy Brief no. 10.Copenhagen, Denmark: CGIAR Research Program on Climate Change, Agriculture and Food Security (CCAFS).

Khandker, V., Vasant P. Gandhi, and Nicky Johnson, 2020. Gender Perspective in Water Management: The Involvement of Women in Participatory Water Institutions of Eastern India. Water, 12, 196; doi:10.3390/w12010196.

Leach, Fiona E. 2003. Practising gender analysis in education. *Oxfam.* ISBN 0-85598-493-7.
Ochola, Washington O., Sanginga, Pascal C., Bekalo, Isaac, 2010. Managing Natural Resources for Development in Africa. *IDRC.* ISBN 9966-792-09-0.
World Bank, FAO & IFAD, 2015. Gender in climate-smart agriculture: module 18 for gender in agriculture source-book. Agriculture global practice. Washington, D.C.: World Bank Group.
World Bank, 2001. Engendering Development-Through Gender Equality in Rights, Resources, and Voice. Washington, DC: World Bank.

Annexure I

Harvard Tool 1: The Activity Profile

Activities	Men	Women
Productive Activities: Agriculture: activity 1 activity 2, etc. Income generating: activity 1 activity 2, etc. Employment: activity 1 activity 2, etc. Other:		
Reproductive Activities Water related: activity 1 activity 2, etc. Fuel related: Food preparation: Childcare: Health related: Cleaning and repair: Market related: Other:		

(*Source:* FAO SEAGA field tool kit. Gender Analysis for Sustainable Livelihoods & A Guide to Gender-Analysis Framework, OXFAM Publication)

Annexure II

Harvard Tool 2: Access and Control Profile – Resources and Benefits

	Access		**Control**	
	Women	**Men**	**Women**	**Men**
Resources				
Land				
Equipment				
Labour				
Cash Education/training, etc.				
Other				
Benefits				
Outside income				
Asset ownership				
Basic needs (food, clothing, shelter etc)				
Education				
Political power/prestige				
Other				

Annexure III

Harvard Tool 3: Influencing Factors

Influencing Factors	Constraints	Opportunities
Community norms and social hierarchy		
Demographic factors		
Institutional structures		
Economic factors		
Political factors		
Legal parameters		
Training		
Attitude of community to development workers		

Annexure IV

SEAGA Development Context Tools: Trend Lines

S.No.	Years	Events	Intensity of Events
1.	1990		
2.	1995		
3.	2000		
4.	2005		
5.	2010		
6	2015		
7.	2020		

Annexure V

SEAGA Livelihood Analysis Tools: Daily Activity Clocks

Time	Women	Men
0 to 1 am		
1 to 2 am		
2 to 3 am		
3 to 4 am		
4 to 5 am		
5 to 6 am		
6 to 7 am		
7 to 8 am		
8 to 9 am		
9 to 10 am		
10 to 11 am		
11 to 12 noon		
12 to 1 pm		
1 to 2 pm		

2 to 3 pm
3 to 4 pm
4 to 5 pm
5 to 6 pm
6 to 7 pm
7 to 8 pm
8 to 9 pm
9 to10 pm
10 to 11 pm
11 to 12 pm

Annexure VI

SEAGA Livelihood Analysis Tools: Resources Picture Cards

S.No.	Resource	Access			Control		
		Male alone	Female alone	Joint	Male alone	Female alone	Joint
1.	Land						
	(i) Family land						
	(ii) Lease land						
2.	Capital						
	(i) Family income						
	(ii) Credit from bank						
3.	Water						
4.	Seeds/seeding materials						
5.	Labour (manual)						
6.	Manures and fertilizers etc.						
7.	Machines/equipments						
8.	Farm produce						
9.	Food						
10.	Technologies						
11.	Trainings						
12.	Extension services						
13.	Market information						
14.	Farm profit						
15.	Co-operatives societies						

18

Gender Budgeting: A Tool for Women Empowerment

Smitha S. and Sulaja O.R.*

1. Introduction

Women empowerment is critical to the process of community development as well as for economic growth. Gender budgeting is recognized as a distinctive technique of women empowerment that directly upholds women development through allocation of budgetary fund. It aims at bringing gender equality through allocation of public funds and affirmative action for under-privileged sections. In fact, the term gender budgeting has been explained differently by various authors. A comprehensive definition states "gender budgeting as a dissection of the government funds to establish its gender differential impacts and to translate gender commitments into budgetary commitments" (Sharp, Rhonda, 1999). It can be viewed as a mainstream application of gender in the budgeting process. It ensures gender assessment of the budget, including reorganization of revenue and expenditure to incorporate gender perspective at all levels of the budget process. Gender budgeting therefore deals with gender specific formulation of laws, policies, plans, schemes and programs by allocation of resources; implementation and execution; monitoring, review and its impact assessment. It analyses the government budget from a gender perspective to assess how it addresses the needs of women in the areas of health, education, employment etc. It uses the budget as an entry point and apply the gender lens to the entire process of programme implementation.

Public budgets have different effects on women and men, and describe the unequal distribution of power in society in terms of economic disparities, differences in living conditions and ascribed social roles. Gender budgeting aims to visualize the gender impact of budgets and to transmute them into an instrument for increasing gender equality. Gender budgets basically includes all types of governments at the national, regional and local levels. The terms gender-

**Author contact: smitha.s@kau.in*

responsive, gender aware or gender-sensitive budget and women's budget may be used as synonyms instead of gender budget(ing). Gender budgeting does not mean a discrete budget for women, but advocates an analytical tool that scrutinizes the capacity of government budgets to divulge its gender differentiated impact and priorities for associated programmes/schemes that benefit women. Thus gender budgeting relates not only with public expenditures but also with the gender differentiated impact of revenue mobilization by the government as illustrated in Figure 1.

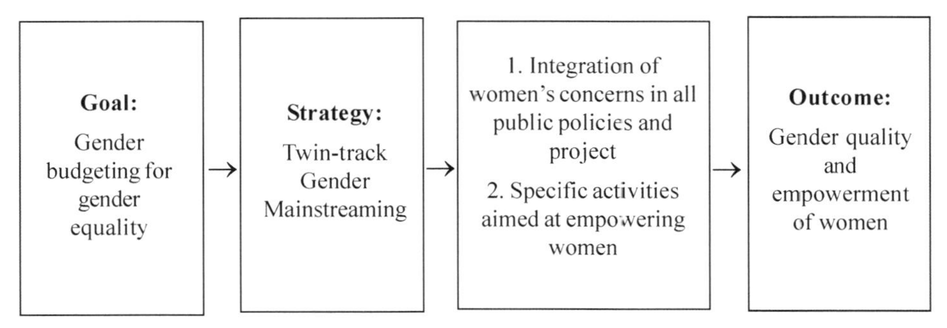

Figure 1. Gender equality through gender budgeting
Source: Ministry of Women and Child Development

2. Why Gender budgeting (GB)?

Gender Budgeting is universally accepted as a powerful tool in achieving development objectives and act as an indicator of commitment to the stated policy of the government. The budgetary policy of the government has a major role in achieving the objectives of gender equality and growth through fiscal and monetary policies. It forms a measure for resource mobilization and affirmative action for the under-privileged sections. Women stand apart as one segment of the population that warrants special attention due to their vulnerability and lack of access to state resources. Thus, gender responsive budgets can augment in achieving the objectives of gender equality, human development and economic efficiency. The purpose of gender budgeting exercise is to assess quantum and adequacy of allocation of resources for women and establish the extent to which gender commitments are translated in to the budgetary commitments. This exercise facilitates increase in accountability, transparency and participation of the community.

Globally there exist significant gaps and inequalities between women and men in terms of political and economic participation, remuneration, distribution of economic assets, and involvement in unpaid care and domestic work. All member states of the Council of Europe are signatory parties to the Convention of the United Nations on the Elimination of all forms of Discrimination against Women

(CEDAW), which requires state partners to take all necessary steps to ensure gender equality. Further to the commitments under the CEDAW Convention, the 1995 Beijing Platform for Action was also endorsed by all most all UN member states. The world community has agreed that the national governments must reallocate and mobilise additional resources for the empowerment of women.

3. Development of Gender Responsive Budgeting (GRB)

Australia, Canada, South Africa and the United Kingdom are considered pioneers in Gender Responsive Budgeting (GRB). The Australian attempt to GRB is considered as the first initiative to analyze government budgets from a gender perspective. It dates back to 1984 when Federal and state governments in Australia introduced Women's Budget Statements (WBS) to include gender in economic and social policy. Every year since 1989, the British Women's Budget Group (WBG), consisting of experts from universities, unions and NGOs, has published comments on the national budget. Their focus was on assessing taxes and transfers, but they also referred constantly to the importance of "engendering" economic policies. In 1993 the Women's International League for Peace and Freedom (WILPF) established a gender budgeting initiative in Canada. Gender budgeting was given additional impetus by the Fourth World Conference on Women, held in Beijing in 1995.The *Beijing Platform for Action* called for ensuring the consideration of a gender perspective and women's needs in budgetary policies and programs. South Africa initiated formation of gender sensitive budget in 1995, through a participatory process of involving parliamentarians and NGOs. The Commonwealth initiative to integrate gender into national budgetary processes was started in 1997 in four countries other than South Africa such as Fiji, St Kitts and Nevis, Barbados and Sri Lanka. Several other nations have also adoptedmeasures to engender their national budget (Canada, UK, Mozambique, Namibia, Tanzania and Uganda). Gender budgeting is now widespread and currently being attempted in more than 100 countries. Within their gender budgeting efforts, some countries have adopted or modified fiscal policies and programs to ensure that the budget supports terminating gender gaps and promotes women's advancement.

3.1. Evolution of Gender Budgeting (GB) over the Five year plans

Central Social Welfare Board (1953) was set up to endorse welfare work through voluntary organizations, charitable trusts and philanthropic agencies under the First Five Year Plan (1951-1956). The second Five Year Plan (1956-1960) saw the development of *Mahila mandal* for grass roots work among women. There were provision for women's education, prenatal and child health services, supplementary feeding for children, nursing and expectant mothers under the third, fourth and interim plans (1961-74). A major shift in the approach towards women, from welfare to development marked the fifth plan (1974-1978). The multifaceted approach with tripartite focus on health, education and employment was included under the sixth plan (1980-85). It was during this plan period, women's development was accepted as a separate economic agenda. Inclusion of a separate chapter on women and children in the plan document started for the first time from the sixth five-year plan onwards. The seventh plan (1985-1990) aimed to bring women into the mainstream of national development. It saw the establishment of the Department of Women and Child Development within the Ministry of Human Resource Development of Government of India.

The eighth plan (1992-1997) earmarked a paradigm shift, from development to empowerment of women. It highlighted a gender perspective and the need to guarantee a definite flow of funds to the developmental sectors for women. However, it was during the ninth plan (1997-2002). Women Component Plan was adopted as one of the major initiatives. The plan directed central and state governments to earmark and allot funds not less than 30% of the total for the women related sectors. National policy of empowerment of women (2001) also envisaged introduction of gender perspective in the budgeting process. Tenth plan (2002-2007) envisioned gender-just and gender-sensitive budget and established the separate Ministry of Women and Child Development. The Department of Women and Child Development (DWCD) continued to act as the Government of India's nodal agency for Gender Budgeting since 2004 and commissioned studies on gender analysis of budgets from 1993-94 to 2002-03. The eleventh Five Year plan (2007-2012) mentioned GB and gender based outcome assessment. It underlined the importance of gender audits of public expenditure, programmes and policies at national, state and district levels. The plan envisaged strengthening of the gender budget cells (GBC) established in various Ministries and Departments.The 12th Five Year Plan (2012-2017) focused to strengthen and empower the GBCs and carry out national level gender outcome assessments through spatial mapping of gender gaps and resource gaps.

3.2. Gender budgeting (GB) in India

The constitution of the Republic of India guarantees equality for both men and women. However, ground reality differs in almost all spheres. Studies confirmed that women do not enjoy equality with men economically, politically or socially. In 2004 the Ministry of Women and Child Development (MWCD) identified GB as a significant tool for women's empowerment, and as a way of addressing the pragmatic inequality. The MWCD approved *budgeting for gender equity* as a mission statement, and outlined a strategic framework of activities to implement this mission and to disseminate it to all the Ministries and Departments of

Government of India. The central government introduced gender budgeting in 2005-06 and has institutionalized it by introducing a Gender Budget Statement (GBS) ever since. Accordingly Ministry of Finance had directed all the ministries to establish GBCs by January 2005, and submit report highlighting budgetary allocations for women. Thus the first Gender Budget Statement (GBS) appeared in the Union Budget from 2005-06. The timeline depicting the evolution of the concept Gender Budgeting in India is presented as Table 1.

Table 1. Timeline depicting the evolution of the concept Gender Budgeting in India

2001	• Special mention by the Finance Minister of India in his Budget Speech.
	• Study on Gender Related Economic Policy Issues by National Institute of Public Finance and Policy (NIPFP). (Commissioned by the then DWCD).
	• The second interim report of the NIPFP (August 2001), analyzed the Union Budget 2001-02 from a gender perspective, for the first time.
2002	• Expenditure on women was elicited from the Union Budgets in the succeeding years and the broad results were reflected in the Annual Reports of the DWCD.
	A step forward in the direction of Gender Analysis of the Union Budget
2003	• In January 2003, the Cabinet Secretary to the Government of India recommended that ministries/departments should include a chapter on gender issues in their annual reports.
2004	• In December 2004, the Department of Economic Affairs of the Ministry of Finance, issued instructions to all Ministries/Departments to establish a 'Gender Budgeting Cell' by 1st January, 2005.
2005	• Every year since 2005-2006, the expenditure division of the Ministry of Finance has published a note on gender budgeting as part of the budget circular. This is compiled and recorded in the form of Statement 20 as a part of the Expenditure Budget Document Volume 1. This GB Statement comprises two parts- Part A and Part B.
	• Part A reflects Women Specific Schemes, i.e. those which have 100% allocation for women.
	• Part B reflects Pro Women Schemes, i.e. those where at least 30% of the allocation is for women.
2007	• On 8th March 2007, the Ministry of Finance published a Charter for Gender Budget Cells (GBCs), describing its functions and composition.
2010	• The Planning Commission of the Government of India, has elucidated that, Gender Responsive Budgeting or Gender Budgeting will replace the Women Component Plan
2012	• The Planning Commission of the Government of India had directed the State Finance Departments to set up Gender Budget Cells on the lines of the Charter for Gender Budget cells issued by the Ministry of Finance to hasten the process of Gender Budgeting.

Source: Ministry of Women and Child Development

Gender Budgeting is based on the modern idea that budgeting is not simply an accounting or bookkeeping exercise. Budgeting is rather a key measure in the planning and implementation process. GB initiatives facilitate coherence between planning, budgeting and gender equality goals by intervening across the planning and budgeting cycle. The Government of India has been a pioneer in adopting GB at the national, state and regional levels. In cognizance of these the budgeting process followed the laid down policies and its commitment to promote gender equality, rather than policies being determined by budget provisions. The major functions served by GB are as follows:

- Identifying the felt needs of women and reprioritizing and/or increasing expenditure to meet these needs;
- Supporting gender mainstreaming in macro economics;
- Strengthening civil society participation in economic policy making;
- Enhancing the linkages between economic and social policy outcomes;
- Tracking public expenditure against gender and development policy commitments;and
- Contributing to the attainment of the Millennium Development Goals (MDGs).

4. Objectives of gender budgeting

4.1. Gender equality

The fundamental objective of gender budgeting refers to refining budgets and related policies with a view to promoting gender equality as an integral part of human rights. Gender budgeting highlights the gender specific effects of budgets and often raises awareness of the implicit dimensions of discrimination against women.

4.2. Accountability

Gender budgets act as a mechanism for *translating government's gender equality commitments into budgetary commitments*. It is a crucial tool for monitoring gender mainstreaming activities.

4.3. Transparency and participation

Gender budgeting improves the transparency and democratising budgetary processes as well as budget policy in general through stakeholder participation. Gender-responsive budget initiatives can contribute to the increased public consultation practices in budget preparation and monitoring of outcomes and impacts, especially by ensuring that women are not excluded from the process.

4.4. Good governance

Gender inequalities results in significant losses in social cohesion, economic efficiency and human development. Budgeting with a gender perspective is an important strategy in the pursuit of equitable citizenship and a fair distribution of resources, helping to eliminate inequality and reduce poverty. Thus, gender-sensitive budgets are tools not only focussing to improve effective economic and financial management, but also the overall governance.

5. Institutional mechanisms and practices for Gender Budgeting

Gender Budget Cell is an institutional mechanism to facilitate the integration of gender analysis into the government budget. Gender Budgeting Cells (GBCs) have been set up with effect from 8 March 2007 in various Ministries with the objective of influencing and effecting changes in the respective Ministry's policies and programmes. It ensures that public resources through the Ministry's budget are allocated and managed accordingly.

6. Tracking Gender Budgeting and capacity building initiatives

When the gender budget statement (GBS) was first introduced in the union budget in 2005–06, it was called the Statement 19. This was later revised as Statement 20 in 2006–07 which continued as such till the 2016–17 budget. Since the budget of 2017–18, it came to be known as Statement 13. The GBS captures the total quantum of resources earmarked for women in a fiscal year.This is a reporting mechanism that can be used by the Ministries/Departments to review their programmes from the perspective of gender lens and is an important tool for presenting data on the allocations for women. The allocations of Gender Budget are reflected in two parts. The first part of the statement called Part A includes schemes with 100% allocation for women while Part B of the statement includes schemes/programmes with 30% to 99% allocation for women (Table 1).

6.1. Capacity building initiatives

A positive trend of organizing pre-budget consultations by the Ministry of Finance over the past couple of years are aimed to ensure the women participation in the budget process. Emphasis has been on the strengthening of key institutions and making adequate investments for the schemes that address gender concerns.While the Ministry of Finance has been instrumental in institutionalizing the gender budgeting process in Union Ministries/Departments, the Ministry of Women and Child development (MWCD) has served as the nodal agency for women that supported the process. The MWCD has been engaged in conducting a number of trainings, workshops, and one to one interactions/discussions. It has also been engaged in the development of resource material like Gender Budgeting Handbook and Gender Manuals for training the officers of Ministries and Departments.

A case in point where the Gender Budgeting concerns corrected the course of women policy in Indian context is in the implementation of MGNREA. In 2005 when the NREGA was formulated, Gender Budgeting was at a nascent stage and the Ministry of Rural Development did not have a Gender Budget Cell. Therefore the Department of Women and Child Development had to intervene to effect a gender inclusive correction to the clause which reserved the right to work to all adult individuals at the time of formulation of the Act. The intervention led to the revision in the clause that mandated for at least one-third of the beneficiaries under NREGA as women. This timely intervention would not have been warranted if the Ministry had a Gender Budget Cell at the time.

6.2. State initiatives in Gender Budgeting

Several State Governments have implemented Gender Budgeting with significant success. States such as Karnataka, Kerala, Gujarat, Rajasthan, Madhya Pradesh, Chhattisgarh and many others have taken significant steps to institutionalize Gender Budgeting to address gender gaps as presented in Table 2. However, it is of concern to note that there are many more states/ UT in India that are yet to formalise and implement the process.

Table 2. Adoption of Gender Budgeting in states/UTs in India

Early Adopters	Subsequent Adopters	Recent Adopters
Odisha (2004-05)	Madhya Pradesh (2007-08)	Rajasthan (2011)
Tripura (2005-06)	Jammu &Kashmir (2007-08)	Andaman & Nicobar Islands (2012)
Uttar Pradesh (2005)	Arunachal Pradesh (2007-08)	Punjab (2012)
Karnataka (2006-07)	Chhattisgarh (2007-08)	Maharashtra (2013)
Gujarat (2006)	Uttarakhand (2007-08)	Dadra and Nagar Haveli (2011-12)
Lakshadweep (2006-07)	Himachal Pradesh (2008)	Jharkhand (2015-16)
West Bengal (2005-06)	Assam (2008- 09)	
	Bihar (2008- 09)	
	Nagaland (2009)	
	Kerala (2010- 11)	

Source: Ministry of women and Child Development -Annual Report (2018-19)

7. Tools for Gender Budgeting

Gender Budgeting as a tool for gender mainstreaming uses the Budget as an entry point to apply a gender focus to the entire policy process. It is concerned with gender sensitive formulation of legislation, policies, plans, programmes and schemes with appropriate allocation of resources. Also it follows a gendered perspective in the implementation and execution; monitoring, review, audit and impact assessment of programmes and schemes that are followed-up with corrective action to address gender disparities. The most common tool used to implement Gender Budgeting that has been followed in many countries is the Five Step Framework (Budlender, 2002).

7.1. Five-Step Framework for Gender Budgeting

It is important to understand Gender Budgeting as a continuous process which is not just about the Budget nor as just a one-time activity. The steps followed in the popular five step framework are as follows:

Step 1: Analysis of the situation of women, men, girls and boys (and the different sub-groups) in a given segment.

Step 2: An assessment of the degree to which the sector's policy addresses the gender issues and gaps described in the first step.

Step 3: Assessment of the competence of budget allocations to implement the gender-sensitive policies and programmes identified in step 2.

Step 4: Monitoring of the entire process and ensuring the money was expended as planned.

Step 5: Assessing the impact of the policy/ programme and the extent to which the situation defined in step 1 has transformed.

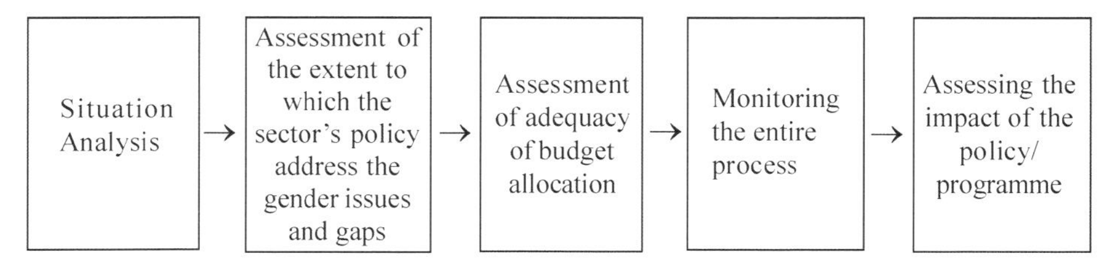

Fig. 2: Five-Step Framework for Gender Budgeting

Box-1

Different Stakeholders in Gender Budgeting

There are a range of different actors who can be involved in Gender budgeting. They have different roles and carry out different activities. Major stakeholders involved are

- Ministry of Finance (Both at the Union and State levels)
- Ministry of Women and Child Development/Social Welfare Department
- Comptroller and Auditor General of India/Local Audit Departments
- Sectoral Ministries like Health, Education, Labour, Agriculture, Power, Roadways, Urban Development etc.
- Researchers, Economists and Statisticians
- Civil society organizations
- Parliamentarians, Budget related committees of both the houses and other representatives of the people at the municipal/panchayat levels
- Development partners/donors etc.
- Media

Table 3. Allocations for women as reflected in the Gender Budget Statement 2005-06 to 2020-21

Year	No. of Ministries/ Departments	Total Magnitude of Gender-related Budget Estimates (Rs. In crores)	Percentage of the total Budget Estimates
2005-2006	9	14379	2.79%
2006-2007	18	28737	5.09%
2007-2008	27	31178	4.5%
2008-2009	27	27662	3.68%
2009-2010	28	56858	5.57%
2010-2011	28	67750	6.11%
2011-2012	29	78251	6.22%
2012-2013	29	88143	5.91%
2013-2014	30	97134	5.83%
2014-2015	36	98030	5.46%
2015-2016	35	81249	4.57%
2016-2017	36	90625	4.59%
2017-2018	31	113327	5.28%
2018-2019	36	115207	4.98%
2019-2020	30	142813	5.29%
2020-21(BE)	34	143462	4.72%

Source: Union Budget (Various years), Gender Budgeting Handbook (2015)

Gender budgeting-Challenges

In spite of the gains achieved by the Gender Budgeting Scheme, the major challenges are regular monitoring, limited enforcement and accountability, limited effective participation of women in field level planning and implementation, prevalence of socio-economic barriers and reliance on women specific schemes and unidimensional focus.These aspects, if not properly addressed will reduce the effectiveness of the scheme to serve as a tool for holistic empowerment.

Conclusion

The annual statement of receipts and expenditures do not qualify to be called a budget. It need to be an instrument of economic policy for fulfilling the social and political obligations. Hence budget reflects the priorities set by the government in allocating resources and are necessarily powerful development documents. It is in this pretext budgeting is universally recognized as a powerful tool in achieving development objectives and an indicator of commitment to the stated policy of the government. Gender Budgeting is a continuous process that is being applied to all levels and stages of the policy development and resource allocation.Gender based assessment in budget aims to render visibility and address any unwanted gender inequalities. Thus it recognises Budget as a powerful tool that can reduce the vulnerability of women and girls and transform their situation.

This will facilitate goal-oriented action and measures to close gender gaps and realise the full potential of both women and men in the economy as well as in other social spheres. Gender budgeting ensure gender mainstreaming in all ministries/departments through allocation of adequate funding coupled with institutionalising full-fledged mechanisms. It supports the principles of good governance, democratic participation, accountability and transparency, as well as effectiveness and efficiency.

References

Annual Report, 2018-19. Ministry of Women & Child Development Department (MWCD), Government of India. Retrieved from http://www.wcd.nic.in/

Budlender, D. 2002. A global assessment of gender responsive budget initiatives. In D. Budlender, D. Elson, G. Hewitt, & T. Mukhopadhyay (Eds.), Gender budgets make cents. Understanding gender responsive budgets (pp. 83–130). The Commonwealth Secretariat.

Budlender, D. 2009. Ten-country overview report. Integrating gender responsive budgeting into the aid effectiveness agenda. UNIFEM.

Dey, Joyashri and Subhabrata Dutta, 2014. "Gender Responsive Budgeting in India: Trends and Analysis." International Journal of Social Science 3 (4): 495-509.

Downes, R., Von Trapp, L., & Nicol, S. 2017. Gender budgeting in OECD countries. OECD Journal on Budgeting, 16(3): 71-107.

Gender Budgeting Handbook for Government of India Ministries and Department, 2007. Ministry of Women & Child Development Department (MWCD), Government of India.

Gender Budgeting Handbook for Government of India Ministries/Departments State Governments/District Officials Research/Practitioners, 2015. Ministry of Women and Child Development, Government of India.

Gender budgeting, 2005. Final report of the Group of specialists on gender budgeting. Directorate General of Human Rights. Council of Europe.

Goel, M M and Suman R. 2009. Gender Budgeting and Women Empowerment in India, Varta. 30. (1&2): 01-11.

Khalifa, R., & Scarparo, S. 2020. Gender Responsive Budgeting: A tool for gender equality. Critical Perspectives on Accounting, 102183.

Lahiri, A., Chakraborty, L., Bhattacharyya, P. N., Bhasin, A., & Mukhopadhyay, H. 2003. Gender budgeting in India. UNIFEM, South Asia and Ministry of Human Resource Development, Government of India, New Delhi.

Ministry of Women and Child Development, Government of India. Retrived from https://wcd.nic.in/gender-budgeting

NITI Aayog Report, 2017. India Three year Action Agenda 2017-18 to 2019-20. Government of India, New Delhi.

Planning Commission, 2012. Report of the Working Group on Women's Agency and Empowerment. XII Five Year Plan, Ministry of Women & Child Development Department (MWCD), Government of India.

Ratho A. 2020. "Gender-Responsive Budgeting in India, Bangladesh and Rwanda: A Comparison," ORF Occasional Paper No. 260. Observer Research Foundation.

Sharp, Rhonda, 2003. Budgeting for equity: Gender budget initiatives within a framework of performance oriented budgeting. New York, UNIFEM.

Sindhu, M. 2020. An Analysis of Gender Budgeting in India. International Journal of Creative Research Thoughts. 8 (2):1199-1207.

Union Budget (various years) Government of India. Accessed from https://www.indiabudget.gov.in/

19

Access and Adoption of Agricultural Technology and Interventions: A Gender Analysis

Anu Susan Sam* and Archana Raghavan Sathyan

1. Introduction

Economic growth and the commercialization of agricultural systems have enhanced the adoption of various technological interventions in the farming systems (Rola-rubzen *et al.*, 2020). The green revolution is one of the successful examples of technology interventions, which led to the agriculture transformation and thereby brought reduction in the hunger and poverty of India (Minten and Barrett, 2008). It conclusively proved that the adoption of these farm technology interventions can increase production, productivity and farm income. The technology interventions consisted of a package of improved crop varieties, irrigation, fertilizers, plant protection chemicals, and agricultural machinery.It also brought to fore that the nation's capacity to maximise its agricultural production potential also relied upon the responsiveness of farmers who were the ultimate decision makers. Thus the availability and accessibility of these technologies to the farmers, and their ability to adopt and utilise these technologies assumed paramount significance in the whole process.

Later decades of twentieth century proved that, along with men, women's role is vital in agricultural production and they are the backbone of the rural economy in many developing countries including India. It is reported that women are responsible for more than 50 per cent of the world's food production and account for about 43 per cent of the total agricultural labour force (Doss, 2018). In India, about 78 per cent of economically active women are engaged in agriculture compared to 63 per cent of men. Moreover,in rural India, almost 50 per cent of rural women are classified as agricultural labourers and 37 per cent as cultivators. According to the agriculture census 2010–11 of India, 12.78 per cent of operational land holdings are operated by women farmers. Although the role played by women is very significant in agricultural production, they tended to lag behind in

**Author contact: anu.susan@kau.in*

adopting various farm technological interventions (Satyavathi *et al.*, 2010; Huyer, 2016). They are often underestimated and unheeded in various development approaches. Gender inequality is found in the access of various agricultural resources such as modern agricultural inputs, technologies, credit and land. They are often not allowed to participate in formal activities like training, extension activities, and official meetings (Lahai *et al.*, 1999). Therefore, the empowerment of women who are engaged in agricultural activities are very important for the overall development of agricultural sector. Sustainable Development Goal 5 (SDG5) of United Nation emphasises women's empowerment and gender equality as their own right and it act as drivers of other SDGs (Yount *et al.*, 2019).

Estimates suggest that agricultural production could have increased by 20–30 per cent, if women farmers had the same accessto resources and services as men. This could lead to the increase in total agricultural output by 2.5–4 per cent in developing countries and there by reducing the number of hungry people in the world by 12–17 per cent (FAO, 2011). Additionally, the productivity and income increase of women could result in their better survival and empowerment. Moreover, the whole family could be more benefited as women spend a larger portion of their income on their children's health, nutrition and education (Mehra, 1997). In fact, the low socio-economic status especially of the rural women has direct relation with their limited availability and accessibility to various resources and improved agricultural technology interventions.These gender inequities have serious implications on agriculture, households and community.

In majority of the developing countries the agricultural systems are gendered and four important areas of gender inequality could be identified. The main gender inequality found in these countries is in the access to key agricultural resources, most critically the access to land.Women rarely own the land on which they work and often have limited decision-making power and control over how to use the land (Krishnan *et al.*, 2016). Ownership of land is an important factor that limit the access to other key agricultural inputs such as fertilizer, pesticides, machinery and infrastructure, improved seeds etc.Another area where gender inequality is visible is in the representation of women in informal and formal agricultural institutions. Men are more engaged in agricultural institutions where as women are discouraged from participating in these institutions. This has resulted in the lack of access to relevant agricultural information about the best practices, as well as the new technological interventions. Roles assigned to men and women is yet another area where gender inequality is rampant in agriculture. Though women and men are engaged in various in agricultural tasks, their roles are different. Women participation is mostly confined to practices such as weeding, harvesting, transplanting, mixing agricultural chemicals, whereas men are more engaged in operating machines, purchase of inputs, application of chemicals etc. There are even communities that bar women from sowing of seeds and related activities based on prevalent social norms.

Access to various technological interventions formed yet another area of gender inequality. Technological interventions are the key enablers that increase the agricultural labour productivity of women and there by hold the potential to reduce the gender inequality in the sector. However, prevalence of widespread gender discriminations limits the access and adoption of agricultural technology by women farmers leading to adverse economic outcomes for women (Seymour *et al.*, 2016).Most of the women farmers especially in the developing nations lack access to various agricultural technology interventions. They are less likely to obtain information on and adopt improved agricultural technologies in comparison to male farmers (Jost *et al.*, 2016). If women are given the opportunity to access to various agricultural technology interventions, they will contribute substantially in the development of the agriculture sector. In order to provide the access to various agriculture technological interventions, the areas of inequalities should be identified properly.This information can be used to mainstream gender in agricultural technology adoption and promote genderequitable outcomes. It is in line with this, the chapter aims to identify the gender inequalities in access to and adoption of various agriculture technology interventions. It also examines the various reasons for gender inequality in the adoption of various agricultural technologies.

2. Gender inequality in access to and adoption of technologies

Developments in technology have positively impacted farmers by providing better means to improve soil fertility, land productivity and thus increasing the overall agricultural output. Women farmers who are more likely to be asset poor and subsistence oriented will be benefited significantly by using farm technologies (World Bank *et al.,* 2009). The gender differences in access to and adoption of various technologies can be divided into three groups viz; (1) technologies that increase agricultural productivity; (2) technologies that help in labour saving and transport; and (3) information and communication technologies (ICTs).

2.1. Productivity improving technologies

One of the approaches for increasing agricultural incomes is to expand and strengthen access to agricultural technology interventions among farmers. Agricultural productivity can be improved in a sustainable manner by combining good agronomic practices and use of productivity-improving technologies such as improved seed varieties, fertilisers and pesticides(Farnworth *et al.*, 2017). The following section deals with the gender inequality in accessing productivity improving technologies.

2.1.1. Improved seed technologies

Improved seed technologies help millions of farmers across the globe in achieving high production targets of food self-sufficiency. Yet many farmers especially women farmers still lack access to improved seed technologies. Reports suggest that these limitations have led to significant yield gaps, even to the tune of 40 per cent and above (Lobell *et al.*, 2009). Hence to increase agriculture productivity, the adoption of modern seed varieties by all is very important (Minten and Barrett, 2008). Women have relatively low rates of adoption of modern crop varieties and this gender gap in adoption is unfavourable for women empowerment in developing countries. It has imposed real costs on societies in terms of untapped potential in agricultural output, food security, and economic growth (Ragasa and Sengupta, 2012). Improved seed technologies are expensive as they are highly dependent on synthetic and technologically-intensive inputs in order to maximize its desirable traits (Krishnan *et al.*, 2016). Due to this, it is difficult to ensure gender equalities in accessing of improved seed varieties as women are financially weak and lack access to credit facilities also.

2.1.2. Fertilizer

Inorganic fertilizers are important for sustaining and improving agricultural production, and thus contributing to income generation (Wu *et al.*, 2019). Gender differences in access to inorganic fertilizer that have important role in agricultural productivity is reported from various parts of the world (Peterman *et al.*, 2014). Across the world male farmers are anticipated to use inorganic fertilizer than are female farmers. According to Farnworth *et al.*, 2017, women farmers who use too little inorganic fertilizers attain lower crop yields and income, which affect the food and nutrition security of their households. The possible reasons for the lower usage of fertilizer by women are high cost, poor transportation network, lack of adequate knowledge and skills in using fertilizer, lack of extension service to enhance technological awareness, climatic condition, risk, the incapability to raise the necessary cash for fertilizer purchases and the lack of information regarding these fertilizers due to lower levels of education etc. (Croppenstedt *et al.*, 2013).

2.1.3. Pesticide

In agriculture, pesticides also play a major role in pest management and is used widely by both male and female farmers. But gender inequality is found in the knowledge on pesticide impacts, pesticide use, practices and protective behaviours (Wang *et al.*, 2017). In developing and underdeveloped countries women are having less knowledge on pesticide use and protective measures. This can be attributed to the dominance of males over various resources in the households. Furthermore, traditionally the male farmers are itinerant which provide more opportunities to get functional knowledge compared to the female farmers who are more confined to the household. Also better literacy rate among male farmers helps them to have better awareness of the impacts of pesticide use and associated health risks. In general, women farmers are illiterate ill-trained, poor and subsistent resulting in indiscriminate pesticide use among them (Gupta *et al.*, 2012).

2.2. Labour saving technologies

Labour saving technologies are often designated as a vital agricultural approach to achieve the economies of scale, off-set the effects of labour shortages, and profitability (Paudel *et al.*, 2020). These technologies often help to raise the productivity and reduce the drudgery of agricultural operations. Labour saving technologies consist of (1) farm mechanisation, (2) agro-processing techniques and (3) transportation facilities. Although various labour-savingtechnologies have huge potential to save time, reduce considerable burden and increase labour productivity of women in general, the use and adoption of these technologies are less among rural women (Carr and Hartl, 2010). In agriculture, compared to women, men adopt new labour saving agricultural technologies at a faster rate across regions due to their inherent advantages.

2.2.1. Farm mechanisation

Farm mechanisation is one of the important factors that contribute towards sustainable agricultural development. It helps in increasing agricultural production by reducing the agricultural cost through better management of external inputs. More significantly, it also helps to reduce the drudgery of various agricultural activities. Women mostly engaged in agricultural operations like seed preparation, digging, weeding, harvesting, separation of pods etc. use traditional tools that involve manual work. Lighter and improved hoes, seed treatment drums, serrated sickles, weeders, and planters are among some of the tools that have been tested to reduce women's work burden and health hazards. In order to overcome the high financial costs involved hiring of agricultural machinery services has become a common practice in various South Asian countries. The important time sensitive agricultural tasks like land preparation, seeding, irrigation, harvesting and post- harvesting operations can be performed effectively with the help of various machines. However, both the availability and accessibility of machines are currently dominated by men, and women have comparatively limited roles in accessing these machines.

Some of the successful farm mechanisation initiativesby Kerala state are Custom Hiring Centres (CHC) and Food Security Army (FSA). As part of mechanisation in Kerala CHCs were established in 14 districts under the State Food Security

Programme during 2008-09. CHCs can efficiently empower farmers to overcome the labour shortage and improve agricultural operation efficiency. Because of CHCs farmers need not spend on costly machines for limited use in their fields and they can get modern machines in good condition, operated by skilled operators. Machineries viz; tractor, power tiller, transplanters, combine harvesters etc. are made available to farmers at a nominal rate.

FSA was developed and implemented by Kerala Agriculture University. FSA is a team of trained/master farmers who take the challenge to revive paddy lands. FSA helps in attaining food and livelihood security and also conserves the rice ecosystem of the state (Alex, 2003). Initially this was started in Vadakkanchery block panchayat of Thrissur. The machines were pooled by the local body so that they can be used as common resources for farmer's training. The FSA functions as a service provider to those farms where the machinery operations are needed. Various trainings on use machinery for paddy cultivation and harvesting are provided to these master farmers. In FSA, women farmers are also taught to use the machinery and to maintain it.

2.2.2. Agro-processing technologies

The agro-processing industry comprisesof all operations from the stage of harvest till the produce reaches the consumer in the desired form, including processes such as packaging, quantity, quality and price. Agro processing technologies help in the conservation and handling of agricultural produce and to convert them as food, feed, fibre, fuel or industrial raw material. These techniques used for the transformation of raw food into various food products are of vital importance in food and nutritional security of a country. The agro processing activities need increased knowledge of techniques and access to specific technologies to improve product quality and quantity. Women are especially more likely to be socially and economically involved in post-harvest activities than men. However, they are often having less access to these agro processing techniques.

Kudambashree of Kerala state is an example of women entrepreneurs who are actively engaged in agro-processing sector. Kudumbashree was launched in 1999 by the State government of Kerala with the active support of National Bank for Agriculture and Rural Development (NABARD) (Sam, 2009). It is a community-based self-help initiative comprising poor women. It has been envisioned as an approach for poverty alleviation primarily focusing on micro finance and micro-enterprise development.Micro enterprises have been promoted on a large scale within the Kudumbashree network. By 2016, there were 13,829 micro enterprises in production, 5316 in services, 422 in trading, and 3922 in sales and marketing.

2.2.3. Transport technologies

Worldwide, women have lesser opportunities than men to use transport technologies (Carr and Hartl, 2010). For instance, rural Indian women have to travel by foot around 2-3 hours to reach the rice processing mills (Ragasa *et al.*, 2014). Furthermore, women often lack cash to pay for transportation fares or to purchase transporting facilities. The common means of transportation especially bicycles and motorbikes are owned more by men. In most of the rural areas, the public commute services are less frequent and more expensive. Moreover, the cultural taboos prevalent in these areas also restrict women from travelling alone in public vehicles (World Bank *et al.*, 2009; Starkey *et al.*, 2002). These suggest the urgent need for the development of gender-specific transport infrastructure or networks especially in rural areas (Carr and Hartl, 2010). Exclusive buses for women run on lady crew, women run call taxies and more reserved compartments in passenger trains that connect rural areas can provide a hassle free women inclusive commutation environment.

2.3. Information and communication technologies (ICTs)

Development in information communication technologies (ICT) offer solutions to numerous challenges in terms of production, marketing and profit faced by the traditional agricultural system. ICTs in agriculture focus on agricultural and rural development through improved information and communication processes. ICT helps in developing the capacity to generate, absorb, disseminate and protect knowledge and exploit it as a powerful tool to derive societal transformation (Meera *et al.*, 2004). It can transform farming sector and can benefit in empowering the farmers particularly the women farmers by providing better access to natural resources, improved agricultural technologies, effective production strategies, markets, banking and financial services (Lokeswari, 2016). ICTs used in agriculture are basically categorised into two types based on the technology use viz; advanced ICT tools and traditional ICT tools. Advanced tools consist of those based on internet and wireless technologies, whereas the traditional ICT tools consist of radio, satellite radio, and television. The rural areas of many developing countries lag behind in the rates of connectivity in access to both internet and mobile phones. Liff *et al.* (2004) report that ICT has made division not only between rural and urban but is also closely linked with the gender division. Generally, women have less access to ICTs and this pattern increases as the technologies become more sophisticated and expensive. Though ICT is considered as an instrument for the promotion of gender equality and the empowerment of women, a "gender divide" has also been identified as less numbers of women are accessing and using ICTs compared with men. These warrant capacity building programs in digital technologies that target women exclusively, so that they too benefit from the technological innovations that break the barriers of distance and time.

Though gender inequality is found in the access to ICTs, attempts are made to increase its access to women. Example of such an attempt is the ICT initiatives for women by M.S. Swaminathan Research Foundation (MSSRF), India (Sulaiman *et al.*, 2011). The important agricultural knowledge is transferred to the farmers through telecentres which are operated by women. They employ a combination of ICT tools to transmit relevant agricultural knowledge to farmers, including telecenters.

3. Reasons for gender inequity in the adoption of agricultural technologies

Due to the lack of access to resources, infrastructure, extension activities, and less education the agricultural technology adoption rates are lower in women farmers (Huyer, 2016). A concise representation of the various reasons for gender inequality that pose impediments inthe adoption of agricultural technologies has been depicted as Fig. 1. An in depth analysis of the various factors that cause gender inequalities in the access and adoption of various technological interventions in Indiais attempted here.

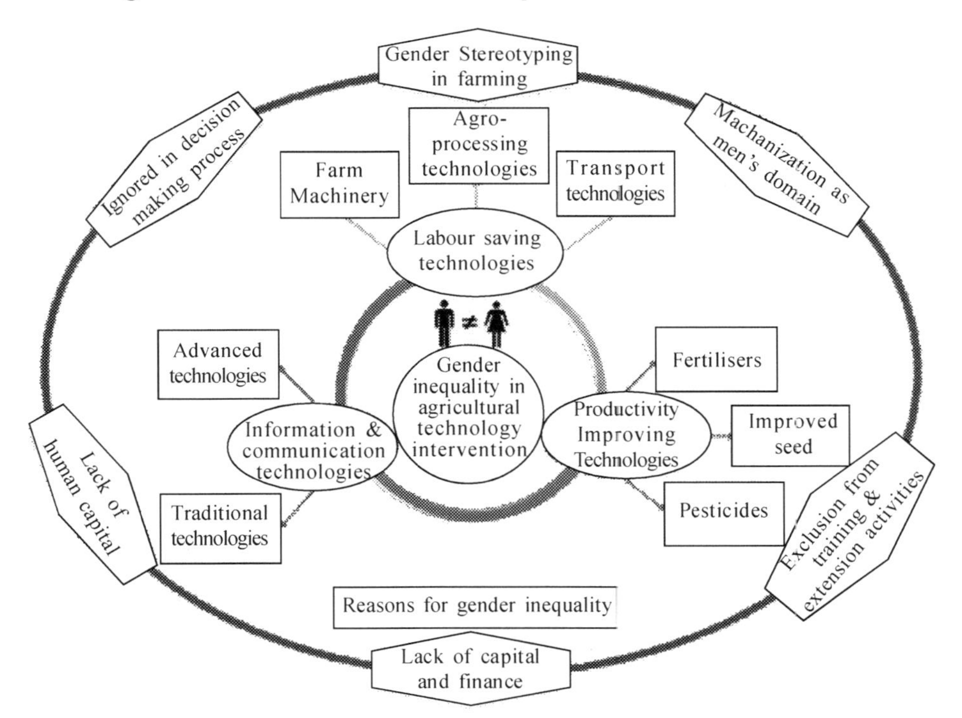

Figure 1. Dimensions of gender inequality in agricultural technological interventions (Author compilation)

3.1. Gender stereotyping in farming: Men as farmers and women as homemakers

In many developing and under developed countries, especially in rural areas the role of men and women are distinguished by prevailing social norms that are skewed against women. Generally, men are considered as the household head and they are expected to work outside their homes. Traditionally, women farmers simultaneously undertook multi-functional rolesthat covered domestic chores, child and family care activities and farm work (Pogoy *et al.*, 2016). However, among these, the more socially acknowledged role has been of a mother and a homemaker. In rural areas, women's status has often been seen as dependent on their roles as mothers and wives, or their confinement to the domestic realm, whereas men are engaged in vocational activities (Rao, 2012). In addition to the role as home maker, rural women also work in farm but women farmers were rarely considered as income contributors to the family. As such, training and extension activities on new farm technologies are often packaged to benefit men. Moreover, the gender of extension agents may hamper the interaction of them with the female farmers in rural India due to social restrictions imposed by cultural taboos. Despite these, there has been ardent efforts to target women in farming through agriculture policy initiatives and exclusive programs for farm women empowerment in recent years. Many successful programs have been reported from across the country that include Poverty Alleviation Mission (Kudumbhasree Mission), Kerala, Mission Shakti ofOdisha and Mahila Sakhtikaran Krishi Yojana by the GOI etc.

3.2. Farm mechanisation considered as men's domain

Mechanisation has often been viewed as an important strategy to raise the productivity and reduce the drudgery of agricultural operations (Fischer *et al.*, 2018). The shortage of agricultural laborers, primarily due to male labour out-migration, has been the driver of farm mechanisation (Paudel *et al.*, 2020). However, most technologies, including farm mechanisation, are traditionally considered the domain of men, and thus women are less involved in technology selection and adoption processes (Devkota *et al.*, 2015). It has been reported that farm mechanisation by women farmers can help in attaining food security of household and it will also help in the revival of agricultural production systems (Alex, 2013). Women farmers who are the household heads seem to be the most disadvantaged in the case of farm mechanisation. Similarly, while the work of women in agriculture is traditionally not as physically demanding as men's tasks, women are often relegated to drudgery.

Mechanisation is considered as an excellent means to increase the productivity of farmers by reducing drudgery. Moreover, it is very necessary to consider ergonomic, comfort and safety aspects while designing farm machinery including hand tools. In a study, conducted by Sam *et al.* (2019), they assessed the safety and efficiency aspects of five coconut climbing devices: sit and climb type (TNAU model), standing type (Chemberi model), KAU coconut palm climber, Kerasureksha and CPCRI model: on ergonomic basis for women climbers. Female climbers showed differed heart rate while operating these five models of coconut climbing devices with a significant difference. Among these, minimum energy expenditure and the oxygen consumption with respect to maximum aerobic capacity were observed for KAU coconut palm climber when compared to other models (Sam *et al.*, 2019). Introduction to new technologies in agricultural operation adopted by farm women leading to the mechanisation will reduce thedrudgery and improve the efficiency (Table 1).

Table 1. Women-friendly improved farm tools and equipment and their advantages

Operation	Improved farm tools and equipment	Advantage over traditional
Ridge making	Hand ridger	It improves the bending posture of the women
Seed treatment	Seed treatment drum	It eludes the bending/squatting posture
Fertilizer	Fertilizer broadcaster	It helps to apply the fertilizer at a uniform rate in a larger area. It also prevents labourers from the direct exposure tofertilizer.
Sowing/Planting	PAU seed drill TNAU paddy drum seeder CRRI rice transplanter	They improve the bending posture that is generally followed in the traditional method. Line sowing can be done in large area and due to this mechanical weeders can be used for weeding.
Weeding	Twin wheel hoe	It avoids the bending/squatting posture and covers more area.
Harvesting	Improved sickle	No need to sharpen the cutting edge thus provides safety to women farmers.

Source: Singh *et al.* (2006)

3.3. Women's role is ignored in decision making process

Low adoption of agricultural technology by women is due to the fact that their needs are often bypassed in the decision-making process. Due to various socio-economic conditions that exist in many patriarchal societies'women are kept away from decision-making process. Despite the multiple roles in which women are engaged such as farm producers, wage earners, care takers of the family, looking after nutrition and manage post-harvestoperations, they are yet to play any decisive role in matters of technology use and adoption. Though women contribute much in on farm and off farm activities, it is unfortunate that in many agrarian societies' women are assigned the exclusive role of a 'worker' rather than a 'decision maker'. Women's contribution to the farming sector in respect of operation and decision making has largely been ignored. Although, women's participation in the decision-making process has a significant impact on their improved status and greater role in society, their involvement in decision making process is low.

3.4. Women's lack of human capital

Development of human capital plays a significant role in the progress of the country as well as in the empowerment of its people. Education is one of the most important means of empowering women with the knowledge, skills, and self-confidence necessary to participate fully in the development process. Education is a practical and economical approach for attaining increased agricultural productivity. Though women in India have made substantial progress since independence, many are still having limited access to education. According to the government of India census 2011, male literacy rate is 82.1% while the literacy rate among females is 65.5%. (Government of India, 2011).Women in the rural areas are not allowed to have education as they are meant to do household chores, which will help them after marriage. Deficits in women's education areconsidered as a key factor constraining women's adoption of new technologies (Fischer *et al.*, 2018).

3.5. Exclusion of women from training and extension activities

In most of the developing and under developed countries, women are actively involved in agriculture and they require assistance to improve farming practices by gaining knowledge on various aspects like technological intervention, modern agriculture practices, marketing activities etc. The gender gap separates the women farmers from accessing the basic information they requirement to enhance production, efficiency, and income. Agricultural research, development, and extension contribute to the development and adoption of new agricultural practices and technologies (Peterman *et al.*, 2010). In addition to that, women are excluded from trainings and agricultural extension services and these activities are traditionally focussed on men (Jiggins *et al.*, 2011). Extension activities are carried out through agricultural organizations and NGOs, but women have lower rates of membership in these organizations than men as they have less access to social capital (Peterman *et al.*, 2010). Though many efforts are taken to include rural farm women in various trainings that are related to agriculture, many of them are able to participate in such trainings and extension activities. Hence, many times women miss out on the new technology and fail to benefit from the new knowledge and skills.

3.6. Women's lack of access to land and finance

Women farmers access to land and capital are seen to be restricted in many developing and underdeveloped countries.In the study of United Nations (2014) they reported that hardly 20 per cent of rural women own the piece of land that they cultivate. Access to land plays a significant role in agriculture productivity or in attaining sustainable resource management. Access to land is a prerequisite for attaining financial aid from various lending institutions. Lack of access to land remains as a major obstacle in the empowerment of women especially economic empowerment. When women have access to land agricultural production and food security also increase. The lack of access to finance along with lack of access to land restricts women from accessing better technology intervention and there by achieving higher yields.

4. Conclusion

The rate of adoption of improved technologies by women farmers is far less than men farmers. In order to promote gender equality and equity in the adoption of various technologies, it is crucial to understand what women want and what their needs are. Along with that it is important to find out the reasons behind this gender inequality so that measures can be taken for reducing this inequality. A careful consideration of gendered needs and implications are essential for increasing the benefits of improved technologies to more people especially women. In order to actualise this, attention should be given from the design of technology till the assessment of its impact on agricultural productivity.

Technology should be designed is in such a way that it should be apt and important for people for whom it is originally target. Female farmers who own land and resources have better chance of adoption of farm technologies. However, there need to be provisions to include women without access to land and other resources also such as custom hire centers and legally tested leased land provisions. Value addition, post-harvest technologies and other avenues that are less land intensive can be popularised to provide them with alternative sources of income. These bring to fore the need to design farm technologies that are important to local needs, requiring minimal energy, reducing drudgery, and allowing management and operation by women. Such technologies can not only help to reduce the workload of women but also prevent displacement ofwomen from their regular employment.

Knowledge and skill owned by women can influence the adoption of various technologies. As such access to education and extension activities serve as cardinal inputs to improve the rate of adoption of various technological interventions. Extension programs should be designed by considering farm women's restraints and needs. Training need of women on technological knowhow, entrepreneurial skills, strategic decision making etc.should form the base of all capacity building initiatives that target women in agriculture (Satyavathi *et al.*, 2010). Reports from different parts across the globe indicate that the extension programs that promote women participation can further empower them and provide better access to human capital, education, information, and resources that are required for the adoption of agricultural technologies (Klugman *et al.*, 2014). More inclusive technology adoption can be achieved by these collective efforts.

References

Alex, J.P. 2013. Powering the Women in Agriculture: Lessons on Women Led Farm Mechanisation in South India. The Journal of Agricultural Education and Extension, 19(5): 487–503. doi.org/10.1080/1389224X.2013.817342

Carr, M., & Hartl, M. 2010. Lightening the load. Labour-saving technologies and practices for rural women. International Fund for Agricultural Development (IFAD) and Practical Action Ltd, UK. Retrieved from http://www.ncbi.nlm.nih.gov/pubmed/23325796

Croppenstedt, A., Goldstein, M. & Rosas, N. 2013. Gender and Agriculture: Inefficiencies, Segregation, and Low Productivity Traps. World Bank Research Observer, 28(1), 79–109. doi.org/10.1093/wbro/lks024

Devkota R., Khadka, K., Gartaula, H., Shrestha, A.,Karki, S., Patel, K., & Chaudhary P., 2015. Gender and labour efficiency in finger millet production in Nepal. A.K. Jemimah Njuki, John R. Parkins (Eds.), Transforming Gender and Food Security in the Global South. Earthscan from Routledge, London and New York, pp. 100-119

Doss, C.R. 2018. Women and agricultural productivity: Reframing the Issues. Development Policy Review, 36(1): 35–50. doi.org/10.1111/dpr.12243

FAO, 2011. FAO policy on gender equality: Attaining food security goals in agriculture and rural development. FAO, Rome.

Farnworth, C. R., Stirling, C., B. Sapkota, T., Jat, M. L., Misiko, M., & Attwood, S. 2017. Gender and inorganic nitrogen: what are the implications of moving towards a more balanced use of nitrogen fertilizer in the tropics? International Journal of Agricultural Sustainability, 15(2): 136–152. doi.org/10.1080/14735903.2017.1295343

Fischer, G., Wittich, S., Malima, G., Sikumba, G., Lukuyu, B., Ngunga, D., & Rugalabam, J. 2018. Gender and mechanization: Exploring the sustainability of mechanized forage chopping in Tanzania. Journal of Rural Studies, 64, 112–122. doi.org/10.1016/j.jrurstud.2018.09.012

Gupta Chetna, D., Gupta Vaibhav, K., Pallavi, N., & Patel Jitendra, R. 2012. Gender differences in knowledge, attitude and practices regarding the pesticide use among farm workers: A questionnaire-based study. Research Journal of Pharmaceutical, Biological and Chemical Sciences, 3(3): 632–639.

Government of India, 2011. Census of India Primary Census Abstract e Odisha e2011. New Delhi: Government of India.

Huyer, S. 2016. Closing the Gender Gap in Agriculture. Gender, Technology and Development, 20(2): 105–116. doi.org/10.1177/0971852416643872

Jiggins, Janice R.K. Samanta and Janice E.O., Awoye, 2011. "Improving Women's Access to Extension Services". Improving Agricultural Extension: A conference Manual. Retrieved November 21,2020 from http://www.fao.org/3/w5830e0b.htm

Jost, C., Kyazze, F., Naab, J., Neelormi, S., Kinyangi, J., Zougmore, R., Kristjanson, P. 2016. Understanding gender dimensions of agriculture and climate change in smallholder farming communities. Climate and Development, 8(2), 133–144. doi.org/10.1080/17565529.2015.1050978

Klugman, J., Hanmer, L., Twigg, S., Hasan, T., McCleary-Sills, J.,& Santamaria. J. 2014. Voice and Agency: Empowering Women and Girls for Shared Prosperity. Washington DC: The World Bank.

Krishnan, P., Raridon, A., Raymond, L., & Subramaniam, M. 2016. Review of the Gender and Social Impacts of Improved Seed Technology in Developing Countries: Policy Implications. Purdue Policy Research Institute (PPRI) Policy Briefs: Vol. 3 (1), Article 1. Retrieved from https://docs.lib.purdue.edu/gpripb/vol3/iss1/1

Lahai, B.A.N., Goldey, P. & Jones, G. E. 1999. The gender of the extension agent and farmers' access to and participation in agricultural extension in Nigeria. The Journal of Agricultural Education and Extension, 6(4): 223–233. doi.org/10.1080/13892240085300051

Liff, S., Shepherd, A., Wajcman, J., Rice, R., & Hargittai, E. 2004. An Evolving Gender Digital Divide? OII Internet Issue Brief No. 2, Retrieved from https://ssrn.com/abstract=1308492 or http://dx.doi.org/10.2139/ssrn.1308492SSRN

Lobell, D.B., Cassman, K.G., & Field, C.B. 2009. Crop Yield Gaps: Their Importance, Magnitudes, and Causes. Annual Review of Environment and Resources, 34(1), 179–204. https://doi.org/10.1146/annurev.environ.041008.093740

Lokeswari K. 2016. A Study of the Use of ICT among Rural Farmers, International Journal of Communication Research, 6(3): 232-238.

Meera, S.N., Jhamtani, A., & Rao, D.U.M. 2004. Information and Communication Technology in Agricultural Development: A Comparative Analysis of Three Projects from India. Retrieved from http://hdl.handle.net/10535/4915

Mehra, R. 1997. Women, Empowerment, and Economic Development. The ANNALS of the American Academy of Political and Social Science, 554(1), 136–149. doi.org/10.1177/0002716297554001009

Minten, B., & Barrett, C.B. 2008. Agricultural Technology, Productivity, and Poverty in Madagascar. World Development, 36(5): 797–822. doi.org/10.1016/j.worlddev. 2007.05.004

Paudel, G.P., Gartaula, H., Rahut, D.B., & Craufurd, P. 2020. Gender differentiated small-scale farm mechanization in Nepal hills: An application of exogenous switching treatment regression. Technology in Society, 61. https://doi.org/10.1016/j.techsoc.2020.101250

Peterman, A., Behrman, J. A, & Quisumbing, A.R. 2010. A review of empirical evidence on gender differences in nonland agricultural inputs, technology, and services in developing countries. IFPRI Discussion Paper 00975. Retrieved from http://ebrary.ifpri.org/utils/getfile/collection/p15738coll2/id/1464/filename/1465.pdf

Peterman, A., Behrman, J.A., & Quisumbing, A.R. 2014. A Review of Empirical Evidence on Gender Differences in Nonland Agricultural Inputs, Technology, and Services in Developing Countries. In Gender in Agriculture (pp. 145–186). doi.org/10.1007/978-94-017-8616-4_7

Pogoy, A.M., Montalbo, I.C., Pañares, Z.A. & Vasquez, B.A. 2016. Role of Women Farmers in Improving Family Living Standard. International Journal of Gender and Women's Studies, 4(1). doi.org/10.15640/ijgws.v4n1a6

Ragasa, C., Sengupta, D., Osorio, M., OurabahHaddad, N., & Mathieson K. 2014. Gender-specific Approaches, Rural Institutions and Technological Innovations. FAO Rome. Retrieved from http://www.fao.org/3/a-i4355e.pdf

Ragasa, C., and Sengupta, D. 2012. Gender and Institutional Dimensions of Agricultural Technology Adoption: A Review of Literature and Synthesis of 35 Case Studies. International Association of Agricultural Economists (IAAE) Triennial Conference. Retrieved from http://ageconsearch.umn.edu/bitstream/126747/2/IAAE.2012. gender.pdf.

Rao, N. 2012. Male "providers" and female "housewives": A Gendered co-performance in rural North India. Development and Change, 43(5), 1025–1048. https://doi.org/10.1111/j.1467-7660.2012.01789.x

Rola-Rubzen, M.F., Paris, T., Hawkins, J. & Sapkota, B. 2020. Improving Gender Participation in Agricultural Technology Adoption in Asia: From Rhetoric to Practical Action. Applied Economic Perspectives and Policy, 42(1): 113–125. doi.org/10.1002/aepp.13011

Sam, A.S. 2009. Household food security and women empowerment: the impacts of Kudumbashree in Kerala state of India. M.Sc thesis submitted at University of Gent, Belgium.

Sam, B., Vahab, H.B. & Regeena, S. 2019. Assessment of ergonomic parameters of coconut climbing devices for women. Current Science, 116(1): 127-133.

Satyavathi, C.T., Bharadwaj, C. & Brahmanand, P.S. 2010. Role of Farm Women in Agriculture. Gender, Technology and Development, 14(3): 441-449. doi.org/10.1177/097185241001400308

Seymour, G., Doss, C., Marenya, P., Meinzen-Dick, R., & Passarelli, S. 2016. Women's Empowerment and the Adoption of Improved Maize Varieties: Evidence from Ethiopia, Kenya, and Tanzania. Agricultural & Applied Economics Association Annual Meeting, 1–30. Boston. Retrieved from http://purl.umn.edu/236164

Singh, S.P., Gite, L.P.J. & Agarwal, N. 2006. Improved Farm Tools and Equipment for Women Workers for Increased Productivity and Reduced Drudgery. Gender, Technology and Development, 10(2): 229–244. https://doi.org/10.1177/097185240601000204

Starkey, P., Ellis, S., Hine, J., & Ternell, A. 2002. Improving Rural Mobility. In World Bank Technical Papers. Retrieved from doi.org/doi:10.1596/0-8213-5185-0

Sulaiman, V.R., Kalaivani, N.J., Mittal, N. & Ramasundaram, P. 2011. ICTs and Empowerment of Indian Rural Women What can we learn from on-going initiatives? CRISP Working Paper 2011-001, available at https://www.empowerwomen.org/en/resources/documents/2013/11/ict-and-empowerment-of-indian-rural-women?lang=en

United Nations, 2014. Improving Access to Finance for the Empowerment of Rural Women in North Africa: Good Practices and Lessons Learned. UN Economic Commission for Africa, Office for North Africa.

Wang, W., Jin, J., He, R., & Gong, H. 2017. Gender differences in pesticide use knowledge, risk awareness and practices in Chinese farmers. Science of the Total Environment, 590–591, 22–28. https://doi.org/10.1016/j.scitotenv.2017.03.053

World Bank, FAO, & IFAD, 2009. Gender in Agriculture Sourcebook.The World Bank, Food and Agriculture Organization, and International Fund for Agricultural Development. Washington, DC. Retrieved from http://siteresources.worldbank.org/INTGENAGRLIVSOUBOOK/Resources/ CompleteBook.pdf.

Wu, Y., Wang, E., & Miao, C. 2019. Fertilizer use in China: The role of agricultural support policies. Sustainability (Switzerland), 11(16). https://doi.org/10.3390/su11164391

Yount, K.M., Cheong, Y.F., Maxwell, L., Heckert, J., Martinez, E.M. & Seymour, G. 2019. Measurement properties of the project-level Women's Empowerment in Agriculture Index. World Development, 124: 104639. doi.org/10.1016/j.worlddev.2019.104639.

20

Extension Toolkits for Gender Responsive Value Chain Development

Aparna Radhakrishnan and Allan Thomas*

The Context

The global food system is dynamic, with newer challenges every day with altering production and consumption subtleties. Roles played by various stakeholders of the global agricultural food system and rural development has changed tremendously in the last few decades. Smallholder farmer stakeholders to multinational companies are affected differently by the present difficulties likethe pandemic, climate change, and other factors. The various stakeholders respond, play roles to the challenges differently, and eventually become the agricultural value chains. The newer challenges affect the vulnerable actors like women, smallholder farmers, etc., who constitute asignificant portion of the value chains. This is specifically so in developing countries like India, where the processes of the 'feminization of agriculture' progresses exponentially. Managing sucha value chain is highly challenging as access and sustainability of various actors differ considerably. Agro-advisory service toolkit for gender-responsive value chain analysis and development aims to select and analyze value chains for opportunities to improve the vulnerable situation and its actors seeking for resilience and reducing gender inequalities. It is now considered a development strategy to promote economic growth. There is an urgent need that thesmallholder farming systems are integrated into the value chains to serve local, national, regional, and global markets.

The chapter presents a brief overview of the gender-responsive value chain concept, factors that limit women's participation in the value chain, interventions to improve the efficiency of value chains in a gendered perspective, and the development of value chain tools for addressing gender inequitiesin the agricultural system, and the challenges faced.

**Author contact: aparna.r@kau.in*

1. The value chain concept

The concept of value chain has gained popularity concerning poverty reduction. A "value chain" describes the complete series of activities required to bring a product or service from conception, through the diverse phases of production (involving a combination of physical transformation and the input of various producer services), distribution to final consumers and final disposal after use (Kaplinsky and Morris 2001, Fonseca *et al.*, 2019). The value chain as a concept defines the full range of activities that firms, farms, and workers do to take a product from its conception to its end use and beyond (Riisgaard *et al.*, 2010, Fonseca *et al.*, 2019). Value chain analysis (VCA) is used to explain market relationships, coordinate the delivery of inputs, improve information flows, and monitor products' quantity and quality. It provides an analytical dimension to the actors involved and provides precision in the interactions, adding value to each units of the production-supply cycle. Studies using the global value chain approach observe different types of value chain governance and the opportunities they provide for technological or functional upgrading of traders and farmers in developing countries (Daviron and Ponte, 2005; Dolan and Humphrey, 2000; Fold, 2002; Muradian and Pelupessy, 2005; Mather and Greenberg, 2003; Poulton *et al.,* 2004, Fonseca *et al.*, 2019) making the work done visible.Rather than the conventional value chains, that rely only on economic aspects, the modern theories analyses the chains from the sustainability angle. Researchers use VCA to understand why particular countries and particular types of enterprise find it challenging to enter specific sectors, why many of the potential benefits of globalization fail to reach the very poor and recognize the implications for value chain development (Mayoux and Mackie, 2008; Fonseca *et al.*, 2019).

Value chains may be men, women-centric, or bothand are called Gender Value Chain (GVC). Globally, women comprise 43% of the total workforce and are involved with 70% of the work in agriculture (Doss and Sofa Team 2007, Fonseca *et al.*, 2019). Thus women-based value chains have immense importance in the agricultural system. Giving priority to men and women, the term gender, which is an overarching, crosscutting sociocultural variable that can be extrapolated to any other variable in sociocultural contexts,has been clubbed with the value chains variable. FAO proposes gender-sensitive value chain mapping as a first step towards making women's work and participation in the value chain visible, including identification of gender-based constraints at each node of the chain (Thitiya *et al.*, 2014; Fonseca *et al.*, 2019).

The chapter delivers a summary of the key lessons emerging from the literature based on the research question: "What are the agricultural extension and advisory services toolkits developed to meet the needs of the present composite and dynamic agricultural setting using a gender-based value chain approach?". It describes how factors such as societal challenges, gendered education differentials, and access and control, and competition in value chain affect men and women who participate and gain in value chains, distinguishing among household, institutional and chain levels of analysis. Any VCA approach should address issues that assist in adaptation planning, analysis of vulnerabilities, hotspots, and risks across a value chain, help women communities adapt, and help develop partnerships for implementing the strategies in which there is mutual benefit. Without understanding the real issues, interventions cannot be made to directly impact/ strengthen the value chain to reduce gender inequalities. Case by case analysis of inequalities and thorough analysis of causal factors are essential to bridge the inequality gap.

2. Tools used for gender value chain analysis

Several evaluations were identified in literature that seek to analyze and address issues of value chain in the women angle (Riisgaard *et al.*, 2010) through mapping gender roles and relations. Different factors influence the opportunity of both men and women in value chains including physical, financial, and human assets and also the benefits depend on how one participates (FAO, 2018). Analysis of gender inequalities and constrains gives an overview of these factors. Factors such as gender division of labor, distribution of benefits, gender dynamic and power relation in the value chain also determine who benefits and how the services are accessed and distributed (Kruijssen *et al.*, 2018). Mainly the evaluation compares individual analytical approaches and learning into action-oriented interventions. Thus, the field practitioners will get some tools they can use while working with different actors along the chain (Rubin and Manfre, 2012). sus group approaches to implementing new technology, facilitating the creation of women-only producer groups, or supporting female entrepreneurs (Riisgaard *et al.*, 2010).

What does a value chain analysis consider? (Life Academy of Vocational Study, 2003)

- What are the financial costs incurred throughout the value chain?
- Where is the most value-added to the value chain?
- Who are the most import actors within the value chain?
- What is the institutional framework of the value chain?
- Where are the bottlenecks in the value chain?
- Where exists the market potential for growth?
- How long is a value chain/ size of a chain?

- What is the possibility for upgrading the value chain?
- What possible synergies exist?

The primary phase of any GVC analysis is the extensive review of literature meant to map the value chain in the region and to understand the different actors at each node of the chain, as well as the linkages between them. This also helps to understand the significant constraints and challenges they are facing in the whole value chain. It also encompassed as an initial attempt for the documentation of women's situations and roles along the chain. The literature review identifies the following tools that could be used for the GVC analysis.

2.1. Tool: Key informant interview on gender aspects of value chains

Key informant interview is one of the most popular and standard method. According to the respondent, checklist, open/closed ended questions can be used. The value chains can be mapped based on the perception and knowledge of the key informants. It also helps identify major actors of the value chain and gives an overall idea of the chain. The key informants may be village heads, researchers, local leaders, implementing partners, state agricultural officers,university scientists, andother actors directly or indirectly involved in agricultural value chains and familiarwithmultiplechain stages. The questions should be in the local language, and it should be simple and straight forward. Sufficient time should be given to understand, comprehend the question and answer. The market trend over the past decade should be asked along with future predictions and key drivers will give an idea about the sustainability and risks involved in the value chain.

2.2. Tool: Explorative rapid market chain assessment

In this tool, primarily the selection of market value chains is done, potential partners are identified and appraisal is carried out. The method helps us get an evidence-based 'first look' into a set of market systems, which are under-performing, and to see quickly why they are not functioning properly (ILO, 2019). The appraisal will conclude in such a way, wheremarket chain actors (farmers, traders, processors, and market agents), meet the leading institution that presents the results of the market chain appraisal and shares actors' different expectations, constraints and capabilities, and facilitates discussion on possible innovations.

2.3. Tool: Indicators based approach to measure empowerment in agricultural value chains

Indicators will be selected that cover aspect of both men and women. Qualitative, quantitative, or both could be used to validate the indicators. Once validation is over, the survey method could measure the indicators, obtaining data from individual households. Equal weightage or weightage based on factor analysis could be used for the data analysis. Data analysis gives the quantitative measure of the index. For example, WEAI (Women Empowerment in Agriculture Index), a survey-based tool to measure women's and men's empowerment and inclusion in agricultural development projects (Malapit *et al.*, 2019), focuses on aspects of empowerment relevant to value chains.

2.4. Tool: Non-Parametric Oaxaca Blinder Decomposition Analysis

The Blinder-Oaxaca decomposition is a statistical method that explains the difference in the means of a dependent variable between two groups by decomposing the gap into that part that is due to differences in the mean values of the independent variable within the groups, on the one hand, and group differences in the effects of the independent variable, on the other hand (Borjas, 2000). In addition, labor and gender economics literature since the mid-2000s has found that women are often in the lowest economic percentiles of income dis-tributions and face barriers in access to income-producing opportunities (Atal *et al.*, 2009; Nopo *et al.*, 2011; World Bank, 2012) the case is same with the women in agriculture. The objective of this tool is to calculate gender wage gaps in agriculture and correlate to other variables like caste, demography, age etc.

2.5. Tool: Time-Use Analysis

This tool provides a quantitative analysis of the time spend per day or per agricultural activity by both men and women. Measuring men's and women's labor burdens could provide interesting insights into how to improve gender balance and labor opportunities for both men and women. The variables on the major productive activities of women are identified and their role in value chain also found out. Accordingly the time spent has to be measured. Many of the studies finding increased workload for women have relied mostly on qualitative information (Lyon *et al.*, 2009; Bolwig and Odeke, 2007).

2.6. Tool: Occupational Segregation Using Duncan Index

Women continue to congregate in sectors and occupations traditionally characterized as "female" that are mostly low-paying jobs. According to the World Bank (2012), removing barriers that prevent women from working in certain occupa-tions would reduce the productivity gap between male and female workers by one-third to one-half, and would increase output per worker by 3–25 percent in some countries. This method is usually used to find gender inequality. The tool estimates gender segregation at each node in the value chain by occupation and can be extended to capture hierarchical segregation by occupation and task (skilled versus non-skilled) depending on available data. Bootstrap techniques could be used to check the impact of classification errors and aggregation on the measurement of occupational segregation.

2.7. Tool: Working Conditions / Access to Work Equality Index

The index provides a quantitative assessment of access to workplace equality. The index is based on three premises: (1) measurement of gender gaps, (2) ease of computation, and (3) a final value bound between 0 (inequality) and 1 (equality) to facilitate comparisons and interpretation. It has two categories: (1) variables that characterize working conditions and (2) variables that describe access to work.

2.8. Tool: Gender-sensitive value chain map

This tool is highly essential as the large part of the gender value chain that requires upgrading is often ignored. The first step in this tool is to mark the chain's main functions and product flow with arrows from left to the right. After that the main segments and interlinkages has to be depicted in the chain. Each production/market segments have to be marked as circles underneath the relevant arrow. The main stakeholders under each market segmenthas to be marked along with marking the support services in the extended value chain. Following this, mark the preliminary information like number of people involved, power relationships and the importance of chain. The factors of enabling environment that facilitate or hinder women's participation or benefits has to be depicted along with the main blockage of upgrading, poorest people involved etc. Finally mark the female stakeholders in a different color showing the power relation involved (Mayoux and Mackie, 2008; Agri-ProFocus, 2014; FAO, 2015; FAO. 2016).

3. Case Study

Case I: Livelihood analysis and value chain mapping of the tribal SHG, Sabari Swasraya Sangam, Nellarachal, Wayanad.

Applying the gender-sensitive value chain map tool, livelihood analysis tool, we analyzed the agricultural value chain that involves biocontrol agent production and mapped women's participation in the differentnodes and functions of the chain. The study examines the economic and social benefits of the SHGs programme implemented by the Krishi Vigyan Kendra, Wayanadduring the year 2004. Eight members of the Sabari SwasrayaSanghom of Nellarachal tribal hamlet were trained inthe application of biotechnology by the Krishi Vigyan Kendra (KVK) under the Kerala Agricultural University (KAU) at Ambalavayal a decade ago.The Self-help groups (SHGs) aim to provide the rural poor, especially women, with savings, credit and insurance, and improve household income security. The fundamental idea behind forming SHGs by KVK Wayanad is to produce sustainable institutions focusing on agriculture and allied activities. It enabled in creating a widespread group of community resource persons for scaling up livelihood interventions generating socio-economic and technical empowerment small, marginal, landless, and socially backward women. The analysis made it clear that as good agriculture practices are gaining momentum in India, this SHG in the Wayanad district of Kerala is creating a success story in production of 13 different varieties of bioagents to support organic farming. Study results indicate that the social and financial assets component of the SHG members has tremendously increased along with the substantial contribution to KVK's income per year.In the biocontrol value chain,the women play a key role in production, processing and marketing aspects that substantially contributed to the enhancement of livelihoods.

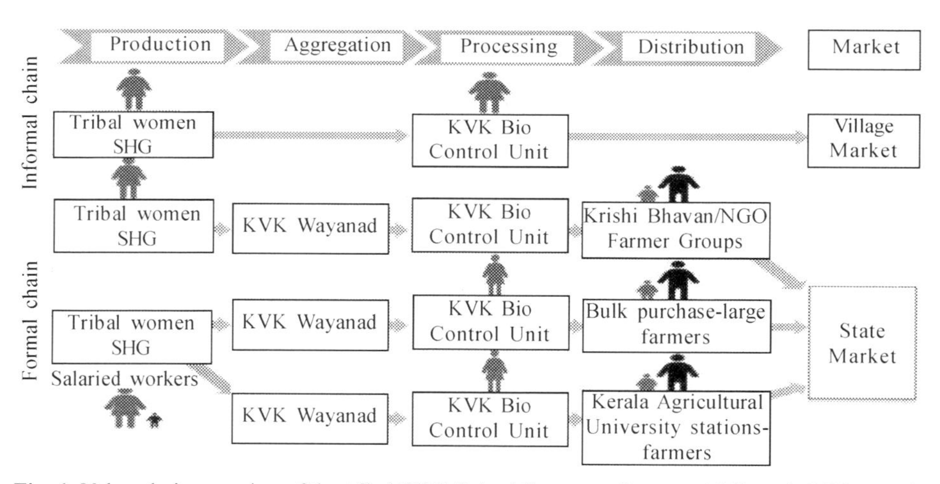

Fig. 1. Value chain mapping of the tribal SHG, Sabari Swasraya Sangam, Nellarachal, Wayanad

Value chain relationships refers to the interaction between the SHG members and with the KVK horizontally and vertically within the chain (Figure 1). External actor competition is the major threar faced by the SHG value chain now. Building trust among the members is important to allow chains to function effectively and sustainably.

4. Factors that limit the participation of women in value chains

Women and men play different roles in value chain like agricultural laborer, performer of agronomic practices, crop protection practices, intercultural operations etc. Women specifically play a major role in production and post-harvest processing which are the key determinants in assessing the final quality of produce (Hoffmann and Roscoe, 2016). Women are more vulnerable and powerless because they have unequal access to information and technologies, they have to pay exorbitant interest to moneylenders and they often do not own their tools of production. Studies also shows that level of education, age of women, land inheritance, distance from house to market Centers, and community responsibilities affect the women participation in value chains (Ola, 2020). Some of the main factors that limit women, as well as men, from benefiting in agriculture value chains are:

4.1. Lack of access to land

Access to land is a fundamental issue because it is a critical asset for food production, a significant determinant of income-earning power, and a major aspect for shelter and community development. It is well documented that women's control over and ownership of land is less than that of men. Both laws and customs often show preference to men owning land over women. Not having land also limits the access of women to many other resources and services, such as producer associations and contract farming opportunities only available to those with land.

4.2. Lack of access to credit

Although there has been much work to improve women's access to microcredit over the years, there are still barriers excluding them from formal credit markets and larger loans required to support larger-scale commercial production and processing. Women's access to credit is constrained by inadequate mobility, low literacy, the lack of assets for collateral, since women seldom have legal ownership of land.

4.3. Social norms

Social norms are unwritten rules that are socially acceptable forms of behavior. In many cultures, the head of the household, which is often the man, is considered the farmer and has access to contracts and agriculture extension despite being heavily involved in farming operations. Thus the participation in value chains gets limited for women.

4.4. Social capital

Social capital comprises the range of relationships, networks, and institutions that allow people to build trust and cooperation (Meinzen et.al, 2014). Across the whole value chain, women face limited access to information, hired labor, technology, assets, and networks (Hoffmann and Roscoe, 2016) thus the social capital component is weak that limit the participation of women in value chains.

4.5. Marketing issues

The increasing awareness of consumers along with market preferences on the food quality standards has aggravated the gender gap in market participation (Asfaw *et al.*, 2010; Oduol *et al.*, 2014). As women have limited knowledge, resources and access to training services (Quisumbing and Pandolfelli, 2010), women generally produce for more localized spot markets and in small volumes than men. When women are involved in marketing of agricultural produce, they usually concentrate at the lower levels of the supply or value chain, in low-cost products (Baden, 1998; Dolan and Sorby, 2003; Oduol *et al.*, 2014) which causes lesser income and profit.

5. Improvement of the efficiency of value chains in a gendered dimension

Value chain interventions aim at creating new value chains, forging and strengthening new links inside a value chain.It also aims to increase the capabilities of target groups to improve the terms of value chain participation, curtailing the possible negative impacts of value chain operations on non-participants and adjacent communities (Riisgaard *et al.*, 2010). Interventions to improve the efficiency of value chains in a gendered dimensionare listed (Coles and Mitchell, 2011).

Process and product upgrading-improving value chain efficiency and quality of produce, for example, the introduction of women-friendly agricultural machinery, drip, and sprinkler irrigation to enhance efficiency.

Horizontal coordination-development of associations among actors within functional 'nodes,'such as forming new fish traders' groups (Walker, 2001) and strengthening producers' groups (Naved, 2000).

Vertical coordination-developing relationships among actors between nodes, for example, farming to a contract (Reynold *et al.*, 2002) and employer's development of pension scheme for agri-processing employees.

Chain upgrading-applying existing skills in a new chain, for example, moving from mixed agriculture to fish farming (Naved, 2000).

'Upgrading' of the enabling environment along with social upgrading, which are applied to actors, but involves changes to policy, law, institutions, support organisations, for example, access to land ownership and provision of credit services for women in smallholder groups.

Access to networks and information is highly essential, it becomes highly limited when the gender-based decision making is skewed with limitations for women. Access to trainings and other skill development interventions are also lacking in women.

6. Ground Realities

Even though gender is receiving increased attention, in-depth analysis is lacking as sex-disaggregated data is scanty in India (Kruijssen *et al.*, 2018) and the analysis is often ignored.The reasons include the lesser number of successful women-based value chains and more dominant male dominated value chains. The competition in the value chain is lesser for men than women due to more access to resources, more willingness of families to devote time and money for men, more advantageous position to take risk, etc. For instance, the chain could exist but is often unnoticed as the economics is only considered for success.

In such cases, improper analysis of GVCs results in unnoticed gender division of labor, which has significant inferences for the way men and women can allocate their time to paid and unpaid work, education, health care, social networks, leisure, and other activities. This affects the genders' decision-making and finally affects the access and control over the assets (Kruijssen *et al.*, 2018). Women often receive lower returns and are disproportionately represented in less-profitable nodes of various value chains (Kruijssen *et al.*, 2018), often excluded from the more profitable part of agricultural value chains. Studies have indicated that in developing countries, women tend to work far longer hours than men. In Asia and Africa, women work as much as 13 hours more per week (FAO, Women, agriculture and food security). Participation of women as skilled labor and managers in the agricultural value chain is weak due to the social perception of women's roles in the community and social dominance of men (FAO, 2015) and also participation in value chain does not imply empowerment always.

7. Conclusion

Value chain analysis is the major step in creating successful value chains. It can be done in many ways,from simple interviews to complex gender sensitive value chain maps. Each approach mentioned in the paper has its own pros and cons and the researchers should use this depending on the time and resources available to undertake the analysis. There is an urgent need to shift from farming system-based approach to value chain based in order to improve the livelihoods of small holder women and men farmers. Gender based value chain analysis also helps policy makers to clearly define the needs of its stakeholders, and formulate gender-specific policy accordingly.

References

Agri-ProFocus, 2014. Gender in value chains toolkit. https:// agriprofocus.com/toolkit.

Asfaw, S. Mithçfer D.Waibel, H. 2010. What impact are EU supermarket standards having on developing countries' export of high value horticultural products? Evidence from Kenya. Journal of International Food and Agribusiness Marketing pp: 252-276 http://dx.doi.org/10.1080/08974431003641398.

Atal, J.P., H. Ñopo and N. Winder, 2009. New Century, Old Disparities. Gender and Ethnic Wage Gaps in Latin America . Research Department Working Paper 109. Washington, DC, United States: Inter-American Development Bank.

Baden, S. 1998. Gender issues in agricultural market liberalization, BRIDGE Report 41, IDS, Sussex.

Bolwig, S., and Odeke, M. 2007. Household Food Security Effects of Certified Organic Export Production in Tropical Africa: A Gendered Analysis. Bennekom, the Netherlands: Export Promotion of Organic Products from Africa.

Borjas, George J. 2000. "Measuring Discrimination". Labor Economics (Second ed.). Boston: Irwin McGraw-Hill. pp. 362–366. ISBN 0-07-231198-3.

Coles, C. and Mitchell, J. 2011, "Gender and agricultural value chains: a review of current knowledge and practice and their policy implications", ESA Working Paper No. 11-05, FAO, Rome.

Daviron, B. and Ponte, S. 2005. The coffee paradox: Global markets, commodity trade and the elusive promise of development. London: Zed Books.

Deborah, Rubin and Cristina, Manfre, 2012. Cultural Practice LLC and MEAS Project. 2012 Technical Note. Applying Gender-Responsive Value-Chain Analysis in EAS.

Dolan, D and Humphrey, J. 2000. Governance and Trade in Fresh Vegetables: The Impact of UK Supermarkets on the African Horticulture Industry, Journal of Development Studies, 37(2): 147-176 Dolan, C. Sorby, K. 2003. Gender and employment in high value agriculture industries.Agricultural and Rural Development working paper No.7, Washington D.C.FAO. 2018. Developing gender-sensitive value chains – Guidelines for practitioners. Rome.

Doss, C. and SOFA Team, 2007. The Role of Women in Agriculture. FAO. Available at: www.fao.org/docrep/013/am307e/am307e00.pdf

Elbehri, Azia and Lee, Maria, 2011. The Role of Women Producer Organizations in Agricultural Value Chains: Practical Lessons from Africa and India. Rome: FAO.

Gibbon, P. 2001. 'Upgrading Primary Production: A Global Commodity Chain Approach', World Development 29(2): 345-363.

FAO, 2015. Gender sensitive value chain analysis for medicinal and aromatic plants in Fayoum. Rome.

FAO, 2016. Developing gender-sensitive value chains – A guiding framework. Rome

FAO, 2018. Developing gender-sensitive value chains – A guiding framework. Rome

Fold, N. 2002. 'Lead Firms and Competition in 'Bi-Polar' Commodity Chains: Grinders and Branders in the Global Cocoa-Chocolate Industry', Journal of Agrarian Change 2(2): 228-247.

Fonseca, CristinoMandinga& Coelho, J. & Soares, Fernando & Correia, Augusta, 2019. Value Chain Analysis: Overview and Context for Development. Direct Research Journal of Agriculture and Food Science. 7: 356-361. 10.26765/DRJAFS13808532.

Hoffmann, N., & Roscoe, A. 2016. Investing in women along agribusiness value chains. Retrieved from https://www.ifc.org/wps/wcm/connect/02c5b53e-420f-4bf4-82bb-6f488ff75810/Women+in+ Agri+VC_Report_FINAL.pdf?MOD=AJPERES& CVID=m0JfSbv.

ILO, 2019. https://www.ilo.org/empent/Projects/the-lab/WCMS_638746/lang—en/index.htm

Kaplinsky, Raphael & Morris, Mike, 2001. A Handbook for Value Chain Research. 113.

Kruijssen F, McDougall CL, van Asseldonk IJM. 2018. Gender and aquaculture value chains: A review of key issues and implications for research. Aquaculture 493: 328–337.

Life Academy of Vocational Study, 2003. Study on Agriculture Value Chain in Mksp Area in Pottangi And Semiliguda Blocks of Odisha. Retrieved from http://lavsodisha.org/downloads/Study_On_Agriculture_Value_Chain.pdf.

Lyon, S., J.A. Bezary, and T. Mutersbaugh, 2009. "Gender equity in fair-trade organic coffee producer organizations: Cases from Mesoamerica."Geforum 41: 93-103.

Malapit, H.C. Ragasa, E.M. Martinez, D. Rubin, G. Seymour, A.R. Quisumbing, 2019. Empowerment in Agricultural Value Chains: Mixed Methods Evidence from the Philippines International Food Policy Research Institute (IFPRI), Washington, DC IFPRI Discussion Paper 1881.

Mather, C. and Greenberg, S. 2003. 'Market Liberalisation in Post-apartheid South Africa: The Restructuring of Citrus Exports after 'Deregulation', Journal of Southern African Studies 29(2): 393-412.

Mayoux, Linda; Mackie, Grania, 2008. A practical guide to mainstreaming gender analysis in value chain development / Linda Mayoux and GraniaMackie ; International Labour Office. Addis Ababa: ILO.

Meinzen-Dick, Ruth & Behrman, Julia & Pandolfelli, Lauren & Peterman, Amber & Quisumbing, Agnes, 2014. Gender and Social Capital for Agricultural Development. 10.1007/978-94-017-8616-4_10.

Muradian, R. and Pelupessy, W. 2005. Governing the coffee chain: The role of voluntary regulatory systems. World Development, 33(12): 2029-2044.

Naved, R. 2000. Intrahousehold impact of the transfer of modern agricultural technology: a gender perspective. https://ispc.cgiar.org/sites/default/files/pdf/278.pdf.

Oduol, Judith Beatrice & Mithöfer, Dagmar & Place, Frank, 2014. Constraints to and Opportunities for Women's Participation in High Value Agricultural Commodity Value Chains in Kenya. 10.13140/2.1.4888.7048.

Ola K.O. 2020. Micro-determinants of Women's Participation in Agricultural Value Chain: Evidence from Rural Households in Nigeria. In: Osabuohien E.S. (eds) The Palgrave Handbook of Agricultural and Rural Development in Africa. Palgrave Macmillan, Cham. https://doi.org/10.1007/978-3-030-41513-6_25

Poulton, C., Gibbon, P., Kydd, J., Larsen, M.N., Osorio, A. and Tschirley, D. 2004. Competition and Coordination in Liberalized African Cotton Market Systems. World Development 32(3).

Quisumbing A.R., Pandolfelli L. 2010. Promising approaches to address the needs of poorfemale farmers: resources, constraints and interventions. World Development 38(4): 581-592.

Reynolds, P.D., Bygrave, W.D., Autio, E., Cox, L.W. and M. Hay, 2002. Global Entrepreneurship Monitor, 2002. Executive Report, Babson College, London Business School and Kauffman Foundation.

Riisgaard, Lone & Fibla, Anna & Ponte, Stefano, 2010. Gender and Value Chain Development. https://www.researchgate.net/publication/263297919_Gender_ and_Value_ Chain_ Development.

Rubin and Manfre, 2012. Applying Gender-Responsive Value-Chain Analysis in EAS. https://www.g-fras.org/en/knowledge/documents/category/17-gender.html? download=122:applying-gender-responsive-value-chain-analysis-in-eas&start=20

Thitiya,Jitmun, John K.M. Kuwornu, Avishek Datta, Anil Kumar Anal, 2019. Farmers' perceptions of milk-collecting centres in Thailand's dairy industry. Development in Practice 29(4): 424-436.

UNDP, 2019. Maninstreaming Gender into ABS Value Chains. Retrieved from https://www.undp.org/content/dam/undp/library/km-qap/undp-gef-bpps-Mainstreaming_Gender_into_ABS_Value_Chains_Toolkit.pdf.

Walker, B.L.E. 2001. Sisterhood and Seine-Nets: Engendering Development and Conservation in Ghana s Marine Fishery . Professional Geographer 53(2): 160-177.
World Bank, 2012. World Development Report 2012 : Gender Equality and Development. World Bank. © World Bank. https://openknowledge.worldbank.org/handle/10986/4391 License: CC BY 3.0 IGO.

21

Empowerment Through Rural Women Collectives

Sulaja O.R. and Smitha S.*

1. Introduction

India Economic Survey (2018) reported that agriculture sector employed more than 50 per cent of the Indian work force. Women accounted for about 33 per cent of cultivators and 47 per cent of agricultural labourers excluding livestock, fisheries and various other ancillary forms of food production in our country (Rao, 2006). There are also reports on gender assessment which indicated that 75 percent of all women workers and 85 percent of rural workers in India are employed in agriculture (Agarwal, 2003). These suggest that the participation of women in the various fields such as agricultural professionals, entrepreneurs, cultivators, and laborers are on a steady increase. Urban migration by rural men for reasons of better economic opportunities has been identified as a major cause that gave this visibility to women as farmers and agricultural labourers. This indicate the emergence of agriculture as a viable livelihood alternative especially among rural women. It has been widely reported that the major difficulty experienced by the farm women has been the lack of control and ownership of the land resources, which rendered them inaccessible to credit, technology and training. Findings of FAO (2011) indicated that, if women farmers had equal property rights, access to scientific knowledge, monetary services, education and training facilities as their male counterparts, agricultural production would have increased to that extent that the number of persons in acute hunger in the world could be reduced by 100-150 million.

A study conducted by Ghosh and Ghosh (2014) on the share of women in agricultural activities across Indian states revealed that majority of women residing throughout the country generate their financial gains through farming and ancillary enterprises. Deviation of this trend could be observed only in Punjab, Kerala, and West Bengal where women were actively engaged in non-farming enterprises related to the well-developed service sector of these states. In the

Author contact: sulaja.or@kau.in

states of Himachal Pradesh, Nagaland, Rajasthan and Bihar more than 80 per cent of women were engaged in agriculture. The study also observed that wages of agricultural labour, income from farming, price volatility of agricultural produces and unpredictable weather conditions had significant influence on the livelihood options of majority of women headed households in India.

2. Women's land rights and empowerment

Women in rural areas are critical agents in bringing about the transformational economic, environmental, and social changes that are needed for long-term growth. However, they face various obstacles, including insufficient access to credit, health care, and education, all of which are compounded by global food and economic crises, as well as climate change. Land entitlements form another major factor that impede progress of women in agriculture sector. The widespread presence of patriarchal traditions in more than 15 countries including India, prevents women from acquiring equal possession rights to inherited land. In India 42 per cent of agriculture labour force constitute women, but their land ownership is less than 2 per cent of area under agriculture. India Human Development Survey (IHDS) report delineated that ownership of farm area is about 57 per cent for rural folks & 9 per cent for urban folks. Further interrogation revealed that 83 per cent of the farming area is inherited by men of the family restricting the ownership of women below 2 per cent. The remaining 15 per cent is being acquired through other means. Lack of legal awareness about inheritance rights among women has been reported as one of the most important obstacle which deter women from claiming land ownership (Mehta, 2018).

Agriculture policies of the Government had immediate and contingent effect on jobs related to farming sector. A land holder in agriculture makes the important decisions relating to resource use and exerts internal control over the agricultural holding. The agricultural holder has authoritative and economic commitment for the holding. Moreover, given the important role played by women in the agricultural labour force globally, capacitating them is significant not only for the overall economic efficiency but for the well-being ofpeople, households, and rural communities, however as well (World Bank, 2019). As multiple stakeholders actively involved in development sector, they hold significant role in launching, endorsing, and fostering multiple types of collective action aimed at guaranteeing fiscal and broader advantages for women by increasing their share in market. However, there is a lack of awareness of the optimal formula which benefit rural women in terms of collective action and the associated 'empowerment' gains. In mixed-sex organisations trying to reinforce sustenance through concerted efforts, gendered power structures usually result in separate and dissimilar results for women. Women have different reasons for concerted efforts than male counterparts, and they bring divergent expertise and standards to groups (Baden, 2013a).

3. Empowerment

Empowerment is a mechanism of increased engagement, decision-making power and influence, and transformative action as a result of increased awareness and capacity building. Women's empowerment includes both individual and collective reform. It improves their natural abilities by providing them with awareness, strength, and experience (Suguna, 2006).

3.1. Strategies for empowerment

Educational, economic, development, collective organization and political approaches have been widely used to enable women's empowerment. Education forms an important part of all strategies for improving a woman's understanding, knowledge, facts, and skills. Knowledge is a prerequisite for addressing oppressive powers in the status quo.The economic approach to empowerment aims to change women's economic status by targeting the factors that trigger gender division of labour, wage inequalities, and women's lack of control over material resources, among other items. Development of women's skills, savings and investment promotion, and increased employment prospects are all emphasized in the economic approach.

The third strategy is the development approach, which describes poverty as a result of people's powerlessness and lack of access to health care, education, and other services. Another perspective holds that women's empowerment necessitates an appreciation of the multiple factors that lead to women's disempowerment. This approach encourages women's collective organization, gender sensitization, gender perspective and strategy, and awareness-raising events. More recently, an organizational approach has been proposed, focused on the idea that organized women will improve gender and social dynamics in public and private lives to benefit women.Finally, political approaches to empowerment agree that women will grow on par with men if politics is free of violence, electoral malpractices, and unethical struggles, among other things, and is focused on values. This will necessitate a greater presence of women in active politics.

Personal and collective empowerment are two distinct forms of empowerment. Individual empowerment is a process of self-esteem, integrity, self-respect, and self-perception. However, issues concerning women's collectivity necessitate collective empowerment. Intervention at the collective level is needed to restructure power relations and reform social values and norms. Collective empowerment aims at transforming collective consciousness, values and attitudes. It is possible to resolve the issue of ensuring improved access to education, expertise, and jobs, as well as material resources and political influence only at the collective level. This requires effective organization among women, mutual assistance and particular amount of sacrifice (Suguna, 2006). This has led to the emergence of Self Help Groups. Joint Liability Groups and other forms of women collectivities into the fore.

3.2. Self-Help Groups

Women's empowerment and involvement in the development process have long been viewed as crucial aspects of change. Gender equality and women's empowerment are generally accepted as critical components of change in all fields. Women in rural areas are being trained to combat the evils of powerlessness and defenselessness faced by them from historical negligence. The idea of Self-Help Groups gained momentum with the efforts of Prof. Mohammed Yunus of Bangladesh who started experimenting with micro-credit and women SHGs in 1976. His strategy of empowering poor women created a quiet revolution in the poverty eradication in Bangladesh. Evidences from a number of countries across the globe suggest that the poor women spend wisely and gained dividends. However, the amount of money they earned has been inadequate. However, there has been empirical evidence that indicated women consistently outperformed men in terms of timely and effective repayment of their loans. Lending loans to women have been proved to be an effective way of guaranteeing that the advantages of accrued financial gain go to the family's overall well-being, notably to the child welfare. In addition women themselves hadthe advantage of the improved position by virtue of the increased income they earned.Based on these SHG can be defined as a small, economically undiversified and affinity based cluster of individuals who have determined to save and contribute to a standard fund to be lent to its members as per the group resolutions. It is a non-profit establishment whose focus is to upgrade members' position as candidates, administrators, and beneficiaries in the political, fiscal, public, and ethnic realms of existence (Gurumoorthy, 2000). The self help group concept got greater visibility, acceptance and reach as a poverty eradication and rural women empowerment tool under the Government of Kerala initiative of Poverty Eradication Mission, popular as the Kudumbasree Mission.

4. Kudumbashree Mission

Kudumbashree Mission plays a critical role in improving the monetary position of the less fortunate women in Kerala State by way of its thrift and credit societies. These societies help them keep money for future use and supply cheap and hassle-free credit. One of the core themes of Kudumbashree Project is economic empowerment, as well as social empowerment of women. The strategy of the Mission is to promote poor people's economic empowerment while also contributing to domestic monetary upliftment. The Mission has a number of fiscal upgradation programmes aimed for the family members of Kudumbashree, especially women and youngsters. Collective farming, livestock farming, and small businesses are encouraged in Kudumbashree programmes. Micro finance has been the powerful tool used for grassroots level fiscal upliftment as it directly improved availability of institutional finance to the weak and poor sections of the society. The reserve money of the women iscollected and combined as a fund which is distributed as loan to the beneficiaries (Kudumbasree story, 2019). Kudumbasree covered about 50% of families in the state of Kerala and the membership accounted to around 39.97 lakhs.Statistics shows that there are 2.58 lakhs Neighbourhood groups, 19,854 Area Development Societies and 1,073 Community Developmemt Societies.

4.1. Organizational structure

Kudumbashree followed the structure of a three-tier system with transparent election process to appoint office bearers, in line with Panchayati Raj rules. The three levels consisted of the a) Neighbourhood Groups (NHGs) – Primary level-Groups of 10-20 women from the same neighbourhood form the foundation of the structure b) Area Development Society (ADS)-Ward level-Federation of NHGs within a ward of the LSG and c) Community Development Society (CDS) -Registered Society as the Federation of ADS within the LSG (Kudumbashree, 2021).

Table 1. Reasons for joining Kudumbashree

S.No.	Reasons	Percentage (N=100)
1.	Availing cash/credit Benefits	67
2.	Developing saving habits	56
3.	Improving standard of living	87
4.	Peer pressure	12
5.	Meeting unexpected demand for cash	10

(*Source:* Saravanaselvi and Pushpa, 2016) (Multiple responses, not to total)

A study was conducted by Saravanaselvi and Pushpa (2016) to investigate the functioning of Kudumbashree and analyzed different dimensions of its interventions. The perceptions of members were collected on why they have joined Kudumbashree as presented in Table 1. Results from the table revealed that majority of the women (87 per cent) joined Kudumbashree with an intention of making improvements in their living standards, followed by 67 per cent availing cash benefits in the form of availability of credit at cheap rate of interest. About 56 per cent of them joined Kudumbashree with an intention of developing saving habits. Ten per cent of the respondents believed that they would meet unexpected demand for cash by their NHG membership. The corresponding two per cent of the respondents were motivated towards Kudumbashree due to peer pressure. The results of the table showed that women had faith in the performance of Kudumbashree and that it will definitely extend a helping hand in times of their need. They also aspired for improved living conditions and financial help without much procedural difficulties.

Economic empowerment of members

The level of economic empowerment was studied with respect to six factors namely income, savings, expenditure, financial management skills, personal belonging and financial security (Table 2).

Table 2. Level of financial empowerment

Indicators of financial empowerment	Level of financial empowerment					Mean score
	Very high	High	Moderate	Low	Very low	
Income		6	72	21	1	2.83
Savings		5	8	68	19	1.99
Expenditure	1	30	38	24	7	2.94
Financial management skills	1	6	30	47	16	2.29
Personal belongings		3	6	18	73	1.39
Financial security	3	7	11	21	58	1.76
Average	1	10	28	33	29	2.20

(*Source:* Minimol and Makesh, 2012)

A perusal of the table pointed out that about three-fourth of participants perceived their income hikes as moderate. But this does not show a corresponding increase in their savings, may be they will be spending more. Majority of the respondents (38%) expressed their expenditure to be moderate while one third of them had high expenditure and one fourth had low expenditure. About half of the members confessed that their financial management skills are low. There is scope for conducting training and educational programmes for Kudumbashree members for improving their monetary decisions. Membership in Kudumbashree could not make remarkable difference in their perception on possession of personal belongings and financial security.

Empowerment Level

The study conducted by Saravanaselvi and Pushpa (2016) among Kudumbashree members had shown that about 55 per cent of respondents opined to have empowerment at medium level. About one fourth of the respondents had high level of empowerment and one fifth of them perceived to have low empowerment level.

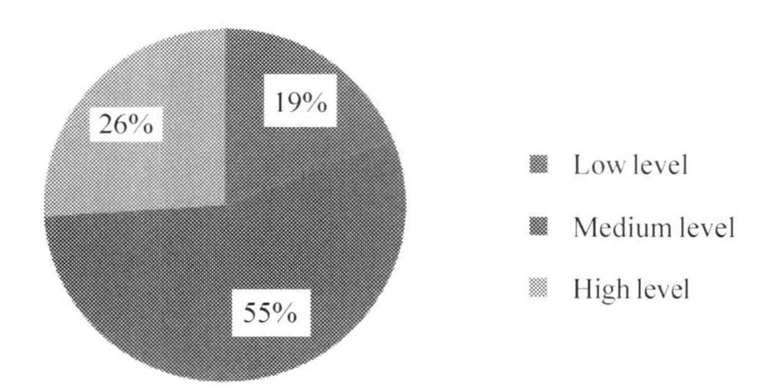

Fig. 1. Level of economic empowerment as perceived by members

Composite figures showed that about 81 per cent of the Kudumbashree members were having medium and high level of empowerment. Their perception is an indicator to the economic freedom and social status they were having after joining Kudumbashree programme.

Since its inception, Kudumbashree Neighbourhood Groups (NHG) have been regarded as powerful instruments of empowerment because they have used a participatory approach to empowerment to produce sustainable results. Social changes are happening in a slow pace, hence the sociological, psychological and economic improvement brought out by Kudumbashree's effort will be more visible over a period of time. It is also possible to draw the conclusion that the Kudumbashree programme would have a more significant impact on the lives of the poor in the coming years.The results are reiterated by the study conducted by Ashalakshmi *et al.* (2017) on Kudumbasree groups. They opined that women empowerment is positively related to membership in SHG/Kudumbasree, availability of credit facility, income level, standard of living (consumption and expenditure), and awareness on financial services. Financial inclusion through membership in SHG/Kudumbasree seems to be highly positively related to availability of credit facility, income level, standard of living (consumption and expenditure) and awareness on financial services.

Banks that were originally reluctant to offer loans to women from economically weaker background have modified their outlook as Kudumbashree efficiently reinforced the Community Based Organizations. Now, banks are competing with one another to draw Kudumbashree NHG accounts by providing special services. Poor women can now easily obtain bank loans with favourable terms and conditions. Kudumbashree members were given the opportunity to learn about banking operations and gained the courage to go to the bank and obtain the banking services. The Kudumbashree project in Kerala has clearly contributed to members' economic, political, socio-cultural, personal & familial empowerment. Despite the various constraints and obstacles, the Kudumbashree project is observed to have the potential to change the state's entire economy (Anupama, 2015).

The World Bank-funded JEEViKA (livelihoods) programme in India has been implemented in Bihar to focus on the livelihoods of rural women with weak economic conditions. Between 2008 and 2018, almost ten million rural women have joined self-help groups as a result of the programme, giving them entry to fiscal and trading platforms to initiate and develop businesses. The achievement of JEEViKA influenced the Indian government's decision to launch the National Rural Livelihoods Mission (NRLM) in 2011. The World Bank is a co-funder of the NRLM, which has expanded this prototype to all 30 states in the country, touching about 45 million women who have raised $1.4 billion and leveraged $20,000 in commercial bank loans. (World Bank, 2019). However, Baden (2013b) had pointed out that there is lack of knowledge about an optimum model for the empowerment of rural women in terms of organized efforts and the resulting "empowerment" benefits. She had conducted research in Ethiopia, Mali, and Tanzania to evaluate the impact of development programmes that supported women's integrated efforts in farm trade sector

Minimol and Makesh (2012) in their study on Self Help Groups reported that the personal empowerment [general] qualities that were identified with the highest level of empowerment were self confidence and mutual respect. It must be accepted that society has come to recognize the unified women's movement as an integral part of the social order. This is evident from the tolerance exhibited by family of women group members when the group meets for weekend evening meetings in one of the members' homes.

Conclusion

The role played by women in the economy of most countries is commendable especially developing and less developed countries. But they hardly enter into organized sectors of employment rather than contributing to 'unorganized', 'unprotected', 'unregistered' or 'informal' sectors. Unpredictable nature and instability of employment in these areas make them vulnerable to financial insecurity and poor standard of living. Women collectives especially in the form of Self Help Groups (SHG) are observed to be viable alternative to address these issues by handholding them to have more assertiveness and bargaining power with outsiders. The impact of Self Help Groups have on members' social empowerment has been admirable. But more viable collective approaches that are good enough to bring empowerment socially and economically among women need to emerge from the diverse contexts in which women operate.

References

Agarwal, B. 2003. Gender and land rights revisited: exploring new prospects via the state, family and market. Journal of Agrarian Change, 3(1 2): 184-224.

Anupama, R. 2015. Role of Kudumbashree in financial inclusion. Available: (Online) https://www.kudumbashree.org/storage//files/jtxru_kshree%20study% 20report46%2040.pdf.

Ashalakshmi R. K., Elsy Johnson and Megha Babu, 2017. Financial Inclusion and Impact on Women Empowerment in Rural area with special reference to Central Travancore Region of Kerala. International Journal of Application or Innovation in Engineering & Management. 6(11): ISSN 2319-4847

Baden, S. 2013a. Women's collective action: Unlocking the potential of agricultural markets. Oxford: Oxfam International.

Baden, S. 2013b. Women's collective action in African agricultural markets: the limits of current development practice for rural women's empowerment, Gender and Development, 21(2): 295-311, DOI: 10.1080/13552074.2013.802882.].

FAO (Food and Agriculture Organization), 2011. Gender differences in assets, prepared by SOFA team, ESA working paper No.11-12, Food and Agriculture Organisation, Rome

Ghosh, Mun Mun and Ghosh, Arindam, 2014. Analysis of Women Participation in Indian Agriculture 1Dr. IOSR Journal Of Humanities And Social Science (IOSR-JHSS) Volume 19, Issue 5, Ver. IV (May, 2014), PP 01-06 e-ISSN: 2279-0837, p-ISSN: 2279-0845. www.iosrjournals.org)

Gurumoorty, T.R. 2000. Self Help Groups empower rural women. Kurukshetra, 48(5): 36-40 https://thekudumbashreestory.info/index.php/programmes/economic-empowerment#.

India Economic Survey, 2018. Availale: (Online) http://mofapp.nic.in:8080/economicsurvey/

Kudumbashree official website, 2021. Available: (Online) https://www.kudumbashree.org/pages/9

Kudumbasree story, 2019. Kudumbasree- National Resource Organization. Department of Local Self Government, Government of Kerala, Thiruvananthapuram, Kerala. Available: https://thekudumbashreestory.info/landing

Mehta, Anupama, 2018. Gender gap in land ownership. National Council of Applied Economic Research (NCAER). Availale: (Online) http://www.ncaer.org/news

Minimol M.C. and Makesh K.G. 2012. Empowering rural women in Kerala: A study on the role of Self Help Groups (SHGs). International Journal of Sociology and Anthropology. 4(9): 270-280 (online at http://www.academicjournals.org/IJSA DOI: 10.5897/IJSA12.003 ISSN 2006-988x ©2012 Academic Journals).

Rao, Krishna E. 2006. Role of women in agriculture: a micro level study. Journal of Global Economy. 2: 107-118.

Saravanaselvi. C.I. and Pushpa, KS. 2016. Sustainability of women empowerment through Kudumbasree. International Journal of Research – Granthaalayah. 5(12): 2017.

Suguna, B. 2006. Empowerment of Rural Women through Self Help Groups. Discovery Publishing House, New Delhi.

World Bank, 2019. Rural Women Collectives in India- Translating Agency into Economic Empowerment.https://www.worldbank.org/en/results/2019/09/10/rural-women-collectives-in-india-translating-agency-into-economic-empowerment.

22

Emotional Intelligence in Gender Empowerment History: Reality and Challenges

*Devi Soumyaja**

1. Introduction

Anybody can become angry-that is easy, but to be angry with the right person and to the right degree and at the right time and for the right purpose, and in the right way-that is not within everybody's power and is not easy.

– Aristotle.

The word emotion is derived from the French word, "emouvoir" which means 'to excite'. Researchers have been interested in emotions since time immemorial. One of the earliest reported studies on emotions was attempted by Aristotle in 4th Century BC wherein he identified 14 distinct emotions. The traditional Indian art and literature talks about 'navarasas', nine emotions namely, joy *(hasya)*, fear *(bhayanaka)*, anger *(raudra)*, love *(shringar)*, courage *(vira)*, sadness *(karuna)*, amazement *(adbhuta)*, disgust *(vibhatsya)* and calmness *(shanti)*. Darwin (1872) theorized that emotions can be innate, evolved, and had a functional purpose. Robert Plutchik in 1980 talked about eight basic emotions, which he grouped into four pairs of polar opposites namely, joy-sadness, anger-fear, trust-distrust, surprise-anticipation. Renowned psychologist Paul Ekman and his colleagues through a cross cultural study in 1992 identified six basic emotions namely, anger, disgust, fear, happiness, sadness, and surprise. The popular animation movie produced by Pixar and released by Disney pictures in the year 2015,explored the emotional adjustment journey of a girl Riley, through the five basic emotions, anger, disgust, fear, happiness, and sadness, adapted from Ekman's work. In a recent study by Cowen and Keltner (2017) researchers identified 27 unique emotions. In yet another study to identify emotions based on facial expressions irrespective of sociocultural influences by Jack *et al.* (2016) just four emotions, happiness, sadness, anger and fear were recognized.

**Author contact: devisoumyaja@gmail.com*

2. The concept

The concept of emotional intelligence (EI) was coined by Salovey and Meyer (1990) and they described it as a form of "social intelligence that involves the ability to monitor one's own and others' feelings and emotions, to discriminate among them, and to use this information to guide one's thinking and action". According to Salovey and Meyer (1990) ability model of emotional intelligence consisted of four basic abilities as follows:

- *Ability to perceive and correctly express their emotions and other people's.*

  *The ability to use emotions in a way that facilitates **thought**.*
- *Capacity to understand emotions, emotional language, and emotional signals.*
- *The ability to manage their emotions in order to achieve **goals**.*

Salovey and Mayer (1990) explained EI as a component of Gardner's perspective of social intelligence. Similar to the so-called 'personal' intelligence proposed by Gardner, EI has been said to include an awareness of the self and others. However the concept was popularised by Goleman (1998) through his popular book, *Emotional Intelligence*.

Faltas (2017) argues that there are three major models of emotional intelligence as detailed below:

- Goleman's EI performance model
- Bar-On's EI competencies model
- Salovey, and Caruso's EI ability model

According to Goleman, EI is a cluster of skills and competencies, which are focused on four capabilities: self-awareness, self-management, relationship management, and social awareness. Bar-On's model of EI consists of five scales viz.self-perception, self-expression, interpersonal, decision-making, and stress management.

Salovey and Caruso's model describes four abilities related to EI emotions as described below.

1. Perceiving emotions: The ability to perceive emotions in oneself and others as well as in objects, art, stories, music, and other stimuli.
2. Facilitating thought: The ability to generate, use, and feel emotion as necessary to communicate feelings or employ them in other cognitive processes
3. Understanding emotions: The ability to understand emotional information, to understand how emotions combine and progress through relationship transitions, and to appreciate such emotional meanings

4. Managing emotions: The ability to be open to feelings, and to modulate them in oneself and others so as to promote personal understanding and growth

3. Research trends in EI

Research has found an association between EI and a broad range of skills such as making decisions or achieving academic success (Cherry *et al.*, 2018). Researchers have shown that our success at work or in life depends 80 per cent on Emotional Intelligence and only 20 per cent on intellect. While our intellect help us to resolve problems, to make the calculations or to process information, EI allows us to be more creative and use our emotions to resolve our problems (Cotrus *et al.*, 2012). According to Cooper (1997) developing and using EI skills offers a set of core capabilities that impact many business issues important to individual and organizational success. The major traits found influenced by EI are personal productivity, leadership development, career success, team performance, empowerment and client relationship management.

In 1995, Times magazine published an issue with emotional intelligence as the cover page and there it was famously noted that, *IQ gets you hired, but EQ gets you promoted*. The most powerful life tools we can equip ourselves with to enable us to succeed in life, is to have the combination of IQ and EQ. In popular television, movies and literature there have been several characters who have been portrayed with emotional intelligence. A very popular character would be Caltech physicist Dr. Sheldon Cooper, from the television series, *The Bigbang Theory*. Sheldon demonstrates what it is like to have a high IQ without a correspondingly high EQ.

4. Gender differences in emotions

"Here is my dilemma... as a woman in a high public position or seeking the presidency as I am, you have to be aware of how people will judge you for being, quote, 'emotional.' And so it's a really delicate balancing act-how you navigate what is still a relatively narrow path-to be yourself, to express yourself, to let your feelings show, but not in a way that triggers all of the negative stereotypes."

– Hillary Rodham Clinton (Clinton, 2016).

Another interesting aspect to be discussed in relation to EI is regarding the gender differences in the perception, and expression of emotions. Small but significant gender differences in emotion expressions have been reported for adults, with women showing greater emotional expressivity, especially for positive emotions and internalizing negative emotions such as sadness (Chaplin, 2015). These differences were highlighted in the popular book, *'Men are from Mars and Women are from Venus* written by John Gray. A contemporary researcher who has extensively researched emotions is neuroscientist Dr. Lisa Feldman Barrett. In her seminal work, published as the book, *How Emotions are Made* discusses several aspects of emotions and also breaks some of the myths and misconceptions about emotions. Barrett *et al.* (2001) study examined sex differences in the complexity and differentiation of people's representations of emotional experience. The study done among females from seven different countries demonstrated that women consistently displayed more complexity and differentiation in their articulations of emotional experiences than did men. Yet another study by Barrett and Moreau (2009) concluded that the stereotype of overly emotional women is linked to the belief that women express emotions as they are emotional creatures and men express emotions as the situation warrants it. Thus they summarized that both men and women express the same amount of emotions but both men and women believe that females are more emotional than males. Hence on an average both men and women are equally emotional.

The belief that women are more emotional than men is one of the strongest gender stereotypes held across cultures (Shields and Shields, 2002). In general, women (including women leaders) are proscribed from displaying high-status, masculine emotions that convey dominance, such as anger (Brescoll & Uhlmann, 2008) and pride (Cheng *et al.*, 2010). Another aspect to be considered here is that different cultural and social contexts require men and women to display emotions that are not necessarily congruent with the displayed rules associated with their gender (Fischer *et al.*, 2013). A metaanalysis done by Brescoll (2016) proposed that gender emotion stereotypes present a set of challenges that are unique to women in advancement to leadership positions. The study also noted that women leaders experience psychological consequences similar to those experienced by people engaging in emotional labour as a result of continuously trying to navigate emotional double binds at work. Hence we could conclude that the gender differences in emotional expression are more to do with cultural and social stereotypes than to do with the physiological differences.

Eagly (1987) was one of the first researchers to talk about gender differences and similarities in social behaviour through social role theory. According to this theory, men are stereotyped with argentic characteristics like confidence, assertiveness, independence, rationality, and decisiveness, whereas women are stereotyped with communal characteristics like concern for others, sensitivity, warmth, helpfulness, and nurturance (Eagly, 1997). These stereotypes led to the assumption that emotional labour comes naturally to women, whereas it requires extra effort from men. The term emotional labour was first used in 1983, by American sociologist Arlie Hochschild in her book, *The Managed Heart*. She described emotional labour as having to "induce or suppress feeling in order to sustain the outward countenance that produces the proper state of mind in others.

Women are expected to perform high degrees of emotional labour whereas men are not over-burdened by these expectations. It can also be noticed that gender representation in jobs have been directly linked to the emotional management required in these jobs. Traditionally tender and caring emotions have been assumed to be naturally occurring/obligatory in women than in men and hence in service professions requiring high emotional labouring like caregiving, catering, cleaning and clerical occupations, women are mostly employed (Guy & Newman, 2004; Pilcher, 2007; Riccucci, 2018).

5. Women and emotions-Cultural challenges

For centuries, movies and literature had celebrated "angry young men" in positive light. However, it's only since the last decade, angry young women were portrayed in a positive light in any forms of art. When Hindi movies celebrated Amitabh Bachan as the 'angry young man', all his heroines were expected to be demure, well-behaved and obedient. This indicates how culture had restricted expression of anger among women. A 2013, Iranian film, "Hush! Girls don't scream" through its name effectively captures how women are suppressed in expression of their legitimate emotions. Soraya Chemaly, an award winning American writer and gender activist, in a Ted talk given in 2018, spoke about the power of women's anger. In her talk, she mentioned that, if you ask women what they fear the most in response to their anger, they don't say violence, they say mockery. We can undoubtedly say that social conditioning to a large extent controls emotional expression among men and women. Young boys are often told not to cry like a girl and to be tough. They are told and trained to renounce the feminine emotionality of sadness or fear and to embrace aggression and anger as markers of real manhood. In a similar fashion girls are told to cross their legs, tame their hair, bite their tongues and swallow their pride. Emotional suppression, often specifically the suppression of anger, can affect an individual's ability to manage pain and recover from illness. Thus, the most important step in using emotions for gender empowerment is to overcome and liberate from emotional suppression.

In 2013 psychologist Susan David introduced a new concept termed as 'emotional agility'. Emotional agility is defined as being flexible with one's thoughts and feelings so that one can respond optimally to everyday situations. It means approaching one's inner experiences mindfully and productively. When we are emotionally agile, we are able to recognize our emotions before reacting, and decide how we'd like to respond, and respond in a way that aligns with our values, and in our best interest. Thus, emotional agility is the way ahead for overcoming emotional suppression.

6. Conclusion

Empowerment is the ability to influence the environment around you for the benefit of all. Personal empowerment happens only when one can control their own emotions. Emotions are experienced by each and every human being irrespective of gender. However, gender differences do exist o in terms of emotional frequency, intensity and management. From the foregoing discussions, we can infer that many of the gender differences are precipitated by the cultural and social stereotypes. Thus the most fundamental step towards empowerment through emotions is to break away from the cultural and gender stereotypes related to emotions. This will not happen overnight and right from childhood, children should be trained to be in touch with their true emotions and to express them freely and effectively, without any gender discrimination. Emotional agility would be a great tool for gender empowerment.

References

Barrett, L. F., & Moreau, Bliss, E. 2009. Affect as a psychological primitive. Advances in experimental social psychology, 41: 167-218.

Barrett, L.F., Gross, J., Christensen, T.C., & Benvenuto, M. 2001. Knowing what you're feeling and knowing what to do about it: Mapping the relation between emotion differentiation and emotion regulation. Cognition & Emotion, 15(6): 713-724.

Brescoll, V.L. 2016. Leading with their hearts? How gender stereotypes of emotion lead to biased evaluations of female leaders. The Leadership Quarterly, 27(3): 415-428.

Brescoll, V.L. and Uhlmann, E.L. 2008. Can an angry woman get ahead? Status conferral, gender, and expression of emotion in the workplace. Psychological science, 19(3): 268-275.

Chaplin, T.M. 2015. Gender and emotion expression: A developmental contextual perspective. Emotion Review, 7(1): 14-21.

Chemaly, S. 2018. Rage Becomes Her. Simon and Schuster.

Cheng, J.T., Tracy, J.L. & Henrich, J. 2010. Pride, personality, and the evolutionary foundations of human social status. Evolution and Human Behavior, 31(5): 334-347.

Cherry, M.G., Fletcher, I., Berridge, D. & O'Sullivan, H. 2018. Do doctors' attachment styles and emotional intelligence influence patients' emotional expressions in primary care consultations? An exploratory study using multilevel analysis. Patient education and counseling, 101(4): 659-664.

Cooper, R.K. 1997. Applying emotional intelligence in the workplace. Training & development, 51(12): 31-39.

Cotru°, A., Stanciu, C. & Bulborea, A.A. 2012. EQ vs. IQ which is most important in the success or failure of a student? Procedia-Social and Behavioral Sciences, 46: 5211-5213.

Cowen, A.S. & Keltner, D. 2017. Self-report captures 27 distinct categories of emotion bridged by continuous gradients. Proceedings of the National Academy of Sciences, 114(38): E7900-E7909.

Eagly, A.H. 1987. Reporting sex differences. American Psychologist, 42(7): 756–757.

Eagly, A.H. 1997. Sex differences in social behavior: comparing social role theory and evolutionary psychology. The American psychologist, 52(12): 1380-1383.

Faltas, I. 2017. Three models of emotional intelligence. Public Policy & Administration: article accessible online at https://www. researchgate. net/publication/314213508.

Fischer, A.H., Eagly, A.H., & Oosterwijk, S. 2013. The meaning of tears: Which sex seems emotional depends on the social context. European Journal of Social Psychology, 43(6): 505-515.

Goleman, D. 1998. Working with emotional intelligence. Bantam.

Guy, M.E., & Newman, M.A. 2004. Women's jobs, men's jobs: Sex segregation and emotional labor. Public administration review, 64(3): 289-298.

Jack, R.E., Sun, W., Delis, I., Garrod, O.G & Schyns, P.G 2016. Four not six: Revealing culturally common facial expressions of emotion. Journal of Experimental Psychology: General, 145(6): 708.

Pilcher, K. 2007. A gendered "managed heart"? An exploration of the gendering of emotional labour, aesthetic labour, and body work in service sector employment. Reinvention: a Journal of Undergraduate Research, 1(1): 1-13.

Riccucci, N.M. 2018. Antecedents of public service motivation: The role of gender. Perspectives on Public Management and Governance, 1(2): 115-126.

Salovey, P. & Mayer, J.D. 1990. Emotional intelligence. Imagination, cognition and personality, 9(3): 185-211.

Shields, S.A. & Shields, S.A. 2002. Speaking from the heart: Gender and the social meaning of emotion. Cambridge University Press.

23

Property Entitlement: Introspection on Women Rights

*Adv. Linipriya Vasan**

1. Introduction

Observing the right to property from the angle of human right, it could be perceived that it is an essential component of freedom and development (Narveson, 2010). In his literature Levy (1995) maintains that, free speech is of little value to a person who has no entitlement on property. He also holds the view that only property can enable the individual to make independent decisions and choices because he is not beholden to anyone and has no need to be subservient. Thus property ownership is basically grounded on considered and intrinsic values of human right. Strategically it assists economic human rights essential for development. Intrinsically it provides every one the right to preserve human dignity and in turn worth to protect human rights. Moreover, every individual should enjoy the legal protection that is perpetually required to preserve his or her human dignity. Access to use and control of productive resources including land are essential to ensure the equality and adequate standard of living among persons. The discussions on woman right to land are somehow interlinked with global food security and also sustainable development of the economy. The crucial obstacles on these rights are inadequate legal standard, ineffective implementation mechanism, and discriminatory cultural values.

In many capitalist societies women are still being vulnerable and subjected to exploitation which emerges from their exclusion from property ownership rights. Indeed the claim from women, for right to own property confirms the proposition laid down by Levy (1995) that the property rights relieves its owner to be subservient to others. However, the deteriorated rule of democracy obviously protect the discriminated and arbitral rules in almost every part of the world. The right to human dignity in terms of social recognition, makes an assertion that without personal property an individual is not a full human being. On this aspect, the worthiness of every civilisation could be judged from the position of

**Author contact: linipriya@gmail.com*

women in the society. Mehta (2012) observed that the legislative format throughout the world has been framed for the benefit of man, and as a result women need to depend on men for her livelihood and existence. The discrimination was so systematically formulated and that placed women at the receiving end (Jaishankar and Haldar, 2004).

The UN Women reports reveal that at least 115 countries recognise the women's property right on equal terms with men. They make a clarification on the point that the process has not been uniform and in many countries and there are significant gaps in the legal framework. (Realising women's Right to Land and other Productive Resources, UDHR, 2013).

2. Indian laws of inheritance

The Indian context is no better as is evident from the oft repeated quote "Her father protect her in childhood, her husband protects her in youth and her son protects her in old age; a women is never fit for independence" from the ancient text, Manusmritiby Manu which served as the guiding principle. This does not imply that she was totally excluded from inheriting property, but the proportional share was far less than that of her counterpart male. In the Indian context, right to property is in the absolute domain of personal laws.Various Dharma Sastras followed custom as the basic principle, and these principles varied within religion or caste. The two main schools of Hindu law were *Mitakshara* and *Dayabhaga*. *Mitakshara*school was further classified into *Benares, Mithila, Dravida* and *Maharashtra*. *Mitakshara* law applies to most parts of India except Bengal. However schools like *Marumakkatayam, Aliyasanthanam and Nambudiri* system were followed in certain part of south India (Section 3 and 17 of the Hindu Succession Act, 1956). Section 3 of the Act defines *Aliyasanthanam* law as *Aliyasantana* law. It indicates the system of law applicable to persons who, if this Act had not been passed, would have been governed by the Madras *Aliyasantana* Act, 1949, or by the customary *Aliyasantana* law with respect to the matters for which provision is made in this Act. The *Marumakkathayam* law applies to whom, if this Act had not been passed, would have been governed by the Madras *Marumakkattayam* Act, 1932; the Travancore Nayar Act; the Travancore Ezhava Act; the Travancore NanjinadVellala Act; the Travancore Kshatriya Act; the Travancore Krishnan vaka *Marumakkathayyee* Act; the Cochin *Marumakkathayam* Act; or the Cochin Nayar Act with respect to the matters for which provision is made in this Act.

Under Mitakshara system, son, grandson and great grandson were the coparceners based on their birth, and females were side lined and were kept outside the system. It recognised the inheritance by succession only on the self-acquired properties. These implied that multiple laws, originated in different social systems, formed the basis of Indian inheritance laws. In fact it created complications in the property laws as the customary practises mostly persuaded the policy of discrimination against women.

The British colonial rule, did not venture to interfere upon the personal laws persisting in our country. Only after the emergence of social reforms, steps to ameliorate the position of women through the process of legislation (the Hindu law of Inheritance act, 1929) had been initiated by the British regime. Another legislation which brought about significant changes in the existing system including all schools of Hindu law, was the Hindu women's right to property Act 1937. It brought about drastic changes in the legal rights, alienation of property, inheritance and adoptions. The 1937 Act enabled women to have equal share to that of a son, but the equal share was only a right to limited estate. The rights where still being incoherent and the anomalies of discrimination where basically not touched by the legislation.

2.1. The Constitutional provisions and safeguards

The kind of ambiguity, inequality and discrimination were gradually reduced, particularly with the adoption of the Constitution and the consequential legislations. It was only after independence the founding fathers of the constitution took note of the discriminatory positions and had taken many positive steps to build up intrinsic equality and protective discrimination. Through the constitutional provisions the principles of equality is enshrined and is one of the basic and fundamental rights of the constitution. But gender disparity is being evident in different legislations especially in the area of property rights. In order to implement the quarterly commitments in the Constitution, the Parliament passed the Hindu succession Act (1956), as a uniform and comprehensive system and applied to the schools of *Mitakshara* and *Dayabhaga*. Remarkable changes were brought about by the legislation, but still retained section 6 of the *Mitakshara* Coparcenary right. This provision is inherently discriminatory as it exclusively dealt with the rights of male members only. Thus though the Hindu Succession Act, 1956 was enacted to overcome the anomalies created by the Hindu Women Rights to Property Act, 1937, the Act was also not free of defects.

The real question focused in this article revolves around the equality of women with respect to the property rights under the Hindu Succession Act. The bill as originally formulated by BN Rau committee, proposed for the abolition of *Mitakshara* survivorship and substituted it with inheritance succession, but the report met with conservative criticisms. The resistance was accurately stated in the house by Sita Ram S. Jajoo from Madhya Pradesh with the statement that" Here we feel the pinch because ittouches our pockets. We male members of this house are in a huge majority. I do not wish that the tyranny of the majority may be imposed on the minority, the female members of this house." Retention of the *Mitaksharacoparcenary* with only male coparceners was the end result of the Hindu succession act. This retention continued the exclusion of female from the inheritance right on ancestral property. The exclusion of daughters from the coparcenary right because of her sex is truly discriminatory and negation of the equality principles in the Constitution. The act also resulted in the elimination of the *Marumakkathayam* and *Aliyasanthanam* communities, in which women enjoyed equal position with full coparceners. The retention of the *Mithakshara* system excluded females from inheriting the ancestral property and she could get the share of the property only when one of the coparceners dies.

Another gender bias is evident in proviso to section 6 of the Hindu Succession Act, which specifies the devolving of interest of the deceased in the *Mitakshara* Coparcenary[1]. It is the devolution by intestate succession if he left surviving female relative in class I of the schedule or male relative, specified in that class, who claims through the female relative. In order to properly understand the gender bias, section 8 of the Act should also be observed. It primarily deal with the devolving of the property of a male Hindu dying intestate[2]. Firstly, the property will be devolved upon the heirs specified in class I of the Schedule. The Class I heirs consists of the mother, widow, son and daughter as the only four primary heirs of the Schedule[3]. The remaining eight persons represent one or another person who would have been a primary heir if he or she would not die before the primary heir. The principle of representation would goup to two degrees in the male line of descent. But in the female line of descent it goes only upto one degree. Accordingly, the son's son's son and son's son's daughter get a share but a daughter's daughter's son and daughter's daughter's daughter were devoid of anything. A further infirmity is that widows of a pre-deceased son and grandson are class I heirs, but at the same time the husbands of a deceased daughter or grand-daughter are not heirs.

The act also included yet another discriminatory provision, which denied the married daughter's right to reside in the residential house of parents unless she would become a widow or would be separated from her husband. The concept of coparcenary has been abolished in Kerala following the recommendations of B.N. Rau Committee (Som, 1994). The recommendations of the committee were supported by P.V. Kane with the statement that "the unification of Hindu

---

[1]Section 6 of the Hindu Succession Act, 1956 deals with the devolution of Interest in the Coparcenery Property

[2]Section 8. General rules of succession in the case of males.Under the 1956 Act.

[3]Section 10 of the Act deals with distribution of property among heirs in class I of the Schedule

Law will be helped by the abolition of the right by birth which is the cornerstone of *Mitakshara*school and which the draft Hindu code seeks to abolish[4].

Legislative amendments were made by the state of Andhra Pradesh, Tamil Nadu, Karnataka and Maharashtra in the Hindu Succession Act. They tried to nullify the existing gender disparity by granting equal status to daughters with sons in the Joint Hindu Family much earlier than the central legislative amendment in 2005. Going further, the enactment of the Kerala Joint Hindu Family System (Abolition) Act, 1975 totally abolished the right to property by birth and thereby putting an end to the Joint Hindu family system that made a remarkable change in the concept of ownership of property in the state. The legislation lays down that the coparceners will hold the property as tenants in common as if the partition has taken place and she or he can hold her separate share. Another notable feature is that the traditional *Mitaksharaco* parcenary and right by birth had been abolished by the legislation. The Kerala legislative pattern resulted in maintaining fairness and harmony in the matrilineal and patrilineal families in Kerala.

Lata Mittals law suit[5] and the Supreme Court decision on cases involving personal property of women deal, created much deliberation on the elemental source of power on land and family wealth in India. The petitioner questioned the Indian cultural inequality as the governing laws for different religion being different in the country. Two discriminatory provisions of the Hindu succession Act 1956 include, one which provides that after the death of the male head of the family, the ancestral property is to be administered by the partnership of male heirs and the women are deprived of any legal right to make claim on the property. Another is the ownership granted to grandson and great grant son who are being still unborn at the time of death of the family-head is considered as partner, at the same time mother is eliminated from the partnership. Unfortunately the Supreme Court restricted itself to make any decision on these points and directed the respondents to put back the belonging of the petitioners to the place where they were thrown out. The decree even went to direct the police officers to ensure the peaceful and undisturbed life of the petitioners[6].

In the year 2005 the Hindu Succession Act was amended to rectify the status of women, where by equal rights were granted to daughters with son. The 174th Law Commission Report played a significant role in creating equal treatment

---

[4]174 th Law Commission Report; Property Rights of Women Proposed Reforms under the Hindu Law

[5]LAWS(SC)-1987-1-45

[6]Supreme court of India (30 Jan 1987).

for both the genders through the amendment. The Amendment Bill which was introduced in 2004 primarily focused on two important areas which shown gender disparity. First one is the introduction of a new section in the legislation, instead of section 6, in order to provide justifiable and equal rights to daughters with sons. The second one is on the omission of section 23 of the legislation, which makes the daughter disentitled to ask for the partition of the dwelling house occupied by the intestate family. Together with these, certain other provisions of the Act were also amended to make the constitutional right of equality a reality.

The amendment deleted the provision excluding the rights of agricultural land from the preview of the Act, and included provisions which removed discrimination against women, deprived of any entitlement or interest on agricultural land. With the removal of the provision women got equal rights on agricultural land as that of men. Yet another land mark change was the deletion of section 6 of the Act and the insertion of a new section, which provided that daughter become the coparcener in the property of Joint Hindu Family by her birth and made her the equal owner of property with son. But obviously the rights of the women will vary depending upon her status in the family whether married, unmarried, deserted etc. It will also vary based on the kind of property, hereditary or ancestral or self-acquired, dwelling house or matrimonial etc. Thus the coparcenary rights of women and the right to seek partition for her share in joint family property (JFP) were recognized only through the amendment of 2005. Another significant change was brought about by the omission of section 23 of the Act, thereby legalizing the right of women to seek for partition of intestate dwelling house.

2.2 Impact of the Amendment Act (2005)

In effect the Hindu Succession Amendment Act of 2005 created history on the property rights of women in the country. It had taken away differential status of women and given her the equal rights with sons. She has been recognized as a coparcener by virtue of the provision and even entitled to get the status of the *Karta* also if she is the senior most member in the family. While interpreting the provision, the judiciary had made it clear that daughter born on or after the amendment date can claim the interest over the ancestral property. The question on the prospective and retrospective nature of the amendment was decided by the Division Benchof the Bombay High Court[7] and they observed that the Act should be applied prospectivelyand daughters born on or after the date of amendment will only be considered as coparceners and those who born prior to the amendment will inherit the coparcenary only after the death of the owner.

[7]Ms.VaishaliSatishGanorkar& Another vs StishKeshaoraoGanorkar& others, 30 January 2012.

[8] Bombay High Court (14 Aug, 2014).

The judgment of the Bombay Division Bench in Vaishali S. Ganorkar case was challenged before the Supreme Court in the Badrinarayan Shankar Bhandri & Ors. and Omprakash Shankar[8] case based on the *in curiume* of the Supreme Court judgment in Ganduri Koteshwaramma case. However, the Supreme Court made it clear that section 6 of the Hindu Succession Act 2005 is retrospective in operation and it applies to daughters born prior to 17th June 1958 or thereafter, especially between 17 June 1956 and 8 September 2005. The condition specified was that they were alive on 9 September 2005 that is on the date when the Amendment Act of 2005 came into force. Admittedly, the amended Section 6 applies to daughters born on or after 9 September 2005. The court also pointed out the demand of social justice and long felt social need, that women should be considered as equal in social and economic sphere. The court also made a categorical statement that the Amendment Act has specifically used the term "on and from", which make it clear that the settled rights of coparcenary property will not get affected by the amended provisions and hence heirs of the daughter died before the date of amendment cannot claim the coparcenary right.

Again due to the unequal social attitude towards women, in 2020 the question concerning the interpretation of section 6 of the Hindu Succession Act 2005 has been referred by the Supreme Court to a larger bench in order to clarify the conflicting verdicts rendered by the Division Bench of the Supreme Court in Prakashan & Ors. V. Phulavati & Ors[9] and Danamma v. Amar &Ors[10]. There were series of decisions pronounced by different high court on the application of Section 6 of the Hindu Succession Act[11]. The Division bench of the Supreme Court in Prakash v. Phulavathi[12], answered three questions. Most important was on the application of the amendment even if the respondent's father died after the Act came into existence. Second question was the application of the

[9](2016) 2 SCC 36

[10](2018) 3 SCC 343

[11]Lokmani & Ors. v. Mahadevamma & Ors., [S.L.P.(C) No.6840 of 2016, in this case the High Court took had taken the stand that inequality has to be removed and the amended provisions are to be given with retrospective effect. In In Balchandra v. Smt. Poonam & Ors. [SLP [C] No.35994/2015], here also the retrospectivity of Section 6 was the issue. In the matter of Sistia Sarada Devi v. Uppaluri Hari Narayana & Ors. [SLP [C] No. 38542/2016], the question was regarding the reopening of partition and claiming redistribution of shares by the daughter. In Girijavva v. Kumar Hanmanta Gouda & Ors. [SLP [C] No.6403/2019], on the operation of section 6, court taken the stand that daughters could not claim any benefit as the father died in 1994. In Smt. V.L. Jayalakshmi v. V.L. Balakrishna & Ors. [SLP [C] No. 14353/2019], in this the issue was on the partition of the ancestral property. The decision of the trial court was modified in the light of Prakash v. Phulavati and the daughters were granted 1/35th share in the ancestral property.

[12](2016) 2 SCC 36

Amendment Act to the partition effectuated without the decree of court. Third on the retrospective application of the Amendment. The Supreme court had rejected the contentions of the respondent that the amendment is a progressive legislation. The court took the stand that even if it is a social legislation, retrospective application is possible only when it is expressly mentioned in the Act. The court laid down that Amendment of the Act could be applied to daughters, whose father was living coparcener as of 9th September 2005, regardless of when the daughter was born.

The decision in the afore mentioned case created lot of ambiguities as the female coparcener could not file the partition suit and entitled to share only if the male coparceners file the partition suit, as the amendment Act does not grant any such rights to the female coparceners. The irreconcilability of the matter was challenged in Danamma v. Amar & Ors. The right of the women coparceners has been elaborately dealt as it is one of the dynamic aspects of the Hindu Succession Act. It was held that the right under the legislation is for gradually recognising the rights of the women and delved in to the technicalities of the case and accepted a unique approach and granted the inheritance right to the coparceners recognising the ulterior motive behind the legislative amendment.

3. Conclusion

The Indian Constitutional jurisprudence embodies the principle of equality. The social as well as the legislative policy on the succession mattes restrict the right of women form equality. The evil of gender disparity is deep rooted in the in the traditional law of inheritance. Particularly in the case of Hindu Joint Family, it comprises a common male ancestor and lineal decedents in the male line. Under the traditional rule of *Mitakshara* and *Dayabhaga* the coparcenary right was patrilineal in nature and females were denied of any right to claim the share. The Hindu Succession Act, 1956 was enacted to make reforms in the rights of female and the scope of this right was further broadened by the 2005 Amendment Act. Social justice demands the substantive equality for women in every aspect of life. As women constitute half of the population, her personal development and the national development requires equal treatment.

The Law Commission in its 174th recommendation had recognised the unequal treatment against women. They recommended that the Amendment should follow the policy laid down by states like Andhra Pradesh, Maharashtra, Karnataka Tamil Nadu and Kerala. The recommendations gave prominence to up gradation of women's right under the Hindu Succession Act.

The Judicial decisions have arrived at conflicting conclusions. Rather than making mere literal interpretation, the judiciary need to glorify the constitutional jurisprudence and should take decisions in order to avoid the gender based discriminations in the society. When the constitution provides for protective discrimination an understandable phenomenon happens on the rights of women. The Supreme Court of India, being the apex Court should take a stand to protect the equality and the Constitutional principles providing the possible extension of the provisions keeping in mind that the suppression of women right to property is the suppression of her voice. Further the legislations shall be framed in a manner which removes all the social vulnerability of women and grant her unequivocal position both in the family and in the society.

References

Jaishankar, K. & Haldar, Debarati, 2004. Manusmriti; A Critique of the Criminal Justice Tenats in the Ancient Hindu Code, I. ERCES Online Q. Review No.3, www.erces.com/journal/articles/archives/v03/v03_05.html (accessed on 12/11/2020)

Levy, Leonard W., 1995. Seasoned Judgments: The American Constitution, Rights, and History. New Brunswick, NJ: Transaction Publishers. p.26.

Mehta, Riju, 2012. Inheritance rights of women: How to protect them and how succession laws vary, accessed fromhttps://economictimes.indiatimes.com/wealth/plan/inheritance-rights-of-women-how-to-protect-them-and-how-succession-laws-vary/ articleshow/ 70407336. cms? utm_ source= contentofinterest & utm_ medium=text& utm_campaign=cppst

Narveson, Jan., 2010. Property and Rights. Social Philosophy and Policy 27(1): pp. 101–134.

Som, Reba, 1994. Jawaharlal Nehru and the Hindu Code: A Victory of Symbol over Substance?Modern Asian Studies. 28(1): 165-194.

Printed in the United States
by Baker & Taylor Publisher Services